Studies in Computational Intelligence

Volume 1087

Series Editor

Janusz Kacprzyk, Polish Academy of Sciences, Warsaw, Poland

The series "Studies in Computational Intelligence" (SCI) publishes new developments and advances in the various areas of computational intelligence—quickly and with a high quality. The intent is to cover the theory, applications, and design methods of computational intelligence, as embedded in the fields of engineering, computer science, physics and life sciences, as well as the methodologies behind them. The series contains monographs, lecture notes and edited volumes in computational intelligence spanning the areas of neural networks, connectionist systems, genetic algorithms, evolutionary computation, artificial intelligence, cellular automata, self-organizing systems, soft computing, fuzzy systems, and hybrid intelligent systems. Of particular value to both the contributors and the readership are the short publication timeframe and the world-wide distribution, which enable both wide and rapid dissemination of research output.

Indexed by SCOPUS, DBLP, WTI Frankfurt eG, zbMATH, SCImago.

All books published in the series are submitted for consideration in Web of Science.

Yuriy P. Kondratenko · Vladik Kreinovich ·
Witold Pedrycz · Arkadii Chikrii ·
Anna M. Gil-Lafuente
Editors

Artificial Intelligence in Control and Decision-making Systems

Dedicated to Professor Janusz Kacprzyk

 Springer

Editors
Yuriy P. Kondratenko
Department of Intelligent Information
Systems
Petro Mohyla Black Sea National
University
Mykolaiv, Ukraine

Institute of Artificial Intelligence Problems
of Ministry of Education and Science
(MES) and National Academy of Sciences
(NAS) of Ukraine
Kyiv, Ukraine

Vladik Kreinovich
Department of Computer Science
University of Texas at El Paso
El Paso, TX, USA

Arkadii Chikrii
Department of Optimization of Controlled
Processes
Glushkov Institute of Cybernetics
of National Academy of Sciences
of Ukraine
Kyiv, Ukraine

Witold Pedrycz
Department of Electrical and Computer
Engineering, Faculty of Engineering
University of Alberta
Edmonton, AB, Canada

Anna M. Gil-Lafuente
Department of Business, Faculty
of Economy and Business
University of Barcelona
Barcelona, Spain

ISSN 1860-949X   ISSN 1860-9503 (electronic)
Studies in Computational Intelligence
ISBN 978-3-031-25761-2   ISBN 978-3-031-25759-9 (eBook)
https://doi.org/10.1007/978-3-031-25759-9

This Springer imprint is published by the registered company Springer Nature Switzerland AG
The registered company address is: Gewerbestrasse 11, 6330 Cham, Switzerland

Academician Professor Janusz Kacprzyk

Professor Janusz Kacprzyk is an outstanding scientist in the field of artificial intelligence including fuzzy sets and fuzzy logic, mathematical optimization, decision-making under uncertainty, fuzzy control systems, computational intelligence, intuitionistic fuzzy sets, data analysis, and data mining, with wide applications in databases, management, mobile robotics, and others.

Janusz Kacprzyk is an academician (full member) of the Polish Academy of Sciences. Currently, he is president of the Polish Operational and Systems Research Society and served as the president of the International Fuzzy Systems Association (IFSA) in 2009–2011.

He is a foreign member of the Spanish Royal Academy of Economic and Financial Sciences (2007), Bulgarian Academy of Sciences (2013), Finnish Society of Sciences and Letters (2018), Royal Flemish Academy of Belgium for Science and the Arts (2019), National Academy of Sciences of Ukraine (2021) as well as a member of Academia Europaea, the European Academy of Sciences and Arts and International Academy of Systems and Cybernetic Sciences (2020).

Professor Kacprzyk is a Honorary Senior Member of the Polish Society for Measurement, Automatic Control, and Robotics (since 2014), Permanent Honorary Member of the Mexican Association for Artificial Intelligence (since 2015), Honorary Member of European Society of Fuzzy Logic and Technology (since 2016) and Fellow of multiple professional societies, including IEEE (Institute of Electrical and Electronics Engineers (since 2006), IFSA (since 1997), EurAI/ECCAI (European Association for Artificial Intelligence, since 2014), SMIA (Mexican Society for Artificial Intelligence, since 2015), IET (Institution of Engineering and Technology (since 2017).

Janusz Kacprzyk was born on July 12, 1947. In 1970, he graduated from the Warsaw University of Technology, Department of Electronics. He has received a Ph.D. (1977) in systems analysis and a Dr. Sc., habil. (1990) in Computer Science from the Systems Research Institute, Polish Academy of Sciences (Warsaw, Poland). In 1997, Janusz Kacprzyk became a Full Professor, awarded by the President of the Republic of Poland.

Since 1970 Janusz Kacprzyk has been working at the Systems Research Institute of the Polish Academy of Sciences (SRI PAS) as senior researcher and deputy director of Research. Now, he is a professor of computer science of SRI PAS, a professor of WIT—Warsaw School of Applied Information Technology and Management (since 1998), a Professor at Industrial Research Institute of Automation and Measurements (PIAP) in Warsaw (since 2007), director of the Center of Information Technologies of WIT and SRI PAS (since 2012), Honorary External Professor at the Department of Mathematics of Yli Normal University (since 2006) and Full professor (part-time) at Chongqing Three Gorges University (since 2017), P.R. of China.

Dr. Kacprzyk worked also as a full professor at the Department of Electrical and Computer Engineering, Cracow University of Technology (2010–2014) and at the Department of Physics and Applied Computer Science, AGH University of Science and Technology, Cracow, Poland (2014–2015).

Dr. Kacprzyk has great international experience in teaching and research as he was Visiting Professor at the Hagan School of Business, Iona College, New Rochelle, NY, USA (1981–1983), Department of Computer Science, University of Tennessee, Knoxville, TN, USA (Spring, 1986), Department of Computer Science, University of North Carolina, Charlotte, NC, USA (Spring, 1988), Department of Management and Computer Sciences, University of Trento, Italy (Spring 1994), School of Science and Technology, Nottingham Trent University, Nottingham, UK (2003–2006), Department of Computer Science, Tijuana Institute of Technology, Tijuana, BC, Mexico (since 2006) and (j) RIKEN Brain Research Institute, Wako, Tokyo, Japan (2006–2016).

Professor Kacprzyk is a world-class high-caliber scientist, and his research contribution is confirmed by different Pioneer and similar awards:

- The 2017 Award of HAFSA (Hispanic American Fuzzy Systems Association) for pioneering contributions to the areas of fuzzy logic and systems
- The 2016 Award of the International Neural Network Society—Indian Chapter for Outstanding Contributions to Computational Intelligence
- Medal of ECSC (European Centre of Soft Computing) for contributions to the establishment and activities of the European Centre of Soft Computing, Mieres, Spain, 2015

- Lifetime Achievement Award in Soft Computing, WAC (World Automation Congress), 2014
- IFSA 2013 Award for outstanding academic contributions and lifetime achievement in the field of fuzzy systems, and continuous support of IFSA, IFSA (International Fuzzy Systems Association), 2013
- Distinguished Lecturer, IEEE CIS, 2010–2012
- Medal of the Polish Neural Network Society for exceptional contributions to the advancement of computational intelligence in Poland, 2010
- Pioneer Award for Outstanding Contributions to Granular Computing and Computing with Words, Silicon Valley Chapter IEEE Computational Intelligence Society (Institute of Electrical and Electronics Engineers, Computational Intelligence Society), 2007
- Pioneer Award, IEEE CIS for pioneering works on multistage fuzzy control, in particular, fuzzy dynamic programming, 2006
- 6-th Kaufmann Award and Gold Medal for pioneering works on the use of fuzzy logic in economics and management, FEGI, 2006
- AutoSoft Journal Lifetime Achievement Award in recognition of pioneering and outstanding contributions to the field of soft computing, 2000
- Award of Scientific Secretary of the Polish Academy of Sciences for the book Fuzzy Sets in Systems Analysis (in Polish), PWN, Warsaw, 1987
- Award of Department IV (Engineering Sciences) of the Polish Academy of Sciences for the book Multistage Decision Making under Fuzziness, Verlag TÜV Rheinland, Cologne, 1984
- Medal of the University of Liège, Belgium, 1981.

Besides, Prof. Kacprzyk was awarded as Honorary Professor at Department of Mathematics, Yli Normal University, Xinjiang, P.R. of China (2006) and as Honorary Doctor (Dr. honoris causa) at Széchenyi István University, Győr, Hungary (2014), Óbuda University, Budapest, Hungary (2016), Prof. Asen Zlatarev University, Bourgas, Bulgaria (2017), Lappeenranta University of Technology, Lappeenranta, Finland (2017), and Petro Mohyla Black Sea National University, Mykolaiv, Ukraine (2020).

Dr. Kacprzyk is an Editor-in-Chief of the two international scientific journals and seven book series at Springer, editor of more than 50 books, author of more than 650 publications, including 7 books and more than 100 papers, which are included in Web of Science. He presented more than 150 plenary and invited talks.

The 2022 year marks a remarkable event for the Polish Academy of Science and, more generally, for our entire scientific community, that is the 75-th anniversary of Prof. Janusz Kacprzyk.

On the one hand, Prof. Kacprzyk has been for years the driving force behind the use of artificial intelligent models and methods in the analysis of broadly perceived complex systems in engineering, management, and economics. He is one of the originators of a comprehensive study of fuzzy theory and its implementation in industry and agriculture, and has proposed many original tools and techniques.

On the other hand, as President of the Polish Operational and Systems Research Society and past president of the International Fuzzy Systems Association (IFSA) in 2009–2011, he has greatly contributed to the activities of both societies and their great stature and visibility.

Third, in addition to his illustrious scientific and scholarly career, Prof. Janusz Kacprzyk has been always aware that theory should be complemented with practice. This is true for virtually all fields of science, in particular engineering. The contributions of Prof. J. Kacprzyk in this respect, i.e., at the crossroads of science and engineering have been remarkable too.

Finally, one should also mention the role which Prof. Kacprzyk has played as a teacher and mentor, and also a loyal and supportive friend to so many people, including all of us.

Given that special role that Prof. Janusz Kacprzyk has played in so many fields of activities, from a purely scientific and scholarly level, through organizational activities, to real engineering application, we have decided—as a token of appreciation of the entire community—to publish this special volume *Artificial Intelligence in Control and Decision-making Systems* dedicated to him, his scientific influence and legacy.

Mykolaiv, Ukraine   Yuriy P. Kondratenko
El Paso, TX, USA   Vladik Kreinovich
Edmonton, AB, Canada   Witold Pedrycz
Kyiv, Ukraine   Arkadii Chikrii
Barcelona, Spain   Anna M. Gil-Lafuente
March 2022

Contents

Introduction

This volume is dedicated to Prof. Janusz Kacprzyk, Academician of the Polish Academy of Sciences, President of the Polish Operational and Systems Research Society, and President of the International Fuzzy Systems Association (IFSA) in 2009–2011.

By dedicating this book to Prof. Janusz Kacprzyk, we, the editors, on behalf of the entire research and professional community, wish to present a small token of appreciation for his great scientific and scholarly achievements, long-time service to many scientific and professional communities, notably those involved in automation, cybernetics, control, management and, more specifically, the artificial intelligence including fuzzy sets and fuzzy logic, mathematical optimization, decision-making under uncertainty, fuzzy control systems, computational intelligence, intuitionistic fuzzy sets, data analysis, and data mining, with the wide area of applications including databases, management, and mobile robotics.

At a more personal level, we also wish to thank him for his constant support of what we and many of our collaborators and students have been receiving from him for so many years.

In terms of structure, the 18 chapters of the book, presented by authors from 16 different countries (Australia, Brazil, Canada, Chile, Germany, Hungary, Israel, Italy, P.R. China, R.N. Macedonia, Saudi Arabia, Spain, Turkey, United States, Ukraine, and Vietnam), are grouped into three parts: (1) *Computational Intelligence and Fuzzy Systems*, (2) *Artificial Intelligence Techniques in Modelling and Optimization*, and (3) *Computational Intelligence in Control and Decision Support Processes*.

The chapters provide an easy-to-follow introduction to the topics that are addressed, including the most relevant references, so that anyone interested in them can start their introduction to the topic through these references. At the same time, all of them correspond to different aspects of work-in-progress being carried out in various research groups throughout the world and, therefore, provide information on the state-of-the-art of these topics.

The first part *Computational Intelligence and Fuzzy Systems* includes six contributions:

The chapter "Methods for the Computation of the Interval Hull to Solutions of Interval and Fuzzy Interval Linear Systems", by Weldon A. Lodwick and Luiz Leduino Salles-Neto, is devoted to computing the solution of interval linear systems as an NP-Hard problem. This presentation develops tractable numerical algorithms to approximate the interval hull of the solution of interval and fuzzy interval linear systems. The general fuzzy interval linear system is an interval linear system at each alpha level. For linear problems where the coefficients are fuzzy intervals that are triangular, the proposed approach is tractable—it consists of solving one interval linear system and one real-valued linear system. In the case of fuzzy trapezoidal intervals, there are two interval linear systems to solve rather than a continuum. The basis of the proposed algorithm is a series of interval optimizations. Authors discuss some successful examples and provide access to their Python code.

In the chapter "Fuzzy-Petri-Networks in Supervisory Control of Markov Processes in Robotized FMS and Robotic Systems", Georgi M. Dimirovski, Yuanwei Jing, Jindong Shen, Kun Wang, Dilek Tukel, Figen Ozen, Gorjan Nadzinski, and Dushko Stavrov focus on the techniques of Petri-net and fuzzy-Petri-net formalisms with applications in Robotized Manufacturing Engineering. The concepts of event-driven and time-driven sequences are found to be instrumental for development of representation models as well as of control and communication algorithms that enable system design analysis with regard to higher, task-oriented decision-and-control level. Authors present an overview of Petri-net formalism and a similar overview on the feasible synergies of fuzzy systems and Petri nets. The resulting model representations and task-control algorithms—based on case-study plants in robotized flexible manufacturing systems—provide relevant illustrations.

Ronald R. Yager and Fred Petry, in the chapter "Using Fuzzy Set Approaches for Linguistic Data Summaries", analyze the basics of linguistic summary approaches using fuzzy set techniques. Authors describe the quantification of information contained in a linguistic summary and consider the formulation of summaries involving rich concepts. Efficiency of the suggested mathematical approach is illustrated by examples.

In the chapter "Fuzzy or Neural, Type-1 or Type-2—When Each Is Better: First-Approximation Analysis", Vladik Kreinovich and Olga Kosheleva discuss the need (in many practical situations) to determine the dependence between different quantities based on the empirical data. Several methods exist for solving this problem, including neural techniques and different versions of fuzzy techniques: type-1, type-2, etc. In some cases, some of these techniques work better, in other cases, other methods work better. Usually, practitioners try several techniques and select the one that works best for their problem. This trying often requires a lot of efforts. Authors use the first-approximation model of this situation to provide a priori recommendations about which technique is better.

A. A. Chikrii and G. Ts. Chikrii, in the chapter "Matrix Resolving Functions in Game Dynamic Problems", consider a non-stationary problem of bringing the trajectory of a conflict-controlled process to a given terminal set that varies over time. Special attention is paid to the method of resolving functions. While the main scheme of the method uses scalar resolving functions, the authors propose using matrix functions of a diagonal form with different elements on the diagonal. This idea makes it possible to cover more general game situations. Authors formulate sufficient conditions for the termination of the game in finite time in the class of quasi- and stroboscopic strategies. The proposed approach expands the resolving functions method and thus, increases the efficiency of the resulting applications.

The chapter "Evolving Stacking Neuro-Fuzzy Probabilistic Networks and Their Combined Learning in Online Pattern Recognition Tasks", by Ye. Bodyanskiy and O. Chala, deals with the pattern's classification-recognition problem under conditions of overlapping classes and a small training dataset. The proposed method is based on a modification of the classical probabilistic neural network by D. Specht, improved with combined learning, which includes (a) classical supervised learning, (b) "lazy learning" based on the concept of "Neurons at data points", (c) self-learning based on the rule "Winner takes all", and (d) evolving learning system architecture. Fuzzy modifications of the probabilistic neural network are introduced, where one-dimensional membership functions are used instead of multidimensional activation functions, which reduce the number of adjustable parameters.

The proposed stack systems are characterized by a high learning speed, which allows them to solve problems that arise within the general problem of Data Stream Mining.

The second part *Artificial Intelligence Techniques in Modelling and Optimization* includes six contributions:

In the chapter "Intelligent Information Technology for Structural Optimization of Fuzzy Control and Decision-Making Systems", Oleksiy V. Kozlov, Yuriy P. Kondratenko, and Oleksandr S. Skakodub investigate the intelligent information technology (IIT) for structural optimization of fuzzy systems (FS) based on the evolutionary search for the optimal membership functions. The proposed IIT uses a combination of several different bioinspired evolutionary algorithms, and enables users to find the optimal membership functions for the involved linguistic terms when solving the compromise problems of multi-criteria structural optimization of various FSs. The proposed technology allows to increase FSs efficiency and to reduce the degree of complexity of the needed parametric optimization. The search of the optimal membership functions is illustrated on the example of a FS of the multi-purpose mobile robot (MR) designed to move along inclined and vertical ferromagnetic surfaces. This example uses a combination of three evolutionary algorithms: genetic, artificial immune systems, and biogeographic. The analysis of the obtained results shows that the use of the proposed IIT significantly increases the efficiency of the MR control, while reducing the overall number of to-be-optimized parameters needed to describe the linguistic terms.

Alisson Porto and Fernando Gomide, in the chapter "Neural and Granular Fuzzy Adaptive Modeling", discuss adaptive nonlinear systems modeling focusing on the uninorm-based evolving neuro-fuzzy network and the granular evolving min-max fuzzy approaches. Neuro-fuzzy networks have shown to be highly competitive with multilayer neural networks, but their performance have not been yet contrasted with contemporary adaptive granular modeling methods such as evolving min-max modeling. Neuro-fuzzy network and granular evolving min-max modeling are benchmarked against state-of-the-art evolving modeling methods on the examples of the Box and Jenkins gas furnace and the Mackey-Glass chaotic time series data. The authors show that evolving granular min-max modeling is a powerful modeling approach: in addition to accuracy, it benefits from the transparency of the resulting models.

The chapter "State and Action Abstraction for Search and Reinforcement Learning Algorithms", by Alexander Dockhorn and Rudolf Kruse, emphasizes that decision-making in large and dynamic environments has always been a challenge for AI agents. Given the multitude of available sensors in robotics and the rising complexity of simulated environments, agents have access to plenty of data but need to carefully focus their attention if they want to be successful. While action abstractions reduce the complexity by concentrating on a feasible subset of actions, state abstractions enable the agent to better transfer its knowledge from similar situations. Authors identify and analyze different techniques for learning and using state/action abstractions, and analyze the effect of these techniques on the agents' training and the agents' resulting behavior.

Jaime Gil Aluja and Jean Jacques Askenasy, in the chapter "A Tentative Algorithm for Neurological Disorders", deal with problems that casts a shadow over the lives of so many humans: neurological disorders. This paper concentrates on an important class of such disorders: Parkinson's disease. This chapter has two main objectives: (a) determining, through the degree or level of intensity of the perceived symptoms, the location and the degree or level of affectation of the involved neurons; (b) determining the relations between neuronal areas, between neurons and symptoms, and between different symptoms—and the degrees of these relations. These two aspects of the work help to provide the correct diagnosis in situations which are, at present, often misdiagnosed in the clinical practice. The resulting algorithm provides the correct diagnosis in all such situations with confidence levels of 0.8 or higher.

The chapter "On the Use of Quasi-Sigmoids in Function Approximation Problems with Neural Networks", by Francesco Carlo Morabito, Maurizio Campolo, and Cosimo Ieracitano, introduces a novel type of nonlinear activation functions for the neurons of a Neural Network (NN). The authors describe and analyze backpropagation algorithm based on such quasi-sigmoids functions, and compare the performance of the proposed quasi-sigmoidal networks with the usual sigmoidal networks on several benchmark test cases, including an important problem of electromagnetics. The performance of quasi-sigmoidal networks as function approximators is shown to be generally superior to sigmoidal one—because the new models are more flexible and at the same time need fewer degrees of freedom to achieve the same accuracy.

As a by-product, the proposed activation function allows to carry out data-driven detections of nonlinearity in the data.

In the chapter "Human-Centric Question-Answering System with Linguistic Terms", Nhuan D. To, Marek Z. Reformat, and Ronald R. Yager consider current Question-Answering systems which can automatically answer questions posed by humans in a natural language. Most such systems can only answer questions that do not contain imprecise concepts and that lead to short answers. Authors synthesize a human-centric Question-Answering system capable of answering questions containing user-defined, personalized linguistic terms. The proposed system works with information represented in the form of knowledge graphs. The authors describe the system and present its main components, emphasizing a few extensions that make the system distinctive.

The third part *Computational Intelligence in Control and Decision Support Processes* includes six contributions:

In the chapter "OWA Operators in Pensions", Anton Figuerola-Wischke, Anna M. Gil-Lafuente, and José M. Merigó discuss the public pension system crisis caused by the changing demographics, and its impact on different countries worldwide. This study presents a new method for optimizing forecasts of the average pension amount by using different types of ordered weighted averaging operators: OWA, IOWA, GOWA, IGOWA, POWA, Quasi-OWA. It also accounts for inflation or deflation, providing a more realistic assessment of the average pension size. The main advantage of this approach is the possibility to explicitly take into account the attitudinal character of experts and decision-makers. As an illustration, the authors use their new method to forecast average pension amounts for all regions of Spain.

Martin Leon-Santiesteban, Alicia Delgadillo-Aguirre, Martin I. Huesca-Gastelum, and Ernesto Leon-Castro, in the chapter "Evaluation of the Perception of Public Safety Through Fuzzy and Multi-criteria Approach", consider the multi-criteria ordering in relation to the perception of public safety in the city of Culiacan, Sinaloa, Mexico. The objective of this study is to come up, for each of the four quadrants into which the city is divided, with an understanding of social dynamics, of the people's perception of insecurity, and of the people's opinion of the authorities responsible for providing public security. To collect the data, they use a question-naire sent to citizens of all 46 neighborhoods of the city of Culiacan. To process the collected data, they used ELECTRE-III method and the OWA operator tech-niques. The proposed approach can help in designing security strategies aimed both at improving public safety (and recovering public spaces) and at improving citizens' perception of public safety.

In the chapter "A Multicriteria Hierarchical Approach to Investment Location Choice", Laura Arenas, Manuel Muñoz Palma, Pavel Anselmo Alvarez Carrillo, Ernesto León Castro, and Anna M. Gil-Lafuente analyze, for different financial and macroeconomic criteria, which Latin American countries provide the best return on investment with respect to these criteria. The analysis considers different subgroups of criteria and uses a hierarchical version of the ELECTRE-III method to apply different aggregated rankings using the dimensions of Financial Market, Economic Situation and Growth, Labor Market and Purchasing Power Indicators, Foreign

Commercial Operations, and Fiscal Indicators. The authors' results show, based on different ranking schemes and investor preferences, that Peru and Chile are the best countries in which to invest, and Argentina is the least attractive.

The chapter "Uncertainty in Computer and Decision-Making Sciences: A Bibliometric Overview", by Carlos J. Torres-Vergara, Víctor G. Alfaro-García, and Anna M. Gil-Lafuente, discusses the application of bibliometric tools and techniques to analyze the dynamics of computer-science research publications focused on uncertainty and decision-making. The analysis is divided into key sections including journals, institutions, key scholars, countries, institutions, collaborations, and word frequency. This paper utilizes 1964–2021 publications from the SCOPUS scientific database, identifying 11,122 relevant papers with an average of 19.94 citations per document and an average of 7.2 years. Results provide a numerical description of the evolution and growth of academic research in the fields of uncertainty and decision support processes.

Tamrat D. Chala and László T. Kóczy, in the chapter "Intelligent Traffic Signal Control Using Rule Based Fuzzy System", propose a new hierarchical traffic signal control system based on Mamdani-type fuzzy control. Compared to existing traffic control systems, the new system is more adaptable: it can implement complex control strategies. The new system is also more flexible: it can utilize information about traffic in different directions, it takes into account the number of vehicles waiting at the intersection and their current waiting time. The system also provides a safer and more efficient handling of situations when emergency vehicles are present. All this makes the system more effective than all existing systems for controlling traffic signals.

In the chapter "Generative Adversarial Networks in Cybersecurity: Analysis and Response", Oleksandr S. Striuk and Yuriy P. Kondratenko note that the main challenge for cybersecurity is a timely and adequate response to any type of threat since every day there are more and more malicious scenarios for compromising information technology data. For cyber-attack defense algorithms to be effective, it is important to maintain a high level of awareness of the widest range of state-of-the-art threat types and malicious technologies. Machine learning and AI in general are considered as both a helpful tool and a threat. Generative adversarial networks (GANs) and their modifications can pose a serious threat to the entire field and thus should be properly and thoroughly researched, especially in light of the fact that GANs can be used to improve known attack types so that even AI-based detection system cannot identify them. The purpose of this article is a comprehensive analysis and structuring of existing GAN methodologies, with the subsequent development of approaches for an adequate response to potential threats.

The chapters selected for this book provide an overview of important problems in the areas of artificial intelligence and control and decision-making systems design, and an overview of the advanced computational intelligence techniques that relevant research groups within this area are employing to solve these problems.

Each chapter was reviewed by two or three highly competent reviewers before being accepted for publication in this book.

We would like to express our deep appreciation to all authors for their contributions as well as to reviewers for their timely and interesting comments and suggestions. We certainly look forward to working with all the contributors again in nearby future.

P. S. Vladik, Witold, Arkadii, and Anna Maria would like to express our appreciation of the heroism of Yuriy P. Kondratenko, the book's main editor who continued to lead the work on this book in the middle of the war.

Mykolaiv, Ukraine   Prof. Dr. Sc. Yuriy P. Kondratenko
El Paso, TX, USA   Prof. Dr. Vladik Kreinovich
Edmonton, AB, Canada   Prof. Dr. Habil. Witold Pedrycz
Kyiv, Ukraine   Prof. Dr. Sc. Arkadii Chikrii
Barcelona, Catalonia, Spain   Prof. Dr. Anna M. Gil-Lafuente
March 2022

Computational Intelligence and Fuzzy Systems

Methods for the Computation of the Interval Hull to Solutions of Interval and Fuzzy Interval Linear Systems

Weldon A. Lodwick and Luiz Leduino Salles-Neto

Abstract It is known that computing the solution of interval linear systems is an *NP-Hard* problem. Fuzzy linear systems are even more of a challenge, since there is a linear interval system to solve for each alpha level. This presentation develops tractable numerical algorithms to approximate the interval hull of the solution interval and fuzzy interval linear systems. The general fuzzy interval linear system is an interval linear system at each alpha level. However, linear problems where the coefficients are fuzzy intervals that are triangular, the approach presented here is tractable adding one interval linear system and one real-valued linear system. In the case of fuzzy trapezoidal intervals, there are two interval linear systems to solve rather than a continuum. The basis of the algorithm is a series of interval optimizations. Examples and access to Python code are provided.

1 Introduction

The interval linear system has a long history dating back to at least the W. Oettli and W. Prager paper in 1964 [15] and R. E. Moore's first book on interval analysis in 1966 [10] (also see [11, 13]). There have been a variety of algorithms that have been proposed and over the subsequent years, among them Hansen [2, 3], Rohn [17–20], Neumaier [12]. The book by Kreinovich et. al. [4] analyzes the computations convexity of interval linear systems.

The interval linear system problem is the following. Let be $[b] = [\underline{b}, \overline{b}]$ an interval vector, where $\underline{b} \leq \overline{b}$, and $[A] = [\underline{A}, \overline{A}], \underline{A} \leq \overline{A}$, an $n \times n$ interval matrix. Here the vector and matrix ordering $\leq$ is done component-wise. We are interested in finding the solutions for the interval system $[A]x = [b]$, which is called the *interval linear system problem*. Shary in a series of articles [21–24] defined four cases of solutions

W. A. Lodwick (✉)
University of Colorado Denver, Denver, USA
e-mail: weldon.lodwick@ucdenver.edu

L. L. Salles-Neto
Universidade Federal de São Paulo, São José dos Campos, Sao Paulo, Brazil

© The Author(s), under exclusive license to Springer Nature Switzerland AG 2023
Y. P. Kondratenko et al. (eds.), *Artificial Intelligence in Control and Decision-making Systems*, Studies in Computational Intelligence 1087,
https://doi.org/10.1007/978-3-031-25759-9_1

to interval linear equations. These are the four cases that were studied by Lodwick and Dubois [9] from a constraint interval point of view [7]. The four cases are as follows:

- **Case 1**: x is solution of the interval linear system $[A]x = [b]$ when for all $A \in [A]$ there exists $b \in [b]$ such that $Ax = b$, that is, $[A]x \subseteq [b]$.
- **Case 2**: x is solution of the interval linear system $[A]x = [b]$ when for all $b \in [b]$ there exists $A \in [A]$ such that $Ax = b$, that is, $[A]x \supseteq [b]$.
- **Case 3**: x is solution of the interval linear system $[A]x = [b]$ when $[A]x \subseteq [b]$ and $[A]x \supseteq [b]$ or if $\forall A \in [A]$ and $\forall b \in [b]$, $Ax = b$, that is, $[A]x = [b]$ in the classical set theoretic sense.
- **Case4**: x is solution of the interval linear system $[A]x = [b]$ when there exists $A \in [A]$ and there exists $b \in [b]$ such that $Ax = b$, that is, $[A]x \cap [b] \neq \emptyset$.

The four cases have solutions sets, which may be empty, that we denote Ω_i, for cases $i = 1, \ldots, 4$. Each of the four solution sets of an interval linear system has a unique minimal interval enclosure, the smallest interval box containing the solution set (its interval hull). These interval hulls are denoted $[\Omega_i]$, $i = 1, \ldots 4$.

Moreover, the problem is *inclusion isotonic*. Inclusion isotonic in interval analysis essentialy means that the operations performed on intervals are guaranteed to contain all possible values. In the context of functions we have the following definition, where Π^k is the space of k-tuple $x_1 \times x_2 \times \ldots \times x_k$, $x_j = [\underline{x_j}, \overline{x_j}]$ a real interval.

Definition 1 (*see* [11]) If $F : \mathbb{V} \subseteq \mathbb{I}^m(\mathbb{R}) \to \mathbb{I}^n(\mathbb{R})$ is inclusion isotonic if $x_i \subseteq \hat{x}_i$, then $F(X_1, \ldots, X_m) \subseteq F(\hat{X}_1, \ldots, \hat{X}_m)$.

Remark 2 The Warmus-Sunaga- Moore interval arithmetic, WSMIA [11], sometimes referred to as naive interval arithmetic, and the constraint interval arithmetic (CIA) [7], are examples of interval arithmetic systems that are inclusion isotonic interval extensions of the classical arithmetic of real numbers.

Lodwick and Dubois [9], delineate what each of the interval linear system problems means in the context of CIA and presented methods to find the interval hull solution for each case. This paper presents an optimization approach for each of the four case. It is noted that [16] developed a method for computing the interval hull for Case 1 and focused on its computability of inner solutions (see Sect. 3 for further details). What we have as the method for Case 1, is essentially the Pivkina and Kreinovich approach for Case 1. In addition, algorithms for Cases 2-4 are developed and the transition to fuzzy linear systems, to our knowledge is new. What follows uses a classical 2×2 interval linear system [1] example to illustrate the methods we developed, which we present next.

Example 3 This example is taken from [1].

$$[2, 4]x_1 + [-2, 1]x_2 = [-2, 2]$$
$$[-1, 2]x_1 + [2, 4]x_2 = [-2, 2] \tag{1}$$

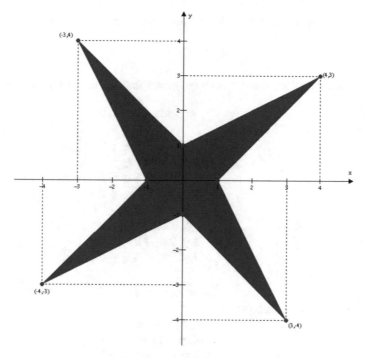

Fig. 1 Case 4 Solution Set for 2 × 2 Example

The depiction of Case 4 corresponding to the interval linear system (1) is given in Fig. 1 (see [1]). From the illustration, it is clear that the interval hull solution to (1) is $[-4, 4] \times [-4, 4]$.

Remark 4 The interval hull is what is called an **outer solution**. For Example 3, we note that the interval $[-1, 1] \times [-1, 1]$ contains points all of which are solutions. This, type of solution is called an **inner solution**. That this so, can be seen from the illustration.

2 Four Cases

2.1 Case 1

This case, solved via linear program, is well-known (see [4]). We restate it here.
    Case 1 for Example 3 is

$$[2, 4]x_1 + [-2, 1]x_2 \subseteq [-2, 2]$$
$$[-1, 2]x_1 + [2, 4]x_2 \subseteq [-2, 2]$$

Remark 5 The symbolic delineation of the general Case 1 problem is

$$\underline{b} \le [A]x \le \overline{b}.$$

Our approach applied to Example 3 generates four linear optimization problems as follows.

$$\underline{x}_i = \min x_i, i = 1, 2$$

Subject to:

$$-2 \le 2x_1 - 2x_2 \le 2$$
$$-2 \le 2x_1 + 1x_2 \le 2$$
$$-2 \le 4x_1 - 2x_2 \le 2$$
$$-2 \le 4x_1 + 1x_2 \le 2$$
$$-2 \le -1x_1 + 2x_2 \le 2$$
$$-2 \le -1x_1 + 4x_2 \le 2$$
$$-2 \le 2x_1 + 2x_2 \le 2$$
$$-2 \le 2x_1 + 4x_2 \le 2$$

$$\overline{x}_i = \max x_i, i = 1, 2$$

Subject to:

$$2 \le 2x_1 - 2x_2 \le 2$$
$$-2 \le 2x_1 + x_2 \le 2$$
$$-2 \le 4x_1 - 2x_2 \le 2$$
$$-2 \le 4x_1 + 1x_2 \le 2$$
$$-2 \le -1x_1 + 2x_2 \le 2$$
$$-2 \le -1x_1 + 4x_2 \le 2$$
$$-2 \le 2x_1 + 2x_2 \le 2$$
$$-2 \le 2x_1 + 4x_2 \le 2$$

The interval hull $[\Omega_1] = [\underline{x}_1, \overline{x}_1] \times [\underline{x}_2, \overline{x}_2]$ for this example is $x_1 \in [-0.5, 0.5]$ and $x_2 \in [-0.5, 0.5]$. The general problem for an $n \times n$ interval linear system $[A]x = [b]$ requires $2n$ linear optimization problem, where each problem has $2n^2$ constraints and is given by

$$m_i = \underline{x}_i = \min x_i$$

Subject to:

$$\underline{b}_i \le \sum_{j=1}^{n} bound(a_{ij})x_j \le \overline{b}_i \qquad\qquad i = 1 \ldots n$$
$$bound(a_{ij}) = \underline{a}_{ij} \text{ or } bound(a_{ij}) = \overline{a}_{ij} \; i = 1 \ldots n, \; j = 1, \ldots, n.$$

$$(2)$$

$$M_i = \overline{x}_i = \max x_i$$

Subject to:

$$\underline{b}_i \le \sum_{j=1}^{n} bound(a_{ij})x_j \le \overline{b}_i \qquad i = 1 \ldots n$$
$$bound(a_{ij}) = \underline{a}_{ij} \text{ or } bound(a_{ij}) = \overline{a}_{ij} \; i = 1 \ldots n, \; j = 1, \ldots, n.$$

(3)

The following theorem holds true.

Theorem 6 *The set $[\Omega_1] = \prod_{i=1}^{n}[m_i, M_i]$ is the interval hull solution for Case 1.*

Proof Note that optimization problems (2) and (3) are continuous problems over compact sets. ∎

Remark 7 In passing, it is noted that the interval hull solution for Case 1 found in [9] Sect. 4.2.2 is not correct. The interval hull solution set $[\Omega_1]$ for Example 3 given by [9] is $x_1 \in [-0.5, 0.5]$ and $x_2 \in [-0.2, 0.2]$. However, the point $(0, 0.5)$, that is, $x_1 = 0$ and $x_2 = 0.5$ is a solution for Case 1 since

$$[2, 4](0) + [-2, 1](0.5) = [-1, 0.5] \subseteq [-2, 2]$$
$$[-1, 2](0) + [2, 4](0.5) = [1, 2] \subseteq [-2, 2].$$

but $(0, 0.5) \notin \Omega_1$ as given by the authors. Their min / max for x_2 is incorrect.

2.2 Case 2

Case 2 for Example 3 is

$$[2, 4]x_1 + [-2, 1]x_2 \supseteq [-2, 2]$$
$$[-1, 2]x_1 + [2, 4]x_2 \supseteq [-2, 2].$$

Remark 8 The symbolic delineation of the general Case 2 problem is

$$\min_{A \in [A], x \in \mathbb{R}} Ax \le \underline{b} \le \overline{b} \le \max_{A \in [A], x \in \mathbb{R}} Ax.$$

Case 2 for Example 3, the problem is infeasible, that is no solution exists. Given the structure of the constraints, it seems that infeasibility will often occur, perhaps, almost always.

Proposition 9 *Consider the interval linear system $[A]x = [b]$, with n rows and n columns. If $0 \in [b]$, $[b] \ne [0, 0]$ and there does not exists a singular matrix $A \in [A]$, such that $Ax = 0$, the solution set $\Omega_2 = \emptyset$.*

Proof Suppose that there exists $x \in \Omega_2$, that is Ω_2 is not empty. Then, there exist $A \in [A]$ such that $Ax = 0$. Since $[b] \ne [0, 0]$, there exists $b^* \in [b]$, $b^* \ne 0$ this implies that $x \neq 0$. Recall that $\min_{A \in [A], x \in \mathbb{R}} Ax \leq \underline{b} \leq \overline{b} \leq \max_{A \in [A], x \in \mathbb{R}} Ax$. Therefore, A is a singular matrix, which is a contradiction. Thus, $\Omega_2 = \emptyset$ as was to be shown. ∎

2.3   Case 3

$$\Omega_3 = \Omega_1 \cap \Omega_2$$

so that

$$[\Omega_3] = [\Omega_1 \cap \Omega_2] \subseteq [\Omega_1] \cap [\Omega_2].$$

Remark 10 The symbolic delineation of the general Case 3 problem is

$$[A]x \subseteq [b] \text{ and } [b] \subseteq [A]x.$$

Case 3 for Example 3 is

$$[\Omega_3] = \emptyset.$$

Proposition 11 *Consider the interval linear system* $[A]x = b$ *with* n *rows and* n *columns. If* $b \neq [0, 0]$ *and* $[A]$ *has one or more rows all of whose coefficients are positive width intervals, then the solution set is empty.*

Proof Suppose that there exists $x \in \Omega_3$. Let kth be the row none of whose coefficients are zero width intervals (real numbers), that is, such that $[a_{kj}]$ is not a real number for all $j = 1, \ldots n$. Since $[b] \neq [0, 0]$, there exists $l \in \{1, 2, \ldots, n\}$ such that $x_l \neq 0$. Let $A^* \in [A]$ be such that $a_{kl}^* = \underline{a}_{kl}$. As $x \in \Omega_3$ we have $A^*x = b$. Let $A^{**} \in [A]$ be such that $a_{ij}^{**} = a_{ij}^*$ for all $i = 1, \ldots n$, $j = 1, \ldots n$, but $a_{kl}^{**} = \overline{a}_{kl}$. Therefore, we have $A^{**}x \neq A^*x$ since $x_l \neq 0$ and $a_{kl}^{**} \neq a_{kl}^*$. So, $A^{**}x \neq b$, which implies that $x \notin \Omega_3$, which is a contradiction. ∎

2.4   Case 4

Case 4 for Example 3 is

$$([2, 4]x_1 + [-2, 1]x_2) \cap [-2, 2] \neq \emptyset$$
$$([-1, 2]x_1 + [2, 4]x_2) \cap [-2, 2] \neq \emptyset$$

Remark 12 The symbolic delineation of the general Case 1 problem is

$$[A]x \cap [b] \neq \emptyset.$$

The interval hull for Case 4 requires the solution to the following two sets of optimization problems.

$$m_i = \underline{x}_i = \min x_i$$

Subject to:

$$\sum_{j=1}^{n} \alpha_{ij} x_j = \beta_i \qquad i = 1 \ldots n$$
$$\underline{a}_{ij} \le \alpha_{ij} \le \overline{a}_{ij} \quad i = 1 \ldots n, j = 1 \ldots n. \tag{4}$$
$$\underline{b}_i \le \beta_i \le \overline{b}_{ij} \qquad i = 1 \ldots n.$$

$$M_i = \overline{x}_i = \max x_i$$

Subject to:

$$\sum_{j=1}^{n} \alpha_{ij} x_j = \beta_i \qquad i = 1 \ldots n$$
$$\underline{a}_{ij} \le \alpha_{ij} \le \overline{a}_{ij} \quad i = 1 \ldots n, j = 1 \ldots n. \tag{5}$$
$$\underline{b}_i \le \beta_i \le \overline{b}_{ij} \qquad i = 1 \ldots n.$$

Example 3 was presented by [9], which needs the solution to the following four optimization problems.

$$m_i = \underline{x}_i = \min x_i, i = 1, 2$$
$$a_{11}x_1 + a_{12}x_2 = b_1$$
$$a_{21}x_1 + a_{22}x_2 = b_2$$
$$2 \le a_{11} \le 4$$
$$-2 \le a_{12} \le 1$$
$$-1 \le a_{21} \le 2$$
$$2 \le a_{22} \le 4$$
$$-2 \le b_1 \le 2$$
$$-2 \le b_2 \le 2$$

$$M_i = \overline{x}_i = \max x_i, i = 1, 2$$
$$a_{11}x_1 + a_{12}x_2 = b_1$$
$$a_{21}x_1 + a_{22}x_2 = b_2$$
$$2 \le a_{11} \le 4$$
$$-2 \le a_{12} \le 1$$
$$-1 \le a_{21} \le 2$$
$$2 \le a_{22} \le 4$$
$$-2 \le b_1 \le 2$$
$$-2 \le b_2 \le 2$$

The interval hull solution for Case 4 is $[\Omega_4] = [-4, 4] \times [-4, 4]$, which corresponds to that given by [9] . The general problem for an $n \times n$ interval linear system $[A]x = [b]$ requires $2n$ linear optimization problem, where each problem has $2n + n^2$ constraints and given by Eqs. (4) and (5).

The following theorem holds true.

Theorem 13 *The set* $[\Omega_4] = \prod_{i=1}^{n}[m_i, M_i]$ *is the interval hull solution for Case 4.*

Proof Note that optimization problems, Eqs. (4) and (5). are continuous problems over compact sets.   ∎

3   The Pivkina/Kreinovich Algorithm

Pivkina and Kreinovich in 2019 (see [16]) proposed a method to compute the solution to a little more general problem to our Case 1. Our Case 1 is a subset of their *tolerance solution* method to linear interval equations. Rohn (see [18–20]) and Shary (see [21–24]) called a solution to Case 1 a *tolerant solution*. Their approach is essentially what we presented for the Case 1 method. Pivkina and Kreinovich prove that to find whether the solution set to Eqs. (2) and (3) is polynomial in time. Moreover, if the solution set is nonempty finding a solution is also polynomial time. Moreover, as can be seen from Eqs. (2) and (3), the solution set is a convex polyhedron. When a tolerance solution exists, there are many values in the solution set. They are primarily interested in finding inner solutions. The remainder of the [16] paper deals with finding "optimal" solutions given various criteria some of which are NP-Hard. Our interest is in computing interval outer solutions (the interval hull) in computationally efficient ways.

4   Fuzzy Linear Systems

This section assumes that the reader is familiar with fuzzy set theory at least at an elementary level. The fuzzy linear system problem

$$\tilde{A}x = \tilde{b},$$

where $\tilde{A}$ and $\tilde{b}$ are fuzzy intervals, requires the solution to the alpha level interval linear system problem, one for each alpha. $\alpha \in [0, 1]$. Typically, one discretizes the alpha levels and interpolates between the succeeding alpha levels. Thus, if there are m discretizations of alpha levels, $m(2n)$ optimizations each with with $2n^2$ constraints for an $n \times n$ fuzzy interval linear system. When all coefficients are triangular fuzzy intervals, one interval linear system problem needs to be done for alpha-level 0 plus one real linear system problem for alpha level 1. If all the coefficients are trapezoidal fuzzy intervals, two interval linear system problems need to be done - one for alpha-level 0 and one for alpha-level 1. In each case one interpolates between alpha level 0 and alpha level 1. Thus, for these two cases, our methods are possible. This approach is applied to all four cases.

4.1 Example

To illustrate this let us assume that the coefficients of the fuzzy interval linear system is composed of triangular fuzzy numbers. While it is assumed that the reader has some familiarity with fuzzy set theory and fuzzy numbers, we mention two concepts. The core of a fuzzy interval (number) is the interval associated with the alpha level one. The support of the fuzzy interval is the closure of alpha level zero. Let us assume that we just have one coefficient that is a triangular fuzzy number.

Example 14 Let the fuzzy linear system

$$\tilde{2}x + 2y = 4$$
$$x - y = 0$$
$$x, y \geq 0$$

where $\tilde{2} = 1/2/4$, a triangular fuzzy number "a little bigger than 2" and illustrated below. Here the notation $1/2/4$ means a triangular number, whose support is $[1, 4]$ and whose core is 2 (see Fig. 2). We have one interval-valued linear system and one real-valued system to solve in the first quadrant. The one interval linear equation is

$$[1, 4]x + 2y = 4,$$
$$x - y = 0, \tag{6}$$
$$x, y \geq 0, \tag{7}$$

and the real valued linear equation is,

$$2x + 2y = 4,$$
$$x - y = 0, \tag{8}$$
$$x, y \geq 0. \tag{9}$$

The solution to the real valued system (Eq. 8) can be seen to be $(1, 1)$. This problem can be solved without use of the algorithms. For the system (6), note that the second equation results in $x = y$, so that we are solving, since x is non-negative,

$$[1, 4]x + y = [1, 4]x + x = 4$$
$$[x, 4x] + [x, x] = [2x, 5x] = 4.$$

Case 1: $[2x, 5x] \subset [4, 4]$, and so $\Omega_1 = \emptyset$.

Case 2: $[2x, 5x] \supset [4, 4]$. $\Rightarrow 2x \leq 4$ or $x \leq 4$, and $5x \geq 4$ or $x \geq \frac{4}{5}$. Thus, the solution is $\left[\frac{4}{5}, 2\right]$.

Case 3: This case is empty since Case 1 is empty.

Case 4: $[2x, 5x] \cap [4, 4] \neq \emptyset$. In this case, since the right side is a real number, the solution is the same as that of Case 2.

Fig. 2 Fuzzy set 1/2/4

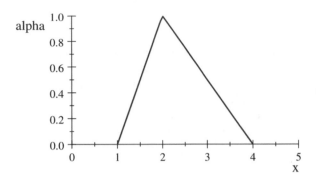

Thus, the fuzzy linear system solution for Case 2 and 4 is obtained as follows. Since the support, alpha level 0, of the fuzzy triangular interval solution is $\left[\frac{4}{5}, 2\right]$ and the alpha level 1 solution is $(1, 1)$, and given the fact that we have a linear problem, the fuzzy interval solution is the triangular fuzzy number $\frac{4}{5}/1/2$.

Example 15 Let $\tilde{A}x = \tilde{b}$ be the 2×2 system

$$\tilde{a}_{11}x_1 + \tilde{a}_{12}x_2 = \tilde{b}_1$$
$$\tilde{a}_{21}x_1 + \tilde{a}_{22}x_2 = \tilde{b}_2$$

where

$$\tilde{a}_{11} = 2/3/4, \tilde{a}_{12} = -2/-\frac{1}{2}/1, \tilde{b}_1 = -2/0/2$$
$$\tilde{a}_{21} = -1/\frac{1}{2}/2, \tilde{a}_{22} = 2/3/4, \tilde{b}_2 = -2/0/2.$$

In this case we have one interval problem

$$[2, 4]x_1 + [-2, 1]x_2 = [-2, 2]$$
$$[-1, 2]x_1 + [2, 4]x_2 = [-2, 2]$$

and one real-valued problem

$$3x_1 - \tfrac{1}{2}x_2 = 0$$
$$\tfrac{1}{2}x_1 + 3x_2 = 0$$

whose solution is $x_1 = 0$, $x_2 = 0$. We have solved the interval problem already.
 Case 1: $x_1 = \left[-\frac{1}{2}, \frac{1}{2}\right]$, $x_2 = \left[-\frac{1}{2}, \frac{1}{2}\right]$. Thus, the fuzzy solutions is, $x_1 = -\frac{1}{2}/0/\frac{1}{2}$, $x_2 = -\frac{1}{2}/0/\frac{1}{2}$.
 Case 2, 3: Empty
 Case 4: $x_1 = [-4, 4]$, $x_2 = [-4, 4]$. Thus, the fuzzy solutions is, $x_1 = -4/0/4$, $x_2 = -4/0/4$.

4.2 General Fuzzy System Case for Triangular and Trapezoidal Fuzzy Interval Coefficients

4.2.1 Triangular Fuzzy Interval

A fuzzy linear system problem all of whose coefficients are triangular fuzzy numbers results in a problem which is solved as follows. Solve the problem where the coefficients are the interval supports of the fuzzy triangular number. The solution of this interval linear system becomes the support of the fuzzy linear system. Then, take the alpha level one of each of the fuzzy triangular linear systems. The solution to his real-valued linear system become the level one value of the solution of the fuzzy interval system. That this is so, comes from the fact that the problem is linear and that the solution is continuous in the values of the support of the coefficients.

4.2.2 Trapezoidal Fuzzy Interval

The trapezoidal fuzzy interval coefficient case is similar to the triangular fuzzy interval case except the alpha-level 1 is now an interval. Thus, there are two interval linear systems to solve, one for the support alpha-level 0 and one for the alpha-level 1. In this case, the support of the fuzzy solution, its alpha-level 0, is the solution of the alpha-level 0 of the interval linear system problem. The alpha level-1 fuzzy solution is the alpha-level 1 of the interval linear system problem. In this case, we have an interval solution for level 0 and an interval solution for level 1 resulting in a trapezoidal fuzzy interval solution to the fuzzy trapezoidal interval linear system.

Example 16 Let $\tilde{A}x = \tilde{b}$ be the 2×2 systems

$$\tilde{a}_{11}x_1 + \tilde{a}_{12}x_2 = \tilde{b}_1$$
$$\tilde{a}_{21}x_1 + \tilde{a}_{22}x_2 = \tilde{b}_2$$

where (here $a/b/c/d$ denotes the trapezoidal interval number whose support is $[a, b]$ and whose core is $[b, c]$).

$$\tilde{a}_{11} = 2/\frac{5}{2}/\frac{7}{2}/4, \tilde{a}_{12} = -2/-\frac{1}{2}/\frac{1}{2}/1, \tilde{b}_1 = -2/-1/1/2$$
$$\tilde{a}_{21} = -1/-\frac{1}{2}/\frac{1}{2}/2, \tilde{a}_{22} = 2/\frac{5}{2}/\frac{7}{2}/4, \tilde{b}_2 = -2/-1/1/2.$$

This yields two interval linear systems

$$[2, 4]x_1 + [-2, 1]x_2 = [-2, 2]$$
$$[-1, 2]x_1 + [2, 4]x_2 = [-2, 2]$$

and

$$[\tfrac{5}{2}, \tfrac{7}{2}]x_1 + [-\tfrac{1}{2}, \tfrac{1}{2}]x_2 = [-1, 1]$$
$$[-\tfrac{1}{2}, \tfrac{1}{2}]x_1 + [\tfrac{5}{2}, \tfrac{7}{2}]x_2 = [-1, 1]$$.

The solution to the first interval linear systems has already been calculated.

Case 1: $x_1 = \left[-\tfrac{1}{2}, \tfrac{1}{2}\right], x_2 = \left[-\tfrac{1}{2}, \tfrac{1}{2}\right]$..

Case 2, 3: Empty Case 4: $x_1 = [-4, 4], x_2 = [-4, 4]$.

The solution to the second interval linear systems his the following:

Case 1: $x_1 = [-0.2857, 0.2857], x_2 = [-0.2857, 0.2857]$..

Case 2, 3: Empty

Case 4: $x_1 = [-0.5, 0.5], x_2 = [-0.5, 0.5]$.

5 Conclusions

This research indicated how to compute the interval hull for the four types of solutions of interval linear system problem. The complexity of the methods using interval optimization is $O(n)$ where each optimization problem has an order $O(n^2)$ number of constraints. Since the fuzzy interval linear system problem involves alpha levels, it can be solved according to our method for each alpha level. If all the coefficients and right-hand sides are trapezoidal fuzzy intervals, only two alpha levels need to be considered, the alpha level zero and alpha level 1.

Dedication to Janusz Kacprzyk The pioneering work of Dr. Janusz Kacprzyk includes major contribution to optimization theory especially in fuzzy optimization and dynamic programming. Dr. Kacprzyk's contribution extends well beyond optimization. It is especially noted that Kacprzyk was one of the first to develop and apply fuzzy optimization and stands among the original developers of the field. While Dr. Kacprzyk's mathematical contributions are not only significant, but wider than optimization, it is also as an Editor for Springer-Verlag that his influence has impacted not only the field, but also given voice to researchers from across the globe that often are not heard. Dr. Kacprzyk is an amazing person. It is with great admiration of the work of Dr. Kacprzyk that we offer this article in dedication to his contributions.

Acknowledgements We would like to thank the Brazilian research agencie FAPESP for their financial support (Grant number 2021/03269-7).

Appendix—Python Code

The software that was used to solve each of the four cases was code by Victor Hugo Godoy Pedrota. However, it should be noted that this software is provided purely for testing out examples and not designed to be production code and it is not guaranteed to be correct. The link to the Python code is https://github.com/vpedrota/Notebooks/tree/main/PULP.

References

1. Barth, W., & Nuding, E. (1974). Optimale losung von intervallgleichungssystemen. *Computing, 2*, 117–125.
2. Hansen, E. (1968). On solving systems of equations using interval arithmetic. *Mathematics of Computation, 22*, 374–384.
3. Hansen, E. (1969). On linear algebraic equations with interval coefficients. In E. Hansen (Ed.) *Topics in interval analysis* (pp. 35–46). Oxford University Press.
4. Kreinovich, V., Lakeyev, A. V., Rohn, J., & Kahl, P. (1998). *Computational complexity and feasibility of data processing and interval computations, Applied optimization, Chapter 10.* Dordrecht, The Netherlands: Kluwer Academic Publishers.
5. Kupriyanova, L. V. (1995). Inner estimation of the united solution set of interval linear algebraic system. *Reliable Computing, 1*(1), 15–31.
6. Lodwick, W. A. (1990). Analysis of structure in fuzzy linear programs. *Fuzzy Sets and Systems, 38*, 15–26.
7. Lodwick, W. A. (1999) Constrained interval arithmetic. *CCM Report,* **138**
8. Lodwick, W. A. (2007) Interval and fuzzy analysis: An unified approach. In P. W. Hawkes (Ed.)*Advances in imagining and electronic physics* (Vol. 148, pp. 75–192). Elsevier Press
9. Lodwick, W. A., & Dubois, D. (2015). Interval linear systems as a necessary step in fuzzy linear systems. *Fuzzy Sets and Systems, 281*, 227–251.
10. Moore, R. E. (1966). *Interval analysis.* Prentice-Hall.
11. Moore, R. E., Kearfott, R. B., & Cloud, M. J. (2009). *Introduction to interval analysis.* Society for Industrial and Applied Mathematics.
12. Neumaier, A. (1984). New techniques for the analysis of linear interval equations. *Linear Algebra, 58*, 273–325.
13. Neumaier, A. (1990). *Interval methods for systems of equations.* Cambridge University Press.
14. Oettli, W. (1965). On the solution set of linear system with inaccurate coefficients. *SIAM Journal on Numerical Analysis, 2*, 115–118.
15. Oettli, W., & Prager, W. (1964). Computability of approximate solution of linear equations with given error bounds for coefficients and right-hand sides. *Numeriche Mathematik, 6*, 405–409.
16. Pivkina, I., & Kreinovich, V. (2019). Finding least expensive tolerance solutions and least expensive tolerance revisions: Algorithms and computational complexity. *International Journal of Intelligent Technologies and Applied Statistics, 12*(2), 151–168.
17. Rohn, J. (1982). An algorithm for solving interval linear systems and inverting interval matrices. *Freiburger Intervall-Berichte, 82*(5), 23–36.
18. Rohn, J. (1986). A note on solving equations of the type $A^I x = b^I$,. *Freiburger Intervall-Berichte, 84*(6), 29–31.
19. Rohn, J. (1986). Solving interval linear systems. *Freiburger Intervall-Berichte, 84*(7), 1–14.
20. Rohn, J. (2003). Solvability of systems of linear interval equations. *SIAM Journal on Matrix Analysis and Applications, 25*(1), 237–245.
21. Shary, S. P. (1991). Optimal solution of interval linear algebraic systems. *Interval Computations, 2*, 7–30.
22. Shary, S. P. (1992). On controlled solution set of interval algebraic systems. *Interval Computations, 4*(6), 66–75.
23. Shary, S. P. (1994). Solving the tolerance problem for interval linear equations. *Interval Computations, 2*, 4–22.
24. Shary, S. P. (2002). A new technique in systems analysis under interval uncertainty and ambiguity. *Reliable Computing, 8*, 321–418.

Fuzzy-Petri-Networks in Supervisory Control of Markov Processes in Robotized FMS and Robotic Systems

Georgi M. Dimirovski, Yuanwei Jing, Jindong Shen, Kun Wang, Dilek Tukel, Figen Ozen, Gorjan Nadzinski, and Dushko Stavrov

Abstract The prospect applications of results and techniques of Petri-net and fuzzy-Petri-net formalisms is surveyed from the points of view of Systems Science and Engineering via discrete-event Markov processes and systemic structures, which are pertinent to applications in Robotized Manufacturing Engineering. The stand point of this analysis is the one of Control and Decision Sciences, and the general mathematical language utilized is the compound one consisted of directed graphs, Markov sequences, and set operations. The concepts of event-driven and time-driven

Honoring Professor Janusz Kacprzik, Academician of Polish Academy of Sciences and several other European Academies, for his many contributions to Computational Intelligence, Fuzzy Systems, Robotics and Operations Research.

G. M. Dimirovski (✉)
Doctoral School FEIT of St Cyril, St Methodius University, 18 Rugjer Boskovic Str, P.O. Box 574, MKD-1000 Skopje, Macedonia
e-mail: dimir@feit.ukim.edu.mk

Y. Jing · J. Shen · K. Wang
College of Information Science and Engineering, Northeastern University, Shenyang 110004, Liaoning, P. R. China
e-mail: jingyuanwei@ise.neu.edu.cn; ywjjing@mail.neu.edu.cn

D. Tukel
Department of Software Engineering, Dogus University, NATO Yolu Cd, No: 265, TR-34775 Umranie/Istanbul, Turkey
e-mail: dtukel@dogus.edu.tr

F. Ozen
Electrical and Electronics Engineering Department, Halic University, 15 Temuz Sehitler Cd. No: 14/12, TR-34060 Eyupsultan/Istanbul, Turkey
e-mail: figenozen@halic.edu.tr

G. Nadzinski · D. Stavrov
Faculty of Electrical Engineering and Information Technologies, St Cyril and St Methodius University in Skopje, Skopje, North Macedonia
e-mail: gorjan@feit.ukim.edu.mk

D. Stavrov
e-mail: d.stavrov@feit.ukim.edu.mk

© The Author(s), under exclusive license to Springer Nature Switzerland AG 2023
Y. P. Kondratenko et al. (eds.), *Artificial Intelligence in Control and Decision-making Systems*, Studies in Computational Intelligence 1087,
https://doi.org/10.1007/978-3-031-25759-9_2

sequences, being both real-world phenomena and mathematical objects, are found to be rather instrumental for development of representation models as well as control and communication algorithms that enable system design analysis with regard to higher, task-oriented decision and control level. Exposition on operating robotic and flexible manufacturing systems, via typical case-studies, follows an overview on the very concept of a system in general based on the system triple-paradigm comprising a conceptual and a representation model as well as real-world object these are aimed to mimic. A brief appropriate overview on Petri-net formalism and a similar overview on the feasible synergies of fuzzy-systems and Petri-nets are given. The resulting model representations and task-control algorithms based on case-study plants in robotized flexible manufacturing systems provide relevant illustrations.

1 Introduction

Science is the measure of all things. ... And we humans measure ideas against science.—Bart Kosko

In industrial manufacturing engineering, either in problems of design and development or in implementation and operation, the underlying nature of the issues involved require to deal with discrete either stochastic or event driven dynamical processes, or both simultaneously. From the view points of control, co-ordination supervision and management decisions, the information flows within the overall system take on the simultaneous signal processes and so are task and control sequences; satisfying mathematically so-called *Markovian* property [1–14]. Markov property guarantees the next state of the system does depend on the previous one only, and not on the entire operating pre-history. Markovian property is assumed satisfied always by systems and networks of concern in this study. Thus realistic treatment of the actual real-world system, its conceptual model and its mathematical representation model(s) is feasible: the triple paradigm of the very system concept [15, 16] is observed (Fig. 1a, b).

There are depicted in Fig. 2a, b two such systems in disciplines of Manufacturing, Mechatronics, and Robotics Engineering. It should be noted, solely their conceptual models, mimicking the respective real-world objects or plants, are seen there and not any of the relevant mathematical models. For, in systems engineering terms, any kind of a system is a composite structure involving at least a quintuple object [7, 18–25], [116, 122] (see Fig. 1): a space of *admissible input variables* (e.g., control and/or management decisions); a space of *feasibly attainable state variables* (e.g., plant process variables); a space of *sustainable output variables* (e.g. measurable process, measured observations and/or performance evaluations); a *functional mapping of current admissible inputs* (controlling actions) and *feasible states into future attainable states* (implying two components of state evolution); and another *functional mapping of current states into sustainable outputs* (system reaction to the forcing inputs and the current state transition).

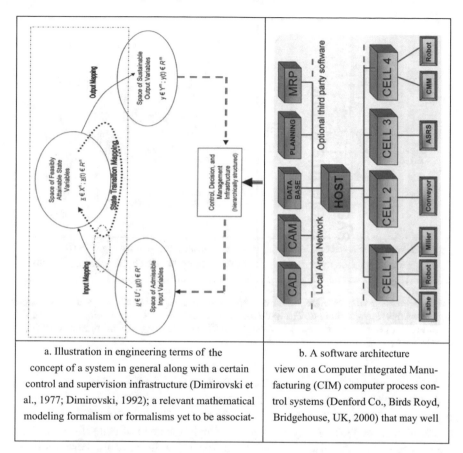

a. Illustration in engineering terms of the concept of a system in general along with a certain control and supervision infrastructure (Dimirovski et al., 1977; Dimirovski, 1992); a relevant mathematical modeling formalism or formalisms yet to be associat-	b. A software architecture view on a Computer Integrated Manufacturing (CIM) computer process control systems (Denford Co., Birds Royd, Bridgehouse, UK, 2000) that may well

Fig. 1 For the purpose of precise definition of the concepts of a system, control, coordination and supervision in compliance with the fundamental natural laws of physics one ought to observe the next two illustrations carefully Fig.1a presents an illustration, in engineering terms, the concept of a system in general along with a certain control and supervision infrastructure [15] (Dimirovski 1992); a relevant mathematical modeling formalism or formalisms yet to be associated accordingly, Dimirovski et al. [17]. In Fig.1b on the other hand depicts a software architecture view on a Computer Integrated Manufacturing (CIM) computer process control systems (Denford Co., Birds Royd, Bridgehouse, UK, 2000) that may well correspond to Fig. 1a in principle

In what follows further down in here, there shall be needed a certain closer insight to both: the (sub)system that is the object of control in the wider sense of the word; and also to the (sub)system that comprises functions of control, management decision and coordinating supervision. That is, deeper insights into the very structure of what is the plant dynamical subsystem and controlling infrastructure dynamical subsystem. However, regardless the rather intuitively clear, tangible concepts of event-driven sequences, related to conditions and events, and time-driven sequences, related to variable changes with respect to time, their implication in systems theory and in systems engineering is rather deep and complex. Also, it should be noted too that

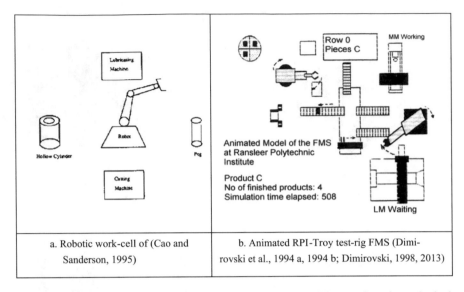

| a. Robotic work-cell of (Cao and Sanderson, 1995) | b. Animated RPI-Troy test-rig FMS (Dimirovski et al., 1994 a, 1994 b; Dimirovski, 1998, 2013) |

Fig. 2 Towards precise definition of the concepts of robotized flexible manufacturing and robotic systems in this research study on Fuzzy-Petri-Network Supervisory Control notice: Fig.2a depicts the structure of Robotic work-cell of [26], while Fig.2b describes on scene of Animated RPI-Troy test-rig FMS [18, 19, 27, 28]

in here the word is about sequences precisely because of the classes of systems and control and decision problems that are in the focus of this study. Lastly but not least, it should be noted that, on one hand, the uncertainty and the involved presence of intuitive concepts and/or qualitative data imply the need of using fuzzy systems based techniques, and discrete event dynamics imply the need for Petri-net based techniques, on the other. This one of the reasons that gave rise to the concepts and techniques involved in fuzzy-Petri-net formalism and justifies its applications.

2   Plant Objects and Processes of Control as Controlled General Systems

Science, thanks to its links with the observation, retains some title to a correspondence theory of truth. Coherence of theory is evidently a lot of ethics.—Willard von Orman Quine

A kind of a summary, within the context discussed above, may well be given by formally defining the very category of a system in engineering terms. It may well be found to be a rather involved composite concept, and yet it must be in consistence with both the natural laws physics (conservation of energy, mass and momentum) and the general systems theory (two mapping processors linking the spaces of input, state and output signals) that possesses a semi-group property, thus belonging to an abstract algebra. Namely, the concept of a system can be understood as a goal-oriented composition of interacting components with a purpose that is subject to certain structural organization and constraints. The formal mathematical definition is even much more involved hence during the past century a number of definitions have been put forward over time [1, 4, 8, 9, 15, 29–41].

For the purpose of the present research endeavor, the system concept can be defined by means of a "tuple" of mathematical objects [1, 3, 9, 15, 16, 30, 37, 39, 42] as described in the system-theoretic equation:

$$\Sigma =< U, \ X, \ Y, \ \varphi, \ \eta, \ T >$$ (1)

The quantities in this set-theoretic Eq. (1) denote:
U—is a set space of admissible input variables;
X—is a set space of feasibly attainable state variables;
Y—is a set space of sustainable output variables;
φ—is a functional mapping (i.e. mechanism) of admissible inputs and attained states to (the next) feasible states

$$\phi : \ U \times X \times T \rightarrow X \times T;$$ (2)

φ—functional mapping of attained states to sustainable outputs,

$$\eta : \ X \times T \rightarrow Y \times T;$$ (3)

T—the set of non-negative numbers representing the time, real-valued in continuous-time and integers in discrete-time cases.

Finally, in order to provide an insight on how PN model representation become complicated when modeling controlled discrete event operation that emerges in the case of more comprehensive FMS plant, the next Fig. 3 depicts the PN-based model for the entire RPI-Troy FMS depicted regardless whether 'Fuzzy" or 'Stochastic' formalism is associated with. On the other hand, notice that Fig. 2b depicts slice-image of the animated simulation sequence of slice-images for this plant in operation, via an appropriate visualization by making use of Visual Basic as means of constructing appropriate animation graphic software [20, 43].

Notice the delicate underlying nature of the internal, state-transition, functional mapping that is embedded into or coupled with the input mapping (more detail is found in references on system theory [3, 15, 38, 39, 41] and theory of systems modeling [1, 9, 37, 42, 44]) due to actual dynamical phenomena involved. Also note that this definition satisfies the fundamental first principles on preservation of energy, mass and momentum [18, 19, 27]. Should in the given case study the parameters and properties of the system do not vary with time, which is the issue of information [15, 45], then the real-valued time set T would not appear explicitly. Moreover, the above framework model of a system is given in terms of set spaces and functional mapping and graphs [1, 20, 39, 43], hence admitting various mathematical

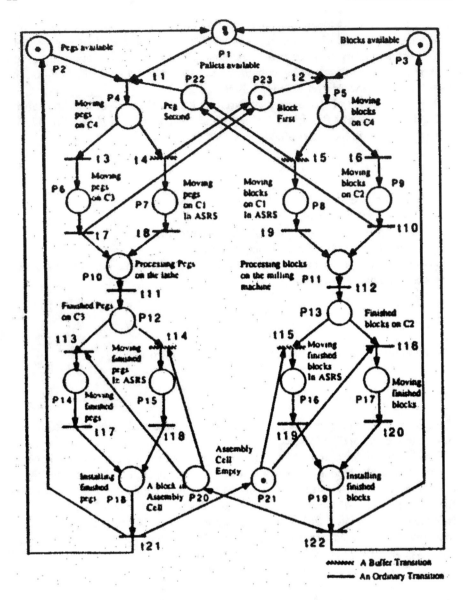

Fig. 3 The PN model representation of the well-known RPI-Troy medium-size FMS, the animated simulation sequence of which is depicted in Fig. 2b [20, 43]

formalisms to be introduced at a later stage; that is, during the needed more detailed deviation of model representation(s) [9, 16, 20, 38, 42, 44, 46–54]. Lastly but not least, for further representation modeling of other attributes or mappings that may be pertinent to a given case study, the compound concept (1) of a general system admits simple inclusion of other mathematical objects or modification of the ones present to capture the features of that particular case study provided these are precisely defined [16, 44, 55].

3 Generalized and IPDI Complex Networks and Systems: Control, Decision and Management

Human spirit does not get satisfied by any kind of material gain.—Lucius Eneus Senecca

In Fig. 2 there are depicted schematics of a robotic peg-cylinder assembly system (a), and a medium-size FMS test-rig of RPI-Troy, NY (b). By their underlying nature of their functioning these both are discrete-event sequential systems and so are their controls. Transients that take place within the individual component machines are taken care of by local servo and/or regulatory controls. On the level of individual machines however local controls can be and are designed solely by using mathematical model representations that provide considerable precision. As a matter of fact the highest precision (and lowest machine intelligence) is associated with machines and robots precisely according to the principle of *Increasing Precision with Decreasing Intelligence* (**IPDI**) for intelligent (in machine sense) engineering systems, a 1988, 1989 discovery due to George N. Saridis [49, 50].

In such compound systemic structures, tasks of control, management decision, and coordinating supervision [16, 46, 56], which are essentially information based [15, 40, 41, 45, 57], are now readily envisaged to reduce to finding the systemic constructs and mechanisms (even man-machine ones), referred to as *control, decision and management* (for short, *controlling*) *infrastructure* in Fig. 1, that will generate a system input space with admissible input variables only. And, the admissible inputs are the solely ones to be applied on the system subject to control and/or management. By and large, this is precisely the essence of industrial and manufacturing systems engineering, and methodologies from control and decision theories, operations research and computational intelligence overlap in order to provide such constructs and mechanisms [18, 27, 30, 35, 43, 52, 53, 55, 58]. These, in turn, are most often implemented by means of hierarchical computer process control architectures, a typical supervisory-level computing process being depicted in Fig. 3. And also, by and large, the fundamental concept employed is the one of task sequence planning and its controlled execution to drive the predominantly discrete-event dynamics of the plant through a sequence of most desired (i.e. optimum in the sense of certain criterion or criteria) states as to perform the desired plan of activities [5, 18–20, 26, 27, 43, 47, 59–63]. In this paper, however, the discussion of the underlying aspects and issues, which are most essential in manufacturing production processes, performed done via examples of a robotic work-cell (RWC) and a flexible manufacturing system (FMS), and task sequence planning and supervisory control problems [16, 18, 19, 27, 35, 47, 59, 64].

It should be noted further that, from the view point of developing an appropriate system engineering design application, the control, decision and management infrastructure can be further hierarchically structured into levels and/or layers to allow for: (a) combined application of compatible formalisms and functional algorithms; (b) integrations of control and supervision functions in appropriate relation to management decision and supervision; and (c) developing intelligent control systems infrastructure according to the IPDI principle (originally introduced and proved by Saridis [49, 50]). For, this outline complies with the underlying fundamental information-based laws of Ashby [29, 57, 65]. Then tools of applied computational intelligence [13, 16, 19, 20, 22, 28, 40, 44, 47, 59, 66–78] for control and decision can be employed to the full [17, 25].

The system controlling infrastructure is emphasized to be a system on its own, describable in terms of a further structured general systems model, a hierarchical and *'larger tuple'* of mathematical objects and mappings as it will be seen in the sequel, implementing functions of control, management decision and coordination supervision. When coupled with the controlled object system these enter into a kind of symbiosis and together create a new system with new spaces of admissible inputs, feasible states and sustainable outputs and functional mapping having properties not present in neither the object system nor the controlling infrastructure system. It is therefore that processes in the object system are forced to follow a different state evolution trajectory, other than without control, as to implement the desired performance and goals [18, 20, 27, 28, 44, 47, 79]. It should be noted, however, controlling infrastructure system has to be hierarchically organized even when it is not designed strictly according to the IPDI principle, and yet alone whet it is. In FMS and robotic systems, on one hand, functions of regulatory and servo controls and of coordination supervision are closely related so that they constitute a control infrastructure level, which is usually referred to as *'lower control level'* and typically comprises two layers implementing these two classes of control and supervision. On the other hand, since management decision involves production planning and optimization, which is based on functions of forecasting and control, and also involves supervision coordination to implement both of these, in the overall controlling infrastructure there exists another upper control level usually referred to as *'higher control level'*. In many respects, functions of forecasting and control constitute the principal essence of this level, so to say—the brain of the higher control level, production planning being its beating heart. It also comprises at least two layers. It may well be stated 'two-level with two-layers' is a necessary condition degree of organization for an well designed overall controlling infrastructure for a complex general system to be controlled.

4   Emulation Simulation of Robotized FMS and Robotic Systems

World is limited and mind is unbounded. Thoughts easily live next to each other, but things painfully collide in space.—Friedrich Schiller

We address briefly in this the issue of developing emulation simulation models and software for analysis and design of FMS and robotic systems since it is the only feasible and efficient approach to handle systems engineering design of such technologically complex systems [2, 5, 25, 35, 39, 41, 46, 47, 49, 55, 59, 69, 80]. In the category context of systems as discussed here in the introduction, the basic system model for emulation simulation of FMS and robotic systems, defined according to Ziegler's DEVS formalism [24, 81], can be established in terms of the following system concept of a *'septimus-tuple'* mathematical structure [16, 20, 47, 51]:

$$M = <X,\ S,\ Y,\ \delta\mu_{\text{int}},\ \delta\mu_{ext},\ \lambda,\ t_a>, \tag{4}$$

The quantities in Eq. (4) denote:

X is a set of external (input) event types;
S is a set of sequential states;
Y is a set of output event types;

The internal transition mapping is

$$\delta\mu_{\text{int}} : S \to S, \tag{5}$$

The external transition mapping is

$$\delta\mu_{ext} : Q \to S \tag{6}$$

Q—is a complete set of states with consumption, that is, a set

$$Q = \{(s,\ e)|\quad s \in S,\ 0 \le e \le t_a(s)\}; \tag{7}$$

The time increment function is

$$t_a : Q \to R^+_{0,\ \infty}; \tag{8}$$

The output mapping is

$$\lambda : S \to Y. \tag{9}$$

At this point of discussing the emulation simulation, it is pointed out that in subsystem model representations of FMS and robotic systems enter into extensive mutual communications according to the overall control strategy, that is under supervision of controlling infrastructure [36, 41, 82]. It should be noted further that the above system representation is fully consistent with the system science theoretic fundamentals of general systems [18, 23, 38], and not solely to the dominating discrete-event dynamics of FMS and robotic system operation. It is also in full consistence with the so-called *object-oriented modeling* (OOM) paradigm in development of application software aimed at emulation simulation [8, 34, 35, 41, 45, 47, 52, 55, 62, 63, 68, 83, 84]. On the other hand, it should be noted that the *discrete event simulation* (DEVS) modeling formalism, because it is aimed at computer emulation of complex systems involving multi-faceted modeling and discrete event simulation [85, 86], also carries the essential information:

- Set of input ports for receiving input events.
- Set of output ports for sending external events.
- Set of state variables and parameters associated.
- Time increment function.
- Internal transition mapping.
- External transition mapping.

Apparently, this information is rather instrumental in building models for FMS and robotic systems employing of control infrastructure with some intelligence according to IPDI principle and protocols handling communications among sub-systems involved [21, 41, 44, 45, 61]. It is also instrumental for model representation and computer simulation of discrete-event processes in FMS as well as of discrete-event dynamic systems in general [28, 81]. Hence, the discrete event representation of FMS and robotic systems in emulation simulation involving processing communications between sub-system objects and with the system environment may be based on two types of actions:

- Receiving of information (of discrete-event type) through input port, namely: *Receive x on input port p.*
- Sending output information (of discrete event type) through output port: *Send y to output port p.*

It should be noted, the hierarchical systemic structure is an inherent property of both conceptually basic computer-aided methodologies of *Booch's object-oriented modeling* (*OOM*) technique [68] and of *Ziegler's discrete-event simulation* (*DEVS*) *formalism* [11]. This property makes Ziegler's DEVS formalism is closed under model coupling, which further endows it with a most important structuring property: namely, *the ability to construct hierarchical models* [9] as followings: The very basic model has inherent hierarchical structure [18, 27]. Therefore the basic model of discrete-event dynamics and discrete communications involved [7, 61, 62, 80, 82] can be embedded into *the DEVS formalism* to form a *multi-couple model* having the following structure:

$$DN =< D, \ \{M_i\}, \ \{I_i\}, \ \{Z_{ij}\}, \ select >, \tag{10}$$

where D is a set of component events with such a property for each i in D to have:

M_i is a 'component basic model';
I_i is a 'set of influences of I';
Z_{ij} is a 'function of i to j output translation';
select is an 'activation function' of the next event.

It should be noted Ziegler's DEVS formalism is a natural derivation from the mathematical systems theory such that it shares and supports the similarity with Booch's object modeling. Both object and system models share the concept of an internal state. The main advantage of this concept is the overcoming of the time basis problem, which is and inherit property of OOM paradigm. Although object-oriented systems feature limited form of communications in relation to modular systems, they can serve as a basis to implement discrete-event and other system modeling formalisms. The actual implementation of hierarchical, modular DEVS formalism over an object-oriented substrate is based on the abstract simulator concept. Since such a scheme is naturally implemented by multiprocessor simulation architectures, models of discrete event systems developed in this form are readily transportable to the distributed parallel simulation [16, 35, 39, 45, 49, 51, 52, 63, 84, 87, 88].

At this point now let consider the PN representation model of the handling discrete event process in a simple FMS in Fig. 4 of two robot-arms operating in a common workspace to perform pick-and-place operations at time to obtain or transfer parts so that no collision occurs; see Fig. 4 where the respective discrete-event operating dynamics [16, 18, 26, 27] and communications among machines [21, 48, 61, 89] are depicted by its PN-based model. The sub-system of the FMS that represents operation of two robot arms [24, 28] servicing different machine tools is summarized in Fig. 4; it is assumed and understood under computer process control [17, 25]. Notice further, one of the arms is transferring semi-products form one machining toll to the buffer while the other one is transferring them from the buffer to the other machining tool. Its PN model in Fig. 4 is fairly obvious representation of the respective operational discrete-event dynamics. This PN model representation of activities of robot arms R1 and R2 can be discussed as follows. Places p_1, p_2, p_3 and transitions t_1, t_2, t_3 model the activities of arm R1, an places p_4, p_5, p_6 and transitions t_4, t_5, t_6 model the activities of arm R2. On the other hand, transitions t_1 and t_4 also represent the concurrent activities of arms R1 and R2. Either of these transitions can fire before or after, or in parallel one with the other.

The controlling infrastructure is taking care of a kind of synchronization in order to avoid collision, and this is accomplished by a rule mechanism of mutual exclusion implemented by the sub-net involving places p_7, p_3, p_6 and transitions t_2, t_3, t_5, t_6. Namely, firing transition t_2 disables transition t_5 assuming the latter is enabled, and vice versa. Thus only one arm can access the workspace at a time. It is also assumed that the buffer storage space has the capacity $K(p) = b$, and this information is captured by means of place p_8. Hence, for instance, when p_8 is empty, then transition t_2 cannot be enabled to fire. This prevents arm R1 form attempting to transfer to the buffer a work-piece when there is no room available. Also, arm R2 is prevented from attempting to access the buffer if there are no parts in it, or place p_9 is empty.

Fig. 4 PN model of operating two robot-arm pick-and-place subsystem [60]

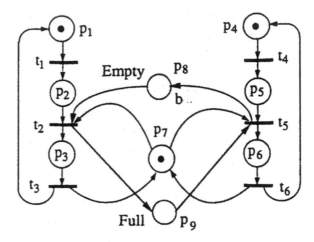

5   Cao and Sanderson's Fuzzy-Petri-Nets for Task-Sequences in FMS and Robotic Systems

There is no much truth in sciences that do not use Mathematics.—Leonardo da Vinci

Let now return our focus back to case studies that are relevant to industrial and manufacturing engineering. In fact, Fig. 2a is taken from Cao and Sanderson [26, 59, 60] shows a simple robotized FMS plant consisted of two sub-system parts: one robot, and two processing machines. In this system, the robot prepares and then inserts the peg cylinder (G) into the hollow cylinder (H) to form a cylinder assembly. In one feasible, complete, and correctly ordered sequence, the robot first picks up the raw peg cylinder, moves it to the cutting machine for cutting, then transfers it to the lubricating machine prior to insertion. On the other hand, Fig. 2b depicts one image-slice of image-sequence from the animated simulation of the medium-size FMS test-rig plant of RPI-Troy, NY, taken from [16, 18, 19, 27, 28, 46]. Its subsystem parts are represented by different symbolic captions. This FMS is consisted of: an input and an in-process storage of work-pieces; output storage of three of finished semi-products; milling-machine MM; lathe-machine LM; two manipulators; and a Gantry-robot with associated conveyors: It is a robotic system of two sub-system parts: one robot, and two processing machines. This FMS plants is aimed at manufacturing four types of parts (semi-products A, B, C, D) which are to be assembled in order to complete a given product. Apparently, the discrete-event dynamics of manufacturing process in this FMS, hence the set of all feasible task sequences, is much more complex.

From the system science and engineering viewpoint, however, in principle, there is no much difference when the higher control level is the one of management decision and coordination supervision, is taken into consideration instead. For, these imply some form of forecasting and control, and provide links to production planning and economical decision making. It should be noted, however, the appropriate formalisms employed will differ considerably always because of the different problem domain.

According to the IPDI principle of Saridis, this level of controlling infrastructure would require higher grade of machine intelligence and lower grade of mathematical precision in representation models and in control strategies employed [16, 18, 19, 27, 30, 43, 90]. This problem is also used to introduce some concepts on fuzzy systems and fuzzy regression model representation of time series for the purpose of possibilistic decision making [22, 53, 66, 71, 75, 91], which is complementary to traditional probabilistic regression and Markov chain decision making [4, 5, 16, 27, 30, 31, 43, 53, 63, 87].

Before going into details underlying this sophisticated methodology let address some practical issues and illustrate what kind of computer assisted decision making techniques is the word about. In this introductory section, in the first place, we focus on the aspect of experts support systems for decision making implementation of forecasting and control that employs Fuzzy-Petri-Net rule chaining (Fig. 5). We illustrate them through a discussion on Zadeh's decision-making in fuzzy environment [22, 66], applied fuzzy regression analysis [75] and possibility theory [71, 73]. Methods for fuzzy regression analysis using the linear possibility systems based on Zadeh's fuzzy-set based theory of possibility [71] have elaborated on in the 1980s; ever since then various types of possibility regression analysis have been elaborated [28], including the consensus of multi-agent systems with dynamic topology [11, 13, 89]. Consensus and synchronization multi-agent systems and multi-networks remains a new one the formulation of which is still being discussed and application still being explored. When thinking about the recently recognized needs for bringing advice from experts on data analysis in, we can hardly miss to see that fuzzy data is one form of expert knowledge. In this context the word is about expert system support, which involves adaptation of actions based on learning iterations. Still, it is a kind of Bayesian decision making, however, operating on some *fuzzy observation space* of data set. Hence fuzzy regression modeling, also involving fuzzy inference and reasoning, are employed instead hence fuzzy, fuzzy-neural or fuzzy-Petri system concepts have to be used one way or another. It ought to be an expert support system implementing the two-level model of decision supervision and control. It is an expert support system that can implement a set of relevant knowledge bases, some of which employ quantitative (i.e., numerical) and some other qualitative computations involving machine intelligence inference and reasoning algorithms (e.g., a fuzzy rule knowledge base) accompanied with an adequate rule-based production system as discussed in previous sections.

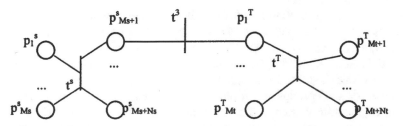

Fig. 5 Representation of the chaining of two arbitrary chained rules in the FPN formalism

6 An Overview of the Background Research and Development Explorations

Human progress has been achieved whenever daring to surpass barriers, cross borders, and exceed limits. The crucial step is the exceeding of limits.—Georgi M. Dimirovski

Let consider again robotic FMS in Fig. 2a, b, and place focus on task-sequence planning as well as observe the controlled sequence execution in industrial manufacturing systems, and in plants of Flexible Manufacturing Systems (FMS), in particular. The objective of task sequence planning *for* a robotic work-cell or flexible manufacturing system is to efficiently represent all feasible and complete task sequences with correct precedence relations and to be able to choose among them. A sequence of the shortest length or other optimality criterion may be selected from these correct sequences.

It has been shown by Cao and Anderson in their previous work [59] that for an assembly system all possible operational sequences, from the initial state to the final state, may be mapped to an ordinary Petri net [36, 37] with specific properties such as *liveliness, reversibility* and *safeness*. Moreover, this Petri-net model guarantees the feasibility constraints based upon the representation. However, all important events or operations, which are considered as *sub-goals*, must be included in all correct sequences following the expected precedence relations. To incorporate these sub-goals and their precedence relations in search for sequences, strong numerical constraints such as generated by the *prime number marking algorithm* on the ordinary Petri-net model generated logically form the plant specifications have to be applied. Then it is possible to obtain a fuzzy-Petri-net representation with *global fuzzy variables* to enable autonomous automated operation of FMS plants leading towards the paradigm of '*Factories of the Future*' (*FoF*) [8, 19–21, 30, 31, 35, 62, 64, 83, 92]. In domain-independent planning area, an approach to reducing the amount of search was to decompose a large problem into sub-goals [35, 51, 58] including independent sub-goals, 'serializable' sub-goals, and non-serializable sub-goals. However, classical plan generation systems [20, 48] are usually not robust and efficient enough to operate in complex robot work-cells or fairly simple FMS plants, and yet alone some more complex ones. Alternative approach that enables both representing and reasoning about domain-related constraints, modeled as sub-goals implicitly, for robotic systems is needed. Then the amount of search necessary to generate the correct sequences would be reduced leading to the feasibility of real-time implementation.

The so-called AND/OR graphs of de Mello and Sanderson [62] have been used in assembly task planning to represent and search all possible assembly sequences. The AND/OR net extends the AND/OR graph representation models [3, 60] to incorporate system mechanisms and devices, and defines an *Internal State Transition* (*IST*) operation which modifies the internal state of an object. The AND/OR net is generated based on the descriptions of objects and all feasible geometric relationships among them, and it is used to plan operations sequences for geometric manipulations including assembly, disassembly, grasping, and robot motions. Recently, Cao and Sanderson showed in [60] that an AND/OR net could be mapped to a Petri net, and this Petri net can then be decomposed to lower level nets while retaining properties of *liveliness, safeness,* and *reversibility*. A further step forward is to expand the domain of the Petri-net mapping as a representation of a robotic work-cell or FMS by defining fuzzy internal states of objects and using transition reasoning rules in the Petri net to impose precedence constraints on key operations. A state of the system is thus composed of a set of membership functions for the completion of the global task on all feasible objects. Transition reasoning rules implement the sequencing constraints required to direct the process to the final global state of the system.

In general, it is prerequisite to define in integrated control and supervision system for FMS plants the generalized concepts of *object, hard-object* and *soft-object* the way it was done in [3, 5, 6, 8, 34, 60, 62] and in [8, 20, 34, 46, 93]. An *object* in the system is defined as a single component, a subassembly of several components, or a complete assembly. A *soft object* is defined to be an object which includes at least one internal state variable. Then a *hard object* is defined to be an object which is not a soft object. An example of soft object is a robot holding a part which is to be measured and cut. If the property of the part the robot holds is unchangeable, then the subassembly of the robot and the part is a hard object. More generally, in an abstract level for instance, an object O consisting of two components, C_1 and C_2, is considered as a soft object if C_1 slides along the surface of C_2. For, the proper handling the triple paradigm of systems engineering—real-world system, its conceptual model (imitation simulator), and its model representation (emulation simulator)—plays a crucial role in systems engineering design applications [3, 16, 20, 27, 47, 49, 62].

The internal state variables of soft objects are described by fuzzy values of tokens. There exist several approaches to map an ordinary Petri-net model into a fuzzy-Petri-net one, some of these being proposed in [3, 19, 20, 26, 83, 88]. It appeared, however, that the prime-number-marking algorithm for mapping an ordinary to a fuzzy Petri-net is superior because it associates the fuzzy-Petri-net models in a manner which guarantees consistent sequencing of operations [26]. In the resulting fuzzy-Petri-net model, each transition has an associated *reasoning rule* and an associated *weighting factor* which evaluates the resulting values in the output places of this transition based on the fuzzy values of tokens in its input places. Transitions which cause the changes of fuzzy values of tokens for objects, are considered by definition to be *key transitions,* carrying most essential information on the operational discrete-event dynamics of the plant, and must be included into all feasible and complete task sequences. The alternative conceptualization employing fuzzy-Petri-net reasoning algorithm over a fuzzy-rule knowledge base, on the other hand, fires solely the key transitions to achieve a desired operating sequence and implements the same end effect [19].

Several articles concerning fuzzy-Petri-nets or similar composite nets encompassing properties synergy of both Petri-net and Fuzzy-system formalisms have been published. The first and most important one is due to Looney [83] who modified the usual Petri net to allow fuzzy rule-based reasoning by propositional logic extended [19, 20, 26, 59]. The resulting net is considered as a new type of a kind of neural network where transition bars serve as the neurons, and the nodes are conditions. Though Valette, Cardoso and Dubois [88] introduced uncertainty and imprecision within Petri net based models in the application of monitoring manufacturing systems. This approach is based on the association of a fuzzy value with the time delay for execution of a transition, which results in attaching a fuzzy date to the transition. In Tsuji and Matsumoto's work [90], a kind of extended Petri net was proposed to model the vague conditions, and the boundedness, liveliness, and reachability for this model of fuzzy inference engines were analyzed. In work [69] fuzzy Petri-networks are used to represent Booelean On/Off production rules of a rule-based system, in other words a Boolean network. An efficient algorithm was also proposed to perform fuzzy reasoning automatically. Similarly, in his investigation of fuzzy-Petri-net reformulation of Saridis' organizing controller, Dimirovski [19, 47] has shown another mapping procedure, also involving a fuzzy-rule-chaining reasoning, that generates an adequate fuzzy-Petri-net model associated to a pre-designed fuzzy-rule production system for a class of robot manipulating tasks. The associated Petri-net representation also possesses the useful properties of liveliness, safeness, and reversibility, and it may be decomposed to lower level nets while retaining these properties.

In more recent works by Cao and Sanderson [26] a definition of the generalized fuzzy Petri net was given which incorporated three types of fuzzy variables: *local fuzzy variables, fuzzy marking variables*, and global fuzzy variables. Property analysis associated with some typical cases of fuzzy Petri nets was given too. Similarly, for a simpler class of robot manipulating tasks, it has been shown clearly by Dimirovski in [19] and Gacovski in [20] that the other mapping procedure that involves the data-driven reasoning algorithm of *fuzzy-rule-chaining* with a *degree-of-fulfillment* and *fuzzy-possibility distributions*, which generates fuzzy-Petri-net model associated to a pre-designed fuzzy-rule production system, also represents a kind of more general Fuzzy-Petri-net and Multi-agent with dynamic topology [33]. Hence these definitions can also be applied to other kinds of applications, including artificial intelligence and knowledge based systems in automated manufacturing. Furthermore, in this way the performance effects predicate/transition nets can be feasible because transition firing can be controlled by imposing conditions on the token values. Nonetheless, the major difference between the predicate/transition net and the fuzzy Petri net, however, is that a fuzzy-Petri-net model can represent more generalized data using fuzzy numbers and complex fuzzy reasoning functions. This is why fuzzy-Petri-net models can cope with the difficult problem of choosing feasible task sequences is to order the correct precedence relationships among all important events or transitions.

The problem emphasized above, which is difficult to handle indeed, in Cao and Sanderson [59] has been solved by using *prime number marking* algorithm in modeling the system, which is based on *Arithmetic Fundamental Theorem*.

Arithmetic Fundamental Theorem of the Arithmetic: If p_i, and q_i are positive primes, and if

$$a = \prod_{i=1}^{n} p_i^{\alpha i} = \prod_{j=1}^{m} q_j^{\beta j} \tag{11}$$

where $1 < p_1 < p_2 < ... < p_{n-1} < p_n$ and $1 < q_1 < q_2 < ... < q_{m-1} < q_m$, then $n = m, p_i = q_i$, and $\alpha_t = \beta_t$ every positive integer has a composition into positive prime factors, which is unique apart from the order of the factors.

In there, the weighting factors as well as the initial tokens and the final tokens are chosen for all soft objects and hard objects so that assigned precedence relationships will be automatically followed. Moreover, all sequences which incorporate any incorrect precedence relationship will be recognized and discarded. The prime token values of soft objects can be interpreted as the degrees of certainty of processing completion for these objects. Their method is based on Cao-Sanderson theorem, and employs *feasible global fuzzy states* obtained during the generation of the fuzzy-Petri-net in the search for feasible, complete, and correctly ordered sequences of operations.

Cao-Sanderson Theorem [60]: Assume the prime number marking algorithm is used to generate all possible sequences from the fuzzy Petri net in which each transition can only be fired at most once for each sequence. These sequences are feasible, complete, and hold correct precedence relationships among key transitions, if and only if the places corresponding to soft objects can only hold the following order of token values: $10^{-L}, 10^{-L} \times P_1, 10^{-L} \times P_1 \times P_2, ..., 10^{-L} \times P_1 \times P_2 \times ... \times P_s$ where $P_1, P_2, ... P_s$ are the first s primes [26].

Corollary 1 For a fuzzy Petri net marked using the prime number marking algorithm, m_t is defined as $Q_t \rightarrow v(Q_t)$, where $v(Q_t) = \{1, \{(1,10^{-L})\}, \{(1,10^{-L} \times P_1 \times P_2)\} ... \{(1, 10^{-L} \times P_1 \times P_2 \times ... P_s)\}\}$.

Corollary 2 Assume each transition can only be fired at most once for one sequence. The search for sequences in the fuzzy Petri net can be halted at a token value $\sigma^i \notin v(Q_t)$, but will not exclude any feasible, complete sequence which has correct precedence relationships among operations.

In effect, Corollary 2 is used to search off-line for all feasible correct sequences which satisfy the basic three properties of the fuzzy-Petri-nets. When a set of enabled transitions are found for the development of partial sequences, those transitions which lead to undesirable token values in the corresponding output places are discarded. In the presentation above, it was assumed that the order of the key transitions is assigned in advance, and only one order is feasible. In practice, it may often occur that more than one partial ordering of feasible key transitions may exist. For instance, suppose t_a and t_b are key transitions, and both sequences $t_a...t_b...$ and $...t_b...t_a...$ may be feasible; the plan representation should include both. As pointed by Cao and Sanderson [26], the problem of multiple partial orderings may be solved in a straightforward manner by enumerating all possible orderings of processing steps and constructing the union of the plans from each set. In general, if there are k feasible assigned key transition sequences, $S_1, S_2, ..., S_k$. For each S_i, the planning strategy above is used to obtain all feasible, complete, correctly ordered sequences represented as $\{plan(S_i)\}$. Then the final complete sequence set is the union $\cup_i{}^k = {}_1\{plan(Si)\}$. If a transition affects two or more soft components, then this transition may be decomposed into two or more transitions, each affects one soft component.

In Dimirovski [47] and Gacovski [20], the same problem has been solved via employing a *fuzzy-rule reasoning* algorithm producing a supervisory guideline with non-negative functions, as appropriate one in the fuzzy-rule-production system and in the associated Petri-net model. Namely, the employed non-negative functions are: of truth along with associated truth-variable, and of fulfillment along with associated degree-of-fulfillment; and a transition function representing the evolution of marking function.

7   Higher Level of Decision and Control: Forecasting, Control and Supervision

Man is measure of all things.—Protagoras

The overall system controlling infrastructure has been emphasized as a hierarchically structured two-level system on its own that implements functions of control, management decision and coordination supervision [7]. It should be noted that it has to be hierarchically organized regardless whether designed strictly according to the IPDI principle or not. In this work we assumed that the IPDI principle has been observed and used in the design of controlling infrastructure. This section is focused on the 'higher control level' system below.

7.1   Fuzzy Regression Based Forecasting and Control Expert Support System

Let now consider into more detail the standard regression models, the difference between the data and the inferred value obtained from the model is taken to be observational error. However, in the *possibilistic* (i.e. possibility reasoning on) fuzzy regression models, it is assumed that the gap between the data and the model appears as an ambiguity in the structure of the system that gives the input and output [12, 66, 76]. Therefore, in fuzzy-regression modeling the coefficient that expresses the system's ambiguity are considered [66]. As in probabilistic statistical decisions, the most used in applications fuzzy regression modeling is based on linear equation whose coefficients are expressed by fuzzy numbers and possibility distributions [44, 78, 86]. Since the fuzzy number for the coefficient gives the possibilities for the coefficient, the system is called a *linear possibility system*. A fuzzy regression model expresses input and output data by a possibility equation.

One interpretation for Zadeh's fuzzy (sub)sets [12, 76, 86] is that membership functions $\mu_F(x)$, that are characterizing and representing fuzzy subsets, can be seen as a possibility distribution. Should some information F from a specialist that is expressed by something like *"about 10"*, it is being replaced by $\mu_F(x)$ and this new expression gives information that represents the possibility. For instance, if $\mu_F(8) = 0.8$, this means the degree of possibility of event occurrence "**8**" is **0.8**. In order to make plausible use of this kind of information F in terms of possibility, possibility distribution functions $\mu_F(x)$ are being employed in model representations. As in the case of probability, if the possibility function $\mu_F(x)$ is known, we can think about establishing the possibility of (fuzzy) event A by means of the fuzzy sets. Possibility functions given in terms of intervals are easily understood, hence consider a case in which the event is also an interval. Namely, let $F = \{X| \ 0 \le x \le 5\}$ and $A = \{x| \ 2 \le x \le 7\}$; then we can think of the possibility of an event A based on the possibility of information $\mu_A(x)$ as equal to 1.00 for A being in F (note A and F overlap in part). This is allowed because $F \cap A \ne \acute{\emptyset}$ ($\acute{\emptyset}$ is the empty set), that is *it is not empty as in the probabilistic setting*. If $A^c = \{x| \ 6 \le x \le 10\}$, which is the complement of A in the respective universal set X, i.e. *Not A*, then apparently the possibility of A^c is $\mu_{A^c}(x) = 0.00$. Note this idea is generally applicable fuzzy sets, and we have the well known definition: When possibility function $\mu_F(x)$ is given, *the possibility measure* $\pi(A)$ is defined

$$\pi(A) = \sup_x \mu_A(x) \wedge \mu_F(x), \tag{12}$$

where operator $\wedge$ represents the logical connective AND, i.e. intersection operation on sets [53, 71, 75]. Then the ambiguous information such as "about 10" is given by a fuzzy number (i.e. interval) and the information obtained from the fuzzy numbers is given as a possibility distribution function that has properties

$$
\begin{aligned}
&(i) \quad \pi_x(\phi) = 0.0, \quad \pi_x(X) = 1.0, \\
&(ii) \ \pi_x(A \cup B) = \pi_x(A) \vee \pi_x(B),
\end{aligned}
\tag{13}
$$

where φ is the empty set and X is the universal set (in the given problem), which are equivalent to those in probability theory and mathematical statistics. From (ii) it follows that a possibility measure, which is monotone, can be associated as $A \subset B \rightarrow \pi_x(A) \le \pi_x(B)$. As in probability and statistics, where various types of probability distribution functions are observed, there is a variety of possibility distribution functions depending on the problem domain of applications.

A *fuzzy number* A of *L-R* type is expressed as $A = (\alpha, c)$, and its *membership function* is defined as

$$\mu_A(x) = L((\chi - \alpha)/c); \ c > 0, \tag{14}$$

where $L(x)$ is called a *reference function* and has the properties: (i) $L(x) = L(-x)$, (ii) $L(0) = 1$, and (iii) $L(x)$ is a strictly decreasing function in the interval $[0, \infty)$, and where α expresses the *centre* and the c the *spread* or *width* of the fuzzy number, which are symmetrical. That is, have symmetrical membership functions (grades of belonging to) and widely used in applications.

The operations with fuzzy numbers are defined by means of Zadeh's extension principle [10, 11, 14] because the function for x is extended to a function of fuzzy numbers (or any kind of fuzzy subsets). It is expressed by the definition: When in a given function $y = f(x_1, x_2, ..., x_n)$ the inputs $x_1, x_2, ..., x_n$ are replaced by the respective fuzzy subsets $A_1, A_2, ..., A_n$, then the fuzzy output is defined by

$$\mu_Y(y) = \sup_{\{x_1,...x_n | y = f(x_1,...x_n)\}} \mu_{A_1}(x_1) \wedge ... \wedge \mu_{A_n}(x_n) \tag{15}$$

This means finding x on the input set $\{x_1, x_2, ..., x_n | y = f(x_1, x_2, ..., x_n)\}$ such that it maximizes the intersection of membership functions $\mu_{A_1}(x_1) \wedge ... \wedge \mu_{A_n}(x_n)$, after y in the output set is fixed. Standard operations for symmetrical L-R fuzzy numbers (for simplicity, fuzzy numbers), are carried out simply according to the following formulae:

$$addition : (\alpha_1, c_1)_L + (\alpha_2, c_2)_L = (\alpha_1 + \alpha_2, c_1 + c_2)_L;$$

$$subtraction : (\alpha_1, c_1)_L - (\alpha_2, c_2)_L = (\alpha_1 - \alpha_2, c_1 + c_2)_L$$

$$scalar \ multiplication : \lambda^\circ(\alpha, c) = (\lambda\alpha, |\lambda|_c).$$

These are also employed in actually expressing and computing linear possibility systemic feature, which are defined as

$$Y = A_1 x_1 + A_2 x_2 + \cdots + A_n x_n, \tag{16}$$

where $A_1,..., A_n$ are symmetrical fuzzy numbers, $A_i = (\alpha_i, c_i)_L$ (Fig. 3), determined by membership functions

$$\mu_{A_i}(a_i) = L((a_i - \alpha_i)/c_i). \tag{17}$$

The fundamental theorem, on which fuzzy regression modeling is based, is recalled here without proof. For, it provides the means of determining membership function for the outputs, i.e. the output possibility distribution.

Basic Possibility Theorem [53, 71]. *The membership function for the output of a linear possibility system is as follows:*

$$\mu_Y(y) = L((y - x^T\alpha)/c^T|x|). \tag{18}$$

In other words, this means that the centre for possibility distribution Y of the output (for a given data set) is defined by $\sum \alpha_i, c_i$, and the width is defined by $\sum c_i |x_i|$. In the present context, we consider the '*possibilistic*' fuzzy regression [71] relative to the traditional 'probabilistic' regressions [52].

Therefore notice that when the coefficients represent possibility distributions $A_j = (\alpha_j, c_j)_L$ (fuzzy numbers), for comparison with the independent probability distributions in which the coefficient are normal distributions $N(e_j, \sigma_j^2)$, it has been found that:

in terms of possibility distributions $- (\alpha_1, c_1)_L x_1 + ... + (\alpha_n, c_n)_L x_n = (\alpha^T x, c^T |x|)_L$,

in terms of probability distributions $- N(e_1, \sigma_1^2) x_1 + ... + N(e_n, \sigma_n^2) x_n = N(e^T x, (\sigma^2)^T x^2)$,

where $x^2 = (x_1^2, ...x_n^2)^T$ represents the input data vector that is treated by normal distribution linear regression model. It is clearly seen now that the possibility is calculated on the grounds of possibility measures and corresponds to the probability calculated on the grounds of probability measures.

Now consider the given data set $(y_i, x_{i1}, ..., x_{in}), i = 1, ..., N$. where y_i is an output for the *i-th* sample and x_{ij} is the j-th input or explanatory variable for the i-th sample. The vector for the explanatory variables is expressed as $x_i = (x_{i1}, ..., x_{in})^T$. In the case of standard regression models, the inference between the actual data and the inferred values

$$y_j - \sum a_i x_{ji} = \varepsilon_j, \quad j = 1, 2, ..., N, \tag{19}$$

is interpreted as observational error, and the regression analysis is performed by means of the probability model. The linear possibility regression analysis considered in here, it follows the 'possibilistic' view on linear regression model. The uncertainty in the data is seen as originating in the system itself, and it depends on the possibility distribution of the system coefficients. The following conceptualization of fuzzy regression problem is readily established:

(1) Let the linear possibility system be the model of fuzzy regression, that is

$$Y_i = A_o + A_1 x_{i1} + \cdots + A_n x_{in}, \tag{20}$$

A_i are symmetrical fuzzy numbers $(\alpha_i, c_i)_L$.

(2) Fuzzy coefficients A_i are determined by the degree h to which the given data set (y_i, x_i) is included in the inferred fuzzy number Y_i through the fuzzy reasoning approach (e.g., Zadeh [40, 71]), that is

$$\mu_{Y_i}(y_i) \geq h, \quad i = 1, ..., N, \tag{21}$$

Y_i is the inferred fuzzy number value from (13).

(3) The fuzzy coefficients A_i that minimize the total width of Y_i are determined via minimizing the criterion

$$J(c) = \sum c^t |x_i|, \tag{22}$$

where $c = (c_1, \ldots, c_n)^T$, and $c^T |x_i|$ is the width of Y_i.

Note that the criterion $J(c)$ corresponds to the sum of errors in the conventional classic regression analysis. Hence, the linear possibility regression problem is reduced to finding the $A_i = (\alpha_i, \; c_i)_L$ that minimize the objective function (15) subject to constraint (14). In other words, solving the considered fuzzy regression problem reduces to the following LP problem:

For data set (y_i, x_i), $i = 1, \; 2, \; \ldots, \; N$, find

$$\min_{\varepsilon, \; c} \; J(c) = \sum c^t |x_i|, \tag{23}$$

subject to

$$y_i \leq x_i^T \alpha + |L^{-1}(h)| c^T |x_i|, \tag{24}$$

$$y_i \geq x_i^T \alpha - |L^{-1}(h)| c^T |x_i|, \tag{25}$$

$$c \geq 0. \tag{26}$$

It is important to note that in this LP problem has been derived due to Basic Theorem that guarantees the constraint (14) can be replaced by the above presented three constraint expressions. Thus, then the possibility fuzzy regression model is determined by solving this well-defined LP problem instead.

7.2 Markov Chain Model State Estimation Via the Fuzzy-Petri-Net Reasoning Supervisor

Management decision making processes, regardless the field of application, are naturally arising as essentially discrete-event, stochastic, dynamic processes. However, these always imply some implicit or explicit forecasting and control. Should in the particular field of application a considerable argument supports the assumption of Markovian property [5, 15] of the underlying discrete stochastic nature, it is well known that Markov chain model is most appropriate one [38, 62, 64]. It is therefore that since Derman's seminal papers [15, 27] Markov-chain based approaches have undergone remarkable theoretical and applications oriented developments. Nonetheless, Markovian property has to be verified one way or another which is not a simple task, and this has given rise to the well known first-, second-, and third-order Markov or semi-Markov stochastic process models This is due to the impact of human factors in the discrete-event dynamics of decision processes, and to implicit forecasting embedded [47]. In discrete-event processes, there are some processes that cause different types of changes, and usually cause leap changes. For actual solving of these problems, transition matrices, probability matrices and Markov chains are employed, but Markov-chain models are crucial [9, 16, 35, 55]. A Markov chain [4] is a stochastic process that is characterized by the following features: (i) a finite number of states; (ii) Markovian transitions; (iii) Stationary transition probabilities; (iv) a set of initial probabilities $P(X_0 = i) = q_i$, $\forall i, i \in [0, N]$. Just for the sake of illustration, let suppose an $n \times n$ matrix of transition probabilities, P, is available such that

$$P_{ij} \geq 0, \ \sum_j p_{ij} \leq 1, \tag{27}$$

along with an array

$$g_i = 1 - \sum_j p_{ij} \tag{28}$$

A set of matrices P can describe a Markov chain where the states of the chain model are indicated by n integers. The element p_{ij} gives the transition probability for the random walk from state i to state j. As long as g is not zero the walk will eventually terminate. The probability that the walk will terminate after state i is given by g_i. It is therefore that Markov chains can be used to solve a very useful class of sequence problems in a rather remarkable way. Hence a Markov-chain model is a sequence of random values whose probabilities at a time interval depend upon the value of the number at the previous time. The controlling factor in a Markov-chain model is the transition probability; it is a conditional probability for the system to go to a particular new state, given the current state of the system.

For many problems, such as simulated annealing or task sequence planning and control, it is not clear how to evaluate these probabilities even the initial ones. One way or another, these have to be inferred from the features of the problem domain. Alternatively, a fuzzy-Petri-net supervisor [19, 47] proposed to generate them in terms of fuzzy possibility distributions estimating probabilities of state sequences. In this way fairly efficient estimates of the actual transition probabilities can be determined in terms of 'defuzzified' values, of course, with a clear distinction of differences between probabilities and possibilities [67, 74]. A modified extension of the reasoning algorithm with our fuzzy-rule Petri-net based production systems [20, 47] can assign the finally computed defuzzified values for output variables as probabilities of the initial state and transition probabilities needed sequence of states of the Markov-chain model for decision making process. In the FPS, Petri-net formalism is used for modeling fuzzy-rule production systems via association identification of its elements (*places* and *transitions*) and features (*marking function*) with the basic elements and features of the fuzzy-rule knowledge base (*propositions, degrees-of-tr*uth, and *implication rules*), i.e. the *crucial FKB background*.

The fuzzy-rule, Petri-net based, production system (FPS) in the processing phase of rule-chaining is illustrated in Fig. 6. Note that a fuzzy-rule knowledge base (FKB) is associated with it, and constitutes the source background of the FPS [19], where all fuzzy rules and their respective propositions are precisely defined both in linguistic and in fuzzy-subset representations. Notice that the focus is on the computing process for obtaining the *degree of fulfillment* (DOF) corresponding to some proposition $\alpha(p_1^T)$ from the DOF of another proposition $\alpha(p_{Ms+1}^s)$, i.e. $g\left(p_1^T\right)$ from $g\left(p_{Ms+1}^s\right)$. Then this can be expressed via

$$\underline{b}_{1,i}^s = \tau^s(g(p_{Ms+1}s)\wedge b_{1,i}^s), i = 1, ..., I \tag{29}$$

where $B_1^s = \left\{b_{1,i}^s\right\}$ is the possibility distribution associated with linguistic value B_1^s in proposition $\alpha(p_{Ms+1}^s)$. The evaluation of the DOF of proposition $\alpha(p_{Ms+1}^s)$ is given by

$$g\left(p_1^T\right) = V\left[\tau^s(g(p_{Ms+1}^s)\wedge b_{1,i}^s)\wedge a_{1,i}^T\right], \tag{30}$$

where $a_{1,i}^T$ is the possibility distribution of some linguistic value A in the propositions of the expert systems FKB. The operators V and Λ in Eqs. (18) and (19) denote the well-defined logical connective operations equivalent to union and intersection of sets [70]. For the more general case in which several rules R^1, ..., R^S, ..., R^T, ... perform inference over a variable, and the same variable is in the antecedent part of at least one later rule R^T, typically. Then such a FKB, can be described as follows:

$$R^1 : IF\ X_1^1\ IS\ A_1^1\ AND\ ...THEN\ X_{M1+1}^1\ IS\ B_1^1\ AND\ ...(\tau^1)$$

$$R^s : IF\ X_1^s\ IS\ A_1^s\ AND\ ...THEN\ X_{Ms+1}^s\ IS\ B_1^s\ AND\ ...(\tau^s) \tag{31}$$

$$R^T : IF\ X_1^T\ IS\ A_1^T\ AND\ ...THEN\ X_{MT+1}^T\ IS\ B_1^T\ AND\ ...(\tau^T)$$

with

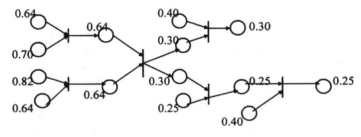

Fig. 6 Directed graph-model representation of the projected Petri-net computing structure for the FPS reasoning process [20, 46]

$$X^1_{M1+1} = X^2_{M2+1} = \ldots = X^s_{Ms+1} = X^T_1. \tag{32}$$

Following a derivation procedure, which is analogous to most of fuzzy based reasoning systems in the literature [20, 72, 73, 83], the obtained DOF for the proposition $p_I{}^T$ is given by formula

$$g(p^T_1) = V\left[V\left[\tau^s(g(p^s_{Ms+1})\wedge b^s_{1,i})\right]\wedge a^T_{1,i}\right], \tag{33}$$

Then the reasoning algorithm of fuzzy-Petri-net support system performs the actual value estimation. The algorithm is consisted of two stages basically: definition of the marking function and production of the DOFs of corresponding propositions, and firing of the active transitions. These stages are sequentially repeated until there are no more active transitions, at in which time the inference process will have ended. Finally, the aggregation—the assignment of a single possibility distribution to each output variable—is performed. Let IP and OP be the sets of input and output places in the FPS, respectively. Consequently, the FPN reasoning algorithm may be summarized as follows:

Step 1—Initially, only the DOFs of the propositions that operate on input variables, that is, those associated with input places, are assumed on the grounds of problem domain. Therefore, the initial marking function will be

$$(p_i) = \{0, \; if \; p_i \notin IP; \; 1, \; if \; p_i \in IP\} \tag{34}$$

Step 2—Next, fire the active transitions. Let t^j be any active transition; that is,

$$t^j \in T \,|\, \forall p_k \in I(t^j), \; M(p_k) = 1. \tag{35}$$

The transition function t_f, which defines the successive marking functions, is defined by an appropriate mapping. The corresponding DOFs obtained are:

$$If \; t^j \in T^R, \; g(p_i) = \wedge g(p_k), \; \forall p_i \in O(t^j) \tag{36}$$

$$If \; t^j \in T^C, \; g(p_i) = V\left[\tau^{rk}(g(p_k)) \; o^{rk} \; \mu_{pk}, p_i\right], \; \forall p_i \in O(t^j). \tag{37}$$

Step 3—Then go back to step 2, while performing

$$\exists t^j \in T \,|\, M(p_i) = 1, \; \forall p_i \in I(t^j). \tag{38}$$

Step 4—For each of output variables X, its associated possibility distribution $\underline{B} = \{\underline{b_i}\}, i = 1,\ldots,I$, is evaluated by

$$b_i = V\tau^r(g(p^r)) \; o^r \; \tau^r(b^r_{n,i}), \tag{39}$$

where P_X is the set of places associated with propositions in which inferences over X are carried out, and it is defined by

$$Px = \left\{ p_n^r \in P \big| \alpha(p_n^r) =" X \ IS \ Bn^{r''} \right\}. \tag{40}$$

Step 5—Terminate reasoning when the markings in (27) are over, and then evaluate defuzzified output variables to estimate Markov state transition probabilities.

The Petri-net computing structure of chained FKB rules via computed degrees-of-fulfillment (DOFs) of propositions in fuzzy rules of the FKB [20, 47] is depicted in Fig. 7 while the resulting conclusion fuzzy subset in Fig. 8. The concluding values by means of two different defuzzification methods are also pointed out.

It should be noted that this fuzzy-Petri-net reasoning algorithm is an on-line data-driven algorithm, and it is executed by the actual data input in the fuzzy-Petri-net supervisory controller. However, the fuzzy-rule production system (such as in Fig. 6) ought to be designed prior to the derivation of its one-to-one equivalent in terms of fuzzy-Petri-net supervisory controller. For the purpose of simulation experiments, a rule base being aimed at estimation of state probabilities in Markov-chain decision models has been designed. Represented in linguistic terms of If-Then rules [53, 73, 75] its construction can be presented as follows:

0. IF X1 = LP\0.12 & X6 = SP\0.10 THEN X2 = ZO, = > dof(A00) = 0.64; dof(A01) = 0.70;
1. IF X1 = LN\0.06 & X3 = SN\0.12 THEN X2 = ZO, = > dof(A10) = 0.82; dof(A11) = 0.64;
2. IF X2 = SN\0.07 & X5 = LN\0.20 THEN X7 = ZO, = > dof(A20) = 0.79; dof(A21) = 0.40;
3. IF X2 = SN\0.15 & X5 = SN\0.25 THEN X4 = ZO, = > dof(A30) = 0.55; dof(A31) = 0.25;
4. IF X4 = SP\0.05 & X3 = SN\0.20 THEN X7 = ZO, = > dof(A40) = 0.85; dof(A41) = 0.40;

The graphical representation of the fuzzy possibility distribution generated for one of the output variable of the above production knowledge base is depicted in the previous figure (Fig. 7), along with the computed dufuzzification values using two of the known methods. These in turn do give the *possibilistic estimation* of the

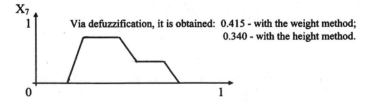

Fig. 7 Possibility distribution and defuzzification values computed for one output variable gives the estimated state transition probabilities of Markov chain model [20, 46]

margin values within which the estimated *state probability of Markov chain model is confined*. The essential factor in a Markov-chain decision models are the transition probabilities, which are governing the sequence of state transitions of Markov. It is a conditional probability for discrete-event system modeled by Markov chain to go to a particular new state, given the current state. Thus, with a clear-cut distinction between fuzzy-possibility and probability distributions, the latter can be reasonably estimated by defuzzified values obtained via fuzzy-Petri-net reasoning support system for application domains satisfying Markovian property.

8 Concluding Summary

So far as the laws of mathematics refer to reality, they are not certain. And so far as they are certain, they do not refer to reality—Albert Einstein

Discrete-event processes and systemic structures, which are pertinent to applications in Industrial Manufacturing Engineering, are surveyed from the points of view of Systems and Control Science as well as Systems Engineering in the prospect of application of results and techniques of Petri-net and fuzzy-Petri-net formalisms. It has been pointed out that their underlying nature is of the type of either Markov or semi-Markov processes in FMS and Robotic plants. The basic mathematical language utilised is the one of set-operations and directed graphs. The notions of event-driven and time-driven sequences, as both real-world phenomena and mathematical objects, are found to be rather instrumental for development of representation models and control and communication algorithms that enable system design analysis with regard to both higher-level task-oriented decision and lower-level control. Furthermore, this way a managing approach and feasible techniques for software implementation within a standard computing platform of distributed, two-level, computer control architecture are provided for.

An overview on the feasible synergies of fuzzy-systems and Petri-nets is presented. Both formalisms are illustrated by means of resulting representation models and task-control algorithms taken from those two case-study plants in flexible manufacturing engineering, which are emulated by employing object-oriented programming [94] and communications [95] among sequential machines [96, 97] in synchronized mode [98]. A particular focus has been placed on presenting the techniques based on two different fuzzy-Petri-net representations formalisms. The one is due to Cao and Sanderson [26, 59, 60]. The other is due to Dimirovski and Gacovski [19, 20, 46–48] where network communications among machines in FMS [95–97] was finally acomplished in collaboration with Jing [21] and his doctoral students Han [89] and Wang [79, 80]. An appropriate outline of similarities and dissimilarities as well as of their respective application potentials and limitations, which have been well established in cooperation with D. Tukel, F. Ozen, G. Nadzhinski and D. Stavrov, are also presented in some detail [25].

Finally, it is pointed out here this work present only a small portion of the existing scientific knowledge, engineering techniques and developed technologies on robotized FMS during last few decades [17]. The interested reader is advised first to study the crucial works by Saridis and Valavanis [49] and Saridis [50] on *Principle of Increasing Precision with Decreasing Intelligence* and by Siljak [38] on his innovated theory of *Dynamic Graphs*.

Acknowledgements The authors would like to acknowledge various invaluable contributions to this long-lasting research endeavors over the decades made by their former doctoral student-collaborators as well as well as postdoctoral fellows, respectively.

References

1. Boyd, S. P., El Ghaoui, L., Feron, E., & Balakrishnan, V. (1994). Linear matrix inequalities in systems and control theory. *SIAM Studies in applied mathematics* (vol. 15). The SIAM.
2. Box, G. E. P., Jenkins, G. M., & Reinsel, G. C. (1995). *Time series analysis: Forecasting and control* (3rd ed.). Prentice Hall.
3. Cheng, D., Ki, H., & Li, Z. (2011). *Analysis and control of boolean networks: A semi-tensor approach.* Springer.
4. Chung, K. L. (1960). *Markov chains with stationary transition probabilities.* Springer-Verlag.
5. Derman, C. (1962). On sequential decisions and Markov chains. *Management Science, 9,* 16–24.
6. Derman, C. (1965). Markovian sequential control processes—Denumerable state space. *Journal of Mathematical Analysis and Applications, 10,* 295–302.
7. Hendeler, J., Tate, A., & Drummond, M. (1990). AI planning: Systems and techniques. *AI Magazine, 3,* 61–77.
8. Lu, J., Zhong, J., Li, L., Ho, D. W. C., & Cao, J. (2015). Synchronization of master-slave probabilistic Boolean networks. *Scientific Reports, 5,* paper 13437.
9. Zhou, M. C., & DiCesare, F. (1993). *Petri-net synthesis for discrete event control of manufacturing systems.* Kluwer Academic Publishers.
10. Willems, J. C. (1972). Dissipative dynamical systems—Part 1: General theory. *Archive for Rational Mechanics and Analysis, 45*(5), 321–351.
11. Chen, Y., LuX, J., & Yu, and D. J. Hill,. (2013). Multi-agent systems with dynamical topologies: Consensus and applications. *IEEE Transactions on Circuits and Systems Magazine, 2913,* 21–34.
12. Li, F., Wu, L., Shi, P., & Lim, C. C. (2015). State estimation and sliding mode control for semi-Markovian jump systems with mismatched uncertainties. *Automatica, 51*(4), 386–393.
13. Yu, J., Shi, P., Xing, W., Chadli, M., & Dimirovski, G. M. (2020). Fuzzy-based dissipative consensus for multi-agent systems with Markov switching topologies. *IFAC PapersOnLine, 53*(2), 4127–4131.
14. Misra, V., Gong, V. B., & Towsley, D. (2000). Fluid-based analysis of a network of AQM routers supporting TCP flows with application to RED. In *Proceedings of the ACM/SIGCOMM 2000,* Stocholm, SW (pp. 151–160). The ACM Press.
15. Dimirovski, G. M., Gough, N. E., & Barnett, S. (1977). Categories in systems and control theory. *International Journal of Systems Science, 8*(9), 1081–1090.
16. Vukobratovic, M. K., & Dimirovski, G. M. (1993). Modeling, simulation and control of robots and robotized FMS. *Facta Universitatis Series Mechanics, Automatic Control & Robotics, 1*(3), 241–280.

17. Dimirovski, G. M., Bilgin Tukel, D., Ozen, F., Jing, Y., Shen, J., Nadzinski, G., & Stavrov, D. (2020). Fuzzy-petri networks in supervisory control stochastic Markov processes in robotized FMS manufacturing plants. *Joint Research Report FMS-03/2020.* IASE-FEIT, SS Cyril and Methodius University, Skopje, MK, DCE-FE, Dogus University, Istanbul, TR, and DEEE, Halic University, Istanbul, TR, ISE-NEU, Shenyang, CN.
18. Dimirovski, G. M., Radojicic, P. C., Markovic, N. B., Iliev, O. L., Gough, N. E., Zakeri, A., & Henry, R. M. (1994). Modeling, control and animated simulation of complex processes in robotized FMS. In G. Buja (Ed.), *Proceedings of IEEE industrial electronics conference* (vol. 2, pp. 1141–1146). The IEEE.
19. Dimirovski, G. M. (1998). Fuzzified petri-nets and their application to the organizing controller. NATO ASI Series F 162In O. Kaynak, L. A. Zadeh, B. Turksen, & I. J. Rudas (Eds.), *Computational intelligence: Soft computing and fuzzy-neuro integration with applications* (pp. 260–282). Springer-Verlag.
20. Gacovski, Z. M. (2002). Intelligent supervisory controllers based on stochastic, fuzzy and parameterized petri nets. Doctoral dissertation, supervisor G. M. Dimirovski. SS Cyril and Methodius University (in Macedonian).
21. Jing, Y.-W., Cheng, B., Dimirovski, G. M., & Sohraby, K. (2001). On leader-follower model of traffic rate control for networks. *Control Theory & Applications, 18*(6), 817–822.
22. Zadeh, L. A. (2008). Is there a need for fuzzy logic? *Information Sciences, 178,* 2751–2779.
23. Shi, P., Boukas, E. K., & Agarwall, R. K. (1999). Control of Markovian jump discrete-time systems with norm bounded uncertainty and unknown delay. *IEEE Transaction on Automatic Control, 44*(11), 2139–2144.
24. Liu, Y., Wang, J., Kao, Y., & Dimirovski, G. M. (2016). Overcoming control complexity of constrained three-link manipulator using sliding-mode control. In I.J. Rudas & S.-F. Su (Eds.), *Proceedings of the 2016 IEEE international conference SMC-2016,* Budapest, HU (pp. 2202–2207). The IEEE; Obuda University.
25. Dimirovski, G. M., Bilgin Tukel, D., Ozen, F., Kolemisevska-Gugulovska, T., & Jing, Y. (2018). Fuzzy-petri networks in supervisory control stochastic Markov processes in robotized FMS manufacturing plants. *Joint Research Report FMS-02/2018.* IASE-FEIT, SS Cyril and Methodius University, Skopje, MK, DCE-FE, Dogus University, Istanbul, TR, and DEEE-FE, Halic Universiy, Istanbul, TR, and ISE-NEU, Shenyang, CN.
26. Cao, T., & Sanderson, A. C. (1995). Task sequence planning using fuzzy Petri nets. *IEEE Transactions on Systems, Man, and Cybernetics, 25*(5), 755–768.
27. Dimirovski, G. M., Iliev, O. L., Vukobratovic, M. K., Gough, N. E., & Henry, R. M. (1994). Modeling and scheduling control of FMS based on stochastic Petri-nets. In G.C. Goodwin & R.J. Evans (Eds.), *Automatic control—World congress 1993* (vol. II, pp. 117–123). Pergamon Press.
28. Dimirovski, G. M. (2013). Vuk and Georgi: An adventure into active systems via mechatronics, robotics and manufacturing engineering. (Invited Lecture). In I.J. Rudas, B. Borovac, J. Fodor, & I. Stajner-Papuga (Eds.), *Proceedings of the 11th international conference on intelligent systems and informatics SISY2013—Remembering Miomir K, Vukobratovic,* Subotica, RS (pp 11–19). The IEEE Piscataway; Obuda University.
29. Ashby, R. W. (1956). *Introduction to cybernetics.* Chapman & Hall.
30. Holvoet, T. (1995). Agents and petri-nets. *Petri Net Newsletter, 49,* 3–8.
31. Huvenoit, B., Bourey, J. P., & Craye, E. (1995). Design and implementation methodology based on Petri-net formalism of flexible manufacturing systems control. *Production Planning & Control, 6*(1), 51–64.
32. Iliev, O. L., Dimirovski, G. M., Gacovski, Z. M., Gough, N. E., & Griffits, I. (1996). Contribution to petri-net based, object oriented modeling of communication protocols for intelligent automation. In M. Jamshidi, J. Yuh & P. Dauchez (Eds.), *Intelligent automation and control: Recent trends and developments* (vol. 4, pp. 325–330). TSI Press.
33. Lu, J., Chen, Y., Yu, X., & Hill, D. J. (2013). Multi-agent systems with dynamical topologies: Consensus and applications. *IEEE Circuits and Systems Magazine,* 21–34.

34. Lin, L., Wakabayashi, M., & Adiga, S. (1994). Object oriented modeling and implementation of control software for a robotic flexible manufacturing cell. *Journal of Robotic & Computer Integrated Manufacturing, 11*(1), 1–12.
35. Moody, J. O., & Antsaklis, P. J. (1998). *Supervisory control of discrete event systems using petri nets*. Kluwer Academic Publishers.
36. Murata, T. (1989). Petri nets: Properties, analysis and applications. *Proceedings of the IEEE, 77*(4), 541–580.
37. Peterson, J. L. (1981). *Petri net theory and the modeling of systems*. Prentice Hall.
38. Siljak, D. D. (2008). Dynamic graphs. *Nonlinear Analysis: Hybrid Systems, 2*, 544–567.
39. Zadeh, L. A., & Polak, E. (1969). *System theory*. Academic Press.
40. Zadeh, L. A. (1973). Outline of a new approach to the analysis of complex systems and decision processes. *IEEE Transactions on Systems, Man, and Cybernetics, SMC-3*(1), 28–44.
41. Chen, G., Wang, X. P., & Li, X. (2014). *Fundamentals of complex networks*, 2nd edn. J. Wiley & Sons.
42. Ziegler, B. P. (1984). *Multifaceted modeling and discrete event simulation*. Academic Press.
43. Iliev, O. L. (1997). Object models and two-level control systems with distributed intelligence for a class of FMS. Doctoral dissertation; supervisor G. M. Dimirovski. SS Cyril and Methodius University (in Macedonian).
44. Shen, J., Jing, Y., & Dimirovski, G. M. (2022). Fixed-time congestion tracking control for a class of uncertain TCP/AQM computer and communication networks. *Control, Automation, and Systems, 20*(3), 758–769.
45. Gitt, W. (1987). Information: The third fundamental quantity. *Siemens Review, 56*, 36–41.
46. Dimirovski, G. M., & Gacovski, Z. M. (2001). Research in robotics and flexible automation: What may be feasible in small developing countries. In P. Kopacek (Ed.), *Proceedings of IFAC symposium on robot control SYROCO'00*. Survey Paper S5 (pp. 1–10). Pergamon Elsevier Science.
47. Dimirovski, G. M. (2005). Fuzzy-petri-net reasoning supervisory controller and estimating states of Markov chain models. *Computing and Informatics, 24*(6), 563–576.
48. Dou, X.-M., & Peng, Y.-W. (1992). The communication protocol between CNC and host-computer in FMS and its modeling using Petri nets. *IFIP Transactions B, B-1*, 215–221.
49. Saridis, G. N., & Valavanis, K. P. (1988). Analytical design of intelligent machines. *Automatica, 24*(2), 123–133.
50. Saridis, G. N. (1989). Analytic formulation of the principle of increasing precision with decreasing intelligence for intelligent machines. *Automatica, 25*(3), 461–467.
51. Wilkins, D. E. (1984). Domain-independent planning: Representation and plan generation. *Artificial Intelligence, 22*, 269–301.
52. Yen, J., & Langari, R. (1999). *Fuzzy logic: Intelligence, control, and information*. Prentice Hall.
53. Zimmerman, H.-J. (1991). *Fuzzy sets, decision making and expert systems*. Kluwer.
54. Li, Z., Duan, Z., Chen, G., & Huang, L. (2010). Consensus of multi-agent systems and syncronization of complex networks: A unified viewpoint. *IEEE Transactions on Circuits and Systems I, 57*(1), 213–224.
55. Tunevski, A., Vukobratovic, M., & Dimirovski, G. (2001). Adaptive control of multiple manipulation on dynamical environment. *Proceedings of the Institution of Mechanical Engineers, Part I, Journal of Systems & Control Engineering, 215*, 385–404.
56. Andrievsky, B. R., Matveev, A. S., & Fradkov, A. L. (2010). Control and estimation under information constraints: Toward a unified theory of control, computation and communications. *Avtomatika i Telemekhanika, 71*(4), 34–99. *Automation and Remote Control, 71*(4), 572–633.
57. Ashby, R. W. (1970). Information flow within coordinated systems. In J. Rose (Ed.), *Progress in cybernetics* (pp. 57–60). Gordon & Bearch.
58. Kopacek, P. (Ed.) (2001). *Preprints of the 10th IFAC symposium INCOM'01 on information control problems in manufacturing*. The IFAC and IHRT-TUW.
59. Cao, T., & Sanderson, A. C. (1991). Task sequence planning using fuzzy Petri nets. In *Proceedings of 1991 International conference on systems, man and cybernetics, SMC1991*, Charlottesville, VA (pp. 349–354). The IEEE.

60. Cao, T., & Sanderson, A. C. (1998). AND/OR net representation for robotic task sequence planning. *IEEE Transactions on Systems, Man, and Cybernetics – Pt. C: Applications and Reviews, SMC-28*(2), 755–768.

61. Kosbar, K., & Schneider, K. (1992). Object oriented modeling of communication protocols. In *Proceedings MILCOM'92—Fusing command, control and intelligence* (vol. 1, pp. 68–72). The IEEE.

62. Homem de Mello, L. S., & Sanderson, A. C. (1990). AND/OR graph representation of assembly plans. *IEEE Transactions on Robotics & Automation, RA-6*(2), 188–199.

63. Sacerdoti, E. (1977). *A structure of plans and behavior.* North-Holland.

64. Korf, R. E. (1987). Planning as search: A quantitative approach. *Artificial Intelligence, 33,* 65–68.

65. Ashby, R. W. (1952). *Design for a brain: The origin of adaptive behavior.* J. Wiley.

66. Bellman, R., & Zadeh, L. A. (1970). Decision-making in fuzzy environment. *Management Science, 17,* B-141–164.

67. Bezdek, J. C. (1994). What is computational intelligence? In J. Zurada, R. Marks, & C. Robinson (Eds.), *Computational intelligence: Imitating life* (pp. 1–12). The IEEE Press.

68. Booch, G. (1986). Object-oriented development. *IEEE Transactions on Software Engineering, SE-12*(2), 211–221.

69. Coad, P., & Yourdon, E. (1991). *Object oriented analysis.* Prentice Hall.

70. Zadeh, L. A. (1975). The concept of a linguistic variable and its application to approximate reasoning—Pts. I, II, III. *Information Sciences, 8,* 199–249, 301–357; *9,* 43–80.

71. Zadeh, L. A. (1978). Fuzzy sets as a basis for a theory of possibility/. *Fuzzy Sets and Systems, 1,* 3–28.

72. Zadeh, L.A. (1980). Inference in fuzzy logic. *IEEE Proceedings, 68,* 124–131.

73. Zadeh, L. A. (1991). The calculus of If-Then rules. *IEEE AI Expert, 7,* 23–27.

74. Zadeh, L. A. (1995). Probability theory and fuzzy logic are complementary rather than competitive. *Technometrics, 37*(3), 271–276.

75. Zadeh, L. A. (1996). Fuzzy logic and calculi of fuzzy rules and fuzzy graphs: A précis. *Multiple-Valued Logic, 1,* 1–38.

76. Wu, Z. G., Shi, P., Su, H., & Chu, J. (2013). Stochastic synchronization of Markovian jump neural networks with time-varying delays using sampled data. *IEEE Transaction on Cybernetics, 43*(6), 1796–1806.

77. Shi, P., Zhang, Y., Chadli, M., & Agarwal, R. (2016). Mixed H-inf and passive filtering for discrete fuzzy-neural networks with stochastic jumps and time delays. *IEEE Transactions on Neural Networks and Learning Systems, 27*(4), 903–909.

78. Ren, T., Sun, S., YanJie, X., Zhe, L., Wang, R., & Dimirovski, G. M. (2020). Synchronization for multi-networks with two types of faults: Pinning control effects. *IET Control Theory and Applications, 14*(11), 1407–1507.

79. Wang, K., Liu, Y., Liu, X., Jing, Y., & Dimirovski, G. M. (2020). Adaptive finite-time congestion controller design of TCP/AQM systems based on neural network and funnel control. *Neural Computing and Applications, 32*(13), 9471–9478.

80. Wang, K., Liu, Y., Liu, X., Jing, Y., & Dimirovski, G. M. (2019). Study on TCP/AQM network congestion with adaptive neural network and barrier Lyapunov functions. *Neurocomputing, 363,* 27–34.

81. Ozen, F., Bilgin Tukel, D., & Dimirovski, G. (2017). Synchronized dancing of an industrial manipulator and humans with arbitrary music. *Acta Polytechnica Hungarica—Journal of Applied Sciences, 14*(2).

82. Planque, P. A., Bastide, R., Dourte, L., & Ciberting-Blanc, C. (1993). Design of user driven interfaces using Petri nets and objects. In C. Rolland, F. Bodard, & C. Cauvet (Eds.), *Proceedings of the 5th international conference on advanced information systems engineering,* Paris (pp. 569–585). Springer.

83. Looney, C. G. (1988). Fuzzy Petri nets for rule-based decision making. *IEEE Transactions on Systems, Man & Cybernetics, SMC-18*(10), 755–768.

84. Zadeh, L. A. (1965). Fuzzy sets. *Information & Control, 8,* 338–353.

85. Shi, P., & Shen, Q. (2016). Cooperative control of multi-agent systems with unknown state dependent controlling effects. *IEEE Transactions on Automation Science and Engineering, 64*(2), 133–142.
86. Shi, P., & Shen, Q. (2017). Observer-based leader-following consensus of uncertain nonlinear multi-agent systems. *International Journal of Robust Control and Nonlinear Systems, 27*(17), 3794–3811.
87. Radiya, A., & Sargent, R. G. (1989). ROBS: Rules and object based simulation. Tucson, AZIn M. S. Elzas, T. I. Oren, & B. P. Ziegler (Eds.), *Modeling and methodology: Knowledge systems' paradigms* (pp. 242–256). Society for Computer Simulation International.
88. Valette, R., Cordoso, J., & Dubois, D. (1989). Monitoring manufacturing systems by means of Petri nets with imprecise markings. In *Proceedings of the IEEE international symposium on intelligent control*, Albany, NY (pp. 233–237). The IEEE.
89. Han, Y., Jing, Y., & Dimirovski, G. M. (2020). An improved fruit fly algorithm –unscented-Kalman-filter Echo-state network method for time series prediction of network traffic with noises. *Transaction of the Institute of Measurement and Control, 42*(7), 1281–293.
90. Tsuji, K., & Masumoto, T. (1990). Extended Petri net models for neural networks and fuzzy inference engines. In *Proceedings of the IEEE international symposium on circuits and systems*, New York (pp. 2670–2673). The IEEE.
91. Chen, G., & Duan, Z. S. (2008). Networks synchronizability analysis: A graph-theoretic approach. *Chaos, 18*(3), 037102.
92. Leborg, P., Jombom, P., Skold, M., & Torne, A. (1992). A model for execution task level specifications for intelligent and flexible manufacturing systems. In *Proceedings of international symposium artificial intelligence transfer* (pp. 74–83). The AAAI Press.
93. Joannis, R., & Krieger, M. (1992). Object oriented approach to the specification of manufacturing systems. *IEEE Transactions on Computer Integrated Manufacturing, CIM-5*(2), 133–145.
94. Bennett, M., & Clark, D. (1994). Object oriented modeling of communication protocols. In *Preprints of IEE colloquium on computer modeling of communication systems* (pp. 1/1–8). The IEE.
95. Hollot, C. V., Misra, V., Towsley, D., & Gong, V. B. (2002). Analysis and design of controllers for AQM routers supporting TCP flows. *IEEE Transaction on Automatics Control, 47*(6), 945–959.
96. Hoare, C. A. R. (1985). *Communicating sequential processes*. Prentice Hall.
97. Vladev, G., Kolemisevska-Gugulovska, T., & Tukel, D. (2012). FMS intelligent supervision and control 2: Communication among machines. In K. Atanassov & V. Jotsov (Eds.), *Proceedings 2012 IEE 6th International conference 'intelligent systems*, Sofia, BG. 6–8 September (pp. 291–295). IEEE; IEEE IM/CS/SMC.
98. Yu, W., Chen, G., Lu, J., & Kurths, J. (2009). Synchronization via pinning control in general complex systems. *SIAM Journal of Control and Optimization, 51*(2), 1395–1416.

Using Fuzzy Set Approaches
for Linguistic Data Summaries

Ronald R. Yager and Fred Petry

Abstract The basics of linguistic summary approaches using fuzzy set techniques are described. Following we describe the quantification of the information contained in a linguistic summary and finally we consider the formulation of summaries involving rich concepts.

1 Introduction

In [1, 2] Yager introduced the idea of linguistic summaries as a user-friendly method of summarizing information in a database. Kacprzyk, Bouchon-Meunier and other researchers [3–16] have made considerable use of the idea of linguistic summary. Here we look at a number of ideas related to linguistic summaries. We first present the fundamental ideas involved in linguistic summaries, we next look at the quantification of the information contained in a summary and finally we look at the formulation of summaries involving rich concepts.

2 Linguistic Summaries Basics

Let D be a database, $D = \{d_1, \ldots, d_n\}$ where the d_i are the objects in the database. Then V is some attribute, with domain X, associated with the database objects. For example, if each d_i is a person then V could be their age. So for each d_i we have a value $V(d_i) = b_i$ where $b_i \in X$. Associated with attribute V is a data set $B = [b_1, \ldots, b_n]$,

R. R. Yager
Machine Intelligence Institute, Iona College, New Rochelle, NY, USA
e-mail: yager@panix.com

F. Petry (✉)
Naval Research Laboratory, Stennis Space Center, MS, USA
e-mail: frednavy3@gmail.com

© The Author(s), under exclusive license to Springer Nature Switzerland AG 2023
Y. P. Kondratenko et al. (eds.), *Artificial Intelligence in Control and Decision-making Systems*, Studies in Computational Intelligence 1087,
https://doi.org/10.1007/978-3-031-25759-9_3

a bag, containing the values of V assumed by the objects is the database Y. We emphasize that a bag allows multiple elements with the same value.

A linguistic summary associated with V is a global statement based on the values in B. If V is the attribute age some examples of simple linguistic summaries are

Most people in the database D are about 35 years old
Few people in the database are elderly
Nearly third of the people in the D are middle aged

A simple linguistic summary is then a statement of the form:

Q objects in the database have V is *S*.

S is called the summarizer and *Q* is called the quantity in agreement. Also associated with a linguistic summary is a measure of validity of the summary, τ. τ indicates the truth of statement that **Q** objects have the property that V is *S* based on the data set B.

A fundamental characteristic of this formulation is that the summarizer and quantity in agreement are expressed in linguistic terms. One advantage of the use of linguistic summaries is that they provide statements about the dataset in terms that are very easy for people to comprehend. In [17] Yager showed that linguistic summarizes are closely related to what Zadeh called Z-numbers [18].

Fuzzy subsets can provide formal semantics for terms used in linguistic summaries. This will be used to formalize the summarizer and quantity in agreement as fuzzy sets enable evaluation of the validity of a linguistic summary. This process is based upon determination of the compatibility of the linguistic summary with the data set B. For a given attribute there are numerous various summaries, so using the data set B the validity, τ, of a conjectured linguistic summary can be determined.

2.1   Validation of Linguistic Summaries

For the approach to validating a linguistic summary use will be made of the representation of a linguistic summarizer by a fuzzy subset over the domain of the attribute. Let V be an attribute with value from the domain X and S is some concept associated with this attribute. Then the concept will be represented by a fuzzy subset S on X. So for $x \in X$, $S(x) \in [0, 1]$ is the degree of compatibility of x with the concept **S**. If V is age and **S** is the concept middle age then $S(45)$ indicates the degree to which 45 years old is compatible with the idea of middle age. Even in environments where the underlying domain is non-numeric this approach permits one to obtain numeric values for the membership grade in the fuzzy subset S corresponding to concept **S**. For example if V is the attribute residence city from the domain of cities in the U.S., the concept **S**, "lives near Chicago", can be a fuzzy subset. The second component for linguistic summaries is the quantity in agreement **Q,** where these are linguistic quantifiers [19]. Examples of linguistic quantifiers are terms such as most, few, about half, all. Essentially linguistic quantifiers are fuzzy proportions, an alternative view of these subjects are generalized logical quantifiers. In [19] Zadeh suggested we could represent these linguistic quantifiers as fuzzy subsets of the unit interval. Using this representation the membership grade of any proportion $r \in [0, 1]$ in the fuzzy set Q corresponding to the linguistic quantifier **Q**, Q(r), is a measure of the compatibility of the proportion r with the linguistic quantifier we are representing by the fuzzy subset Q. For example if Q is the fuzzy set corresponding to the quantifier *Most* then Q(0.9) represents the degree to which the proportion 0.9 satisfies the concept *Most*.

2.2 Natural Language Linguistic Quantifiers

In [20] Yager identified three classes of linguistic quantifiers covering most of those used in natural language. (1) Q is said to be monotonically non-decreasing if $r_1 > r_2$ $\Rightarrow Q(r_1) \geq Q(r_2)$, examples of this type of quantifier are *at least 20%, most, all*. (2) A quantifier Q is said to monotonically non-increasing if $r_1 > r_2 \Rightarrow Q(r_1) \leq Q(r_2)$, examples of this type of quantifiers are *at most 30%, few, none*. (3) A quantifier Q is said to be unimodal if there exists two values $a \leq b$ both contained in the unit interval such that for $r < a$, Q is monotonically non-decreasing, for $r > b$, Q is monotonically non-increasing and for $r \in [a, b]$, $Q(r) = 1$, an example of this type of quantifier is *about 0.4*.

Antonyms can also be associated with linguistic quantifiers. If Q is a linguistic quantifier its antonym is also a linguistic quantifier, denoted $\widehat{Q}$, such that $\widehat{Q}(r) = Q(1 - r)$. The operation of taking an antonym is involutionary, that is $\widehat{\widehat{Q}} = Q$. So antonyms come in pairs. Typical examples of antonym pairs are **all-none** and **few-many**. Consider the quantifier *at most 0.2* defined as $Q(r) = 1$ if $r \leq 0.2$ and $Q(r) = 0$ if $r > 0.2$ Its antonym has $\widehat{Q}(r) = 1$ if $r \geq 0.2$ and $\widehat{Q}(1 - r) = 0$ if $r \geq 0.2$. This is equivalent to $\widehat{Q}(r) = 1$ if $r \geq 0.8$ and $\widehat{Q}(r) = 0$ if $r < 0.8$. So the antonym of *at most 0.2* is *at least 0.8*.

One must distinguish between the antonym of a quantifier and its negation. The negation of Q denoted $\overline{Q}$ was defined such that $\overline{Q}(r) = 1 - Q(r)$. So the negation of *at most 0.2* is $\overline{Q}(r) = 0$ if $r \leq 0.2$ and $\overline{Q}(r) = 1$ if $r \geq 0.2$, this corresponds *to at least 0.2*.

After discussing ideas of summarizers and quantity in agreement, now it is possible to describe a methodology used to calculate validity τ of a linguistic summary. Assume $D = [a_1, a_2, ..., a_n]$ is the collection of values that appear in the database for the attribute V. Consider the linguistic summary:

Q items in the database have value for V that are S.

The basic procedure to obtain the validity τ of this summary in the face of the data D is:

(1) For each a_i in D, calculate $S(a_i)$, the degree to which a_i satisfies the summarizer S.
(2) Let $r = \frac{1}{n} \sum_{i=1}^{n} S(a_i)$, the proportion of D that satisfy S.
(3) $\tau = Q(r)$, the grade of membership of r in the proposed quantity in agreement.

2.3 Examples

Assume we have a database consisting of 10 entries. Let D be the collection of ages associated with these entries: D = $<30, 25, 47, 33, 29, 50, 28, 52, 19, 21>$

1. Consider the linguistic summary ***most*** people are ***at least 25***. In this case the summarizer "at least 25" can be expressed simply as $S(x) = 0$ if $x < 25$ and $S(x) = 1$ if $x \geq 25$. We define the quantity in agreement "most" by the fuzzy subset $Q(r) = 0$ if $r < 0.5$ and $Q(r) = (2r - 1)_{1/2}$ if $r \geq 0.5$. For the first eight items in the data collection $S(x) = 1$ while for the remaining two items have $S(x) = 0$. Thus in this case $r = 8/10 = 0.8$ and hence $\tau = Q(0.8) = 0.77$.

2. Consider the linguistic summary "***About half*** *the people have ages are* ***near 30***." Here we define "near thirty" by $S(x) = e^{-(\frac{x-30}{25})^2}$ and we define "about half" as $Q(r) = e^{-(\frac{r-0.5}{0.25})^2}$.
   In this case $r = 3.94/10 = 0.394$ and $\tau = Q(0.394) = 0.95$.

3. Consider the linguistic summary ***most*** *of the people are* ***young***. Here we define young such that $S(x) = 1$ if $x < 20$, $S(x) = -\frac{1}{10} x + 3$ if $20 \leq x \leq 30$ and $S(x) = 0$ if $x > 30$. In this case $r = 0.27$ and using our previous definition of most we get $\tau = Q(0.27) = 0$.

A number of interesting properties can be associated with these linguistic summaries. Consider a proposed summary "**Q** items have V is **S**" and assume the data set D has cardinality n. The associated validity is $\tau = Q\left(\frac{\Sigma_i S(x_i)}{n}\right)$. Now consider the summary

$$\widehat{Q} \text{ items have V is (\underline{not} S)}$$

where $\widehat{Q}$ is the antonym of Q. In this case the measure of validity τ is

$$\tau_1 = \widehat{Q}\left(\frac{\Sigma_i \overline{S}(x_i)}{n}\right) = \widehat{Q}\left(\frac{\Sigma_i (1 - S(x_i))}{n}\right) = \widehat{Q}\left(1 - \frac{\Sigma_i S(x_i)}{n}\right)$$

however, since $\widehat{Q}(1 - r) = Q(r)$ we see that $\tau_1 = \tau$. These two linguistic summaries have the same measure of validity. The prototypical manifestation of this is that for a given piece of data D the summary *Many people are middle-aged* will have the same validity as *Few people are not middle-aged*.

Consider now the summary "Not Q objects are S." In this case its validity τ_2 is

$$\tau_1 = \overline{Q}\left(\frac{\Sigma_i S(x_i)}{n}\right) = 1 - Q\left(\frac{\Sigma_i S(x_i)}{n}\right) = 1 - \tau$$

This statement has validity complementary to our original proposition. From this we see that

$$\tau(S, \ Q) + \tau(S, \overline{Q}) = 1.$$

2.4 Multi-attribute Linguistic Summaries

Thus far we have considered linguistic summaries involving only one attribute. The approach described above can be extended to the case of multiple attributes from a database. We shall first consider linguistic summaries of this form

*Most people in the database are **heavy** and **middle-aged**.*

Assume U and V are two attributes appearing in the database. Let R and S be concepts associated with each of these attributes respectively, then the generic form of the above linguistic summary is

Q objects in the database have U is *R* <u>and</u> V is *S*.

In this case our data consists of a collection of pairs, $D = \, < (a_1, b_1), (a_2, b_2), \ldots,$ $(a_n, b_n) >$ where $U(x_i) = a_i$ is a value of attribute U and $V(x_i) = b_i$ is a value for attribute V. Our procedure for obtaining the validity of the linguistic summary in this case is.

1. For each i calculate $R(a_i)$ and $S(b_i)$.
2. Let $r = \frac{1}{n}\sum_{i=1}^{n}(R(a_i)S(b_i))$
3. $\tau = Q(r)$

The above procedure can be easily generalized to consider any number of attributes. We also can consider linguistic summaries of the forms.

Q objects in the database have U is *R* <u>or</u> V is *S*.

In this case the procedure is the same except in step two the product is replaced by a union operation, t-conorm, such as $Max(R(a_i), S(b_i))$ or $R(a_i) + S(b_i) - R(a_i)S(b_i)$.
Consider now another class of linguistic summaries manifested by statements like

*Most **heavy** people in the database are **middle-aged**.*

This form of linguistic summary is related to the type of association rule discovery that is of great interest in many applications of data mining. The linguistic summary expressed here can be equivalently expressed as

In *most* cases of our data; if a person is **heavy** then they are **middle-aged**.

Here then we have an embedded association rule.

If weight is **heavy** then age is **middle − aged**.

Furthermore we are qualifying our statement of this association rule with the quantifier **most**.

In this case we have as our generic form

Q of the U is **R** objects in the database have V is **S**

In the above we call R the **qualifier** of the summary. The procedure for calculating the validity of this type of linguistic summary has a similar three-step process.

1. For each data pair (a_i, b_i) calculate $R(a_i)$ and $S(b_i)$.
2. Calculate $r = \frac{\sum_{i=1}^{n} R(a_i) S(b_i)}{\sum_{i=1}^{n} R(a_i)}$
3. $\tau = Q(r)$

A fundamental distinction between this and the previous case is in step two, here instead of dividing by n, the number of objects in the database, we divide by the number of objects having R.

We note that we can naturally extend this procedure to handle summaries of the form:

Most **middle-aged** people in the database are **heavy** and live **around Boston**

Few **lower-income** and **middle-aged** people in the database live in the **suburbs**.

Consider the linguistic summary corresponding to the quantified association rule QR are S and the related summary $\widehat{Q}$ R are $\overline{S}$ where $\widehat{Q}$ is the antonym of Q and $\overline{S}$ is the negation of S. The validity of the first summary is $\tau_1 = Q(r)$ where $r = \frac{\sum_{i=1}^{n} R(a_i) S(b_i)}{\sum_{i=1}^{n} R(a_i)}$.

The validity of the second summary is $\tau_2 = \widehat{Q}(r^*)$ where $r^* = \frac{\sum_{i=1}^{n} R(a_i)(1 - S(b_i))}{\sum_{i=1}^{n} R(a_i)} = 1 − r$. However we see that

$$\tau_2 = \widehat{Q}(r*) = Q(1 − r*) = Q(1 − (1 − r)) = Q(r) = \tau_1$$

Hence the above two statements have the same validity for a given data set D. Thus we see that the statement.

Most senior employees have *high* salaries

has the same validity as the statement

*Few senior employees have **not-high*** salaries

3 Information Measures of Summaries with Linguistic Content

One reason for linguistic summaries is providing useful global information about the database. For the usefulness of a summary a significant measure is one indicating the amount of information conveyed in the summary. One natural consideration is that only the degree of validity of a summary measures information about usefulness of the summary. However see otherwise. Consider database of employees and the possible summary

Most employees are over 15years old.

Although this is valid there is not significant information. Indeed this follows an observation about the measure of validity. That is, consider the summaries: Q objects are S_1 and Q objects are S_2 for which Q is a monotonically increasing quantifier. If $S_1 \subseteq S_2$, $S_1(x) \leq S_2(x)$ for all x, then it follows that $\tau_2 \geq \tau_1$. So for simple monotonic increasing quantifiers, the validity of a summary can always be increased by broadening summarizers used in a summary. Likewise if Q_1 and Q_2 are two monotonically increasing quantifiers so that $Q_1 \subseteq \tau Q_2$ then for a fixed S, $\tau(Q_1, S) \leq \tau(Q_2, S)$. As an example the summary "at least 60% of the employees are heavy" will have a smaller degree of validity then the summary "at least 30% of them are heavy." So if the quantifier or summarizer is made broader this increases the validity. On the other hand too much broadening can lead to the situation where content of a summary is vacuous. This then indicates that production of useful information by linguistic summaries requires trade-off between the size quantifiers size and summarizer used and the resulting validity.

3.1 Informativeness of Linguistic Summaries

We now will describe informativeness associated with simple linguistic summaries based on concepts in fuzzy set theory. Let V be an attribute with domain X and F_1, F_2, ..., F_q be a collection of fuzzy subsets corresponding to linguistic concepts associated with this attribute. We can use the specificity of a fuzzy subset [21] that determines the degree to which the fuzzy subset focuses on one element as its representative. Consider that a concept, "35 years old" is more specific then "about thirty-five" which is then more specific then "at least 25." Next we use measures of specificity associated with normal fuzzy subsets over a space X [22].

Let the domain of V be an interval $X = [a, b]$ and F a normal fuzzy subset defined over X. So the specificity is $Sp(F) = 1 - \frac{1}{b-a} \int_a^b F(x)dx$. In this case this is the negation of average membership grade.

This specificity measure has the following properties

(1) $Sp(F) = 1$ if $F(x) = 1$ for exactly one element and $F(x) = 0$ for all other elements
(2) $Sp(F) = 0$ if $F = X$
(3) $Sp(F_1) \geq Sp(F_2)$ if $F_1 \subset F_2$.

For linguistic quantifiers since $X = [0, 1]$ one obtains $Sp(Q) = 1 - \int_0^1 Q(x)\, dx$, the negation of the area of the quantifier. Also if $\overline{Q}$ is the negation of Q then $Sp(\overline{Q}) = 1 - \int_0^1 \overline{Q}(x)\, dx = 1 - \int_0^1 (1 - Q(x))\, dx = \int_0^1 Q(x)\, dx = 1 - Sp(Q)$. However if it is an antonym of Q, $\widehat{Q}(x) = Q(1 - x)$, then $Sp(\widehat{Q}) = Sp(Q)$. So the specificity of antonyms of quantifiers is the same as the specificity of the quantifier, but the negation has a specificity the same as the complement.

Now consider the specificity of a number of prototypical quantifiers starting with non-decreasing quantifiers. Consider the quantifier "at least α", $Q(x) = 0$ if $x < \alpha$ and $Q(x) = 1$ if $x > \alpha$. In this case $Sp(Q) = \alpha$. Another quantifier is $Q(x) = x^\beta$, where $\beta > 0$, and here $Sp(Q) = 1 - \int_0^1 x^\beta dx = \frac{\beta}{1+\beta}$. The specificity increases as β increases.

Specifying "most" as $Q(r) = 0$ if $r < 0.6$ and $Q(r) = (2r - 1)^{1/2}$ if $r \geq 0.6$. For this quantifier we get $Sp(Q) = 2/3$. In general examine the class $Q(r) = 0$ if $r < \alpha$ and $Q(r) = (\frac{r-\alpha}{1-\alpha})^\beta$ if $r \geq \alpha$, letting $\alpha \leq 1$ and $\beta \geq 0$. Then $Sp(Q) = \frac{\beta+\alpha}{\beta+1}$. So we can say $Sp(Q)$ provides a crisp approximation of the quantity in agreement to some extent and refer to this as the focus of a quantifier.

Now examine the decreasing quantifiers, for which $Q(r_1) \geq Q(r_2)$ if $r_1 < r_2$. As before if Q is a decreasing quantifier then the antonym $\widehat{Q}$, ($\widehat{Q}(r) = Q(1 - r)$) is a non-decreasing quantifier. The conclusion is that non-increasing and non-decreasing quantifiers always correspond to antonym pairs. Additionally because $Sp(Q) = Sp(\widehat{Q})$ the pairs have the identical specificity. Specifically observe that "at most α" has as its antonym "at least $(1 - \alpha)$" which has specificity $1 - \alpha$.

An important class of uni-modal quantifiers are the ones centered about a value a and with a spread $2b$, $Q(r) = 0$ if $0 \leq r \leq a - b$, $Q(r) = 1$ if $a - b \leq r \leq a + b$ and $Q(r) = 0$ if $r \geq a + b$, having assumed $a - b \geq 0$ and $a + b \leq 1$. Then $Sp(Q) = 1 - \int_{a-b}^{a+b} dx = 1 - 2b$. Now the focus of a uni-modal quantifier Q can be defined as $FOC(Q) = \frac{\int_0^1 Q(x)\, x\, dx}{\int_0^1 Q(x)\, dx}$.

3.2   Linguistic Summaries and Specificity

Next information associated with a linguistic summary can be discussed using specificity. Consider a typical simple linguistic summary involving a non-decreasing quantifier: Q_1 objects have V is S_1. Based on database values assume we can establish this summary has validity equal to τ_1. This sort of statement then provides more useful information for larger validity. Also informativeness is increased because if S_1 is a narrow fuzzy subset, then S_1 specificity is large. Relative to Q_1 a higher focus is better, a large specificity corresponding to a narrow fuzzy subset. This leads to development of a measure of useful information for this type of linguistic summary

$$I(Q_1, \ S_1) \ = \ \tau_1 Sp(Q_1) \ Sp(S_1)$$

In the situation of a simple non-increasing quantifier, Q_2 objects have V is S_2 which has validity τ_2. This is equivalent to a summary

$$\widehat{Q}_2 \ \text{objects have V is } \overline{S}_2$$

where $\widehat{Q}_2$ is an antonym of Q_2. Because Q_2 is non-increasing quantifier, $\widehat{Q}_2$ is non-decreasing and so a measure of usefulness of summary can use the preceding form $I(Q_2, S_2) = \tau_2 \ Sp(\widehat{Q}_2) \ Sp(\overline{S}_2)$. Also since for antonym $Sp(\widehat{Q}_2) = Sp(Q_2)$, the information can be expressed as $I(Q_2, S_2) = \tau_2 \ Sp(Q_2) \ Sp(\overline{S}_2)$.

Next we examine unimodal statements such as "Q_3 objects have V is S_3 with validity τ_3". More analysis is needed and informativeness is increased by making Q_3 and S_3 specific and τ_3 large, leading to the form $I(Q_3, S_3) = \tau_3 \ Sp(Q_3) \ Sp(S_3)$. However location must be taken in to account. Two summaries to consider are: **About 65%** of the objects are *heavy*; **About 20%** of the objects are *heavy*. More useful information is provided by the first statement. To illustrate this we utilize of the focus of a quantifier. This leads to the following measure of usefulness of a uni-modal quantifier

$$I(Q_3, \ S_3) \ = \ Foc(Q_3) \ Sp(Q_3) \tau Sp(S_3)$$

3.3 Informativeness of Summaries with Association Rules

Next informativeness of summaries involving association rules, such as *Q heavy people are older* are described representing this as **QR** are **S**. The first case is Q is a non-decreasing quantifier. We must consider the effect of R on informativeness. Examine the two propositions

All fifty $-$ fiveyear olds are heavy

All people are heavy

Clearly the second summary provides more information. In general the broader the context of an antecedent, all else being equal, the more informative a summary.

So for the measure of informativeness we use the form

$$Q_1 R_1 \; are \; S_1$$

where Q_1 is non-decreasing.

$$I(Q_1, \; R_1, \; S_1) = \tau_1 Sp(Q_1) \; Sp(S_1) \; (1 - Sp(R_1))$$

So the desirable characteristics are wide antecedents, narrow consequents and large quantities in agreement as well as large validity.

If R_1 is the whole space then the previous statement becomes QX are S_1 an unqualified summary. Here $R = -X$, so the specificity of R is $Sp(X) = 0$ and $(1 - Sp(R)) = 1$. This means the suggested measure reduces to the unqualified measure.

For qualified summaries involving non-increasing or unimodal quantifiers' measure of informativeness can be extended by adding the term $(1 - Sp(R))$ such as for qualified summaries involving non-decreasing quantifiers.

4 Concepts and Hierarchies

For linguistic summaries of the form "Q objects in a database have V is S" where "V is S" is a *property* associated with values in a database. To determine τ, validity of this linguistic summary, the validity t_i of the property must be computed for each element y_i in the database. If $V(y_i)$ is a value of an attribute V for object y_i then $t_i = S(V(y_i))$ is the degree to which object y_i has property S.

Hence t_i is the membership grade of $V(y_i)$ in fuzzy subset S. This can also be viewed as truth of statement: y_i has property V is S. So $r = \frac{1}{n} \sum_{i=}^{n} t_i$, the proportion of objects in Y having the property. So the validity of the statement "Q objects in a database have property V is S" can be determined as $\tau = Q(r)$.

4.1   Rich Concepts

We wish to extend the range of types of properties that can be included in a linguistic summary using those properties based only on a single attribute associated with objects in the database to a class of more sophisticated description of properties associated with objects in the database. These are now called **rich concepts**. So we are interested in statements of form,

Q objects in the database have the property $''concept''$

Some examples of this could be

Many objects in the database are "upwardly mobile"

At least half the people in the database are "potential customers"

Around half people in the database are "affluent"

A fundamental aspect of these concepts is that they are not necessarily a directly observable attribute of the objects but may involve the combination of attributes associated with the objects in the database. That is these concepts can involve a complex aggregation of attributes and other concepts associated with the objects in the database.

One feature required of these concepts is that if Con is a concept then for any object y_i we can obtain the degree to which object y_i satisfies the concept, $Con(y_i)$, such that $Con(y_i) \in [0, 1]$. This property allows determination of validity of linguistic summaries of the form "Q objects in database have property Con." The procedure developed earlier to obtain validity on this linguistic summary involving the property Con can be used as follows:

(1) For each object y_i in the database obtain the degree to which it satisfies the concept Con, $t_i = Con(y_i)$

(2) Then obtain the proportion of elements in the database satisfying the concept Con, $r = \frac{1}{n} \sum_{i=1}^{n} Con(y_i)$.

(3) Lastly the validity of the linguistic summary is $\tau = Q(r)$.

So the central issue is determination of $Con(y_i)$ that we now consider.

Assume our database is such that each object has attributes V_j for $j = 1$ to q and each attribute takes a value in its domain, X_j. Hence $V_j(y_i) \in X_j$ is the value of attribute V_j for object y_i.

Associate with V_j is the statement V_j *is* S_{jk} where S_{jk} is a fuzzy subset over domain X_j representing a linguistic value associated with V_j. Such statements of the form V_j *is* S_{jk} are termed an ***atomic concept*** since truth for object y_i can be directly obtained as $t_i = S_{jk}(V_j(y_i)))$. This atomic concept will often be represented simply as a Con $= <S_{jk}>$ where the associated attribute V_j is implicit.

4.2 Aggregation of Concepts

Next concepts can be constructed from other concepts. Let Agg be an aggregation operator [23] taking values in the unit interval and returning values in the unit interval. So Agg can construct a new concept from other concepts,

$$Con = Agg < Con_1, Con_2, \ldots, Con_m > \tag{1}$$

The validity of Con for a database object y_i is

$$Con(y_i) = Agg(Con_1(y_1), Con_2(y_1), \ldots, Con_m(y_i))$$

Fig. 1 Structure of
composed concept

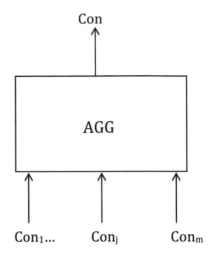

As each $Con_k(y_i) \in [0, 1]$ then $Con(y_i) \in [0, 1]$, so Con is a valid concept.

Con shall be termed the composed concept and Con_j for $j = 1$ to n as the constituent concepts.

In the special case where all the constituent concepts are atomic concepts,

$$Con = Agg(S_{j_1 k_1}, S_{j_2 k_2}, \ldots, S_{j_m k_m}), \text{ then } Con(y_i) = Agg(S_{j_1 k_1}(V_{j_1}(y_i)), \ldots, (S_{j_m k_m}(V_{j_m}(y_i))).$$

While Agg has some formal mathematical properties, the actual algorithmic process of combining the $Con_j(y_i)$, to obtain $Con(y_i)$ often utilizes cognitive/semantic structures underlying intended meaning of the composed concept.

Formula (1) can be illustrated as a module as shown in Fig. 1.

4.3   Hierarchical Framework for Rich Concepts

Using the preceding ideas we now provide a hierarchical framework that can be used for obtaining *rich concepts* from primal properties of the database objects. In the following we shall Fig. 2 useful in our discussion.

In Fig. 2 a box/module indicates a concept composed by aggregation of constituent concepts. An atomic concept is represented by a circle.

The hierarchy starts with a concept of interest, concept 1, defined in terms of the aggregation of other concepts (constituent concepts), here concepts 2 and concept 3. This structuring continues until it ends with an atomic concept. This sort of hierarchy is termed "**rich concept**". Each branch in the hierarchy terminates in an atomic concept V_j is S_{jk} having the associated attribute V_j. We note that a_{ji} the value of the attribute V_j for object y_i, $V_j(y_i)$, is retrieved from the database. Also for object y_i the truth-value of this branch terminating atomic proposition, V_j is S_{jk}, can be obtained as $S_{jk}(a_{ji})$, the membership grade of a_{ji} in the fuzzy subset S_{jk}.

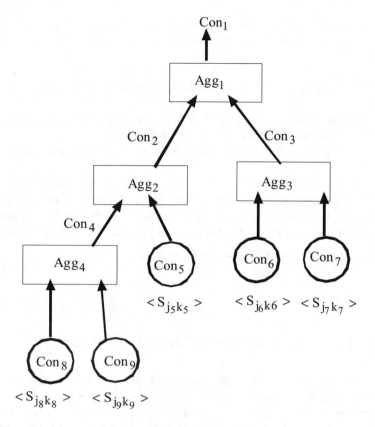

Fig. 2 Hierarchical framework for constructing *rich concepts*

Determining degree of truth of a rich concept for a database object y_i proceeds upward in the hierarchy. Starting from the bottom of the hierarchy and determine the truth of each terminal atomic concept for object y_i. Here these could directly retrieved using the attribute values associated with objects in the database. Then using the hierarchy structure to determine the degree to which y_i satisfies the rich concept defined by the hierarchy is determined. In particular truth of a composed concept related to Agg module in the hierarchy is obtained using the prescribed aggregation of the truth-values of the input constituent concepts.

Thus using the structure of the hierarchical description of the rich concept along with the information contained in the database about with the objects, the y_i, we can obtain, t_i, the degree to which y_i satisfies the rich concept RC. Here then $RC(y_i) = t_i$. Using the collection of these t_i's from each of the elements in the database we are able to determine τ the validity of the linguistic summary "Q objects in the database have property **Rich Concept**."

Let us now consider qualified linguistic summaries that are exemplified by statements of the form **Most tall** people in the database are *young*. This statement has a generic form:

$$Q \text{ of the U is } R \text{ objects in the database have V is } S$$

We see here that both U is R and V is S are examples of atomic concepts. This inspires us to consider linguistic summaries of the form:

Q of the objects in a database having property **Concept-1** also have Property **Concept 2**.

Here then each of Concept 1 and Concept 2 can be defined using a concept hierarchy of the type described earlier which allows the determination of $Con_1(y_i)$ and $Con_2(y_1)$, the truths of **Concept-1** and **Concept-2** for each of the objects in Y. Using this we can the validity τ of the preceding linguistic summary as follows:

(1)   For each y_i calculated $Con_1(y_1)$ and $Con_2(y_1)$

(2) $r = \dfrac{\sum_{i=1}^{n} Con_1(y_i)\,(Con_2(y_i)}{\sum_{i=1}^{n} Con_1(y_i)}$

(3) $\tau = Q(r)$

5   Composing Concepts Using the OWA Operator

Aggregation operators to formulate composed concepts from other constituent concepts include the Ordered Weighted Averaging, OWA, operator [24, 25].

Definition: An OWA operator of dimension n has an associated weighting vector $W = [w_1, \ldots, w_n]$ in which **1.** $w_j \in [0, 1]$ and **2.** $\sum_{j=1}^{n} w_j = 1$ and where $OWA(a_1, \ldots, a_n) = \sum_{j=1}^{n} w_j b_j$ with b_j being the *j*th largest of the a_i. If we let $\rho = \{1, \ldots, n\} \to \{1, \ldots, n\}$ be a bijective index function such $\rho(j)$ is the index of the *j*th largest of a_i then we can express $OWA(a_1, \ldots, a_n) = \sum_{j=1}^{n} w_j\, a_{\rho(j)}$.

By using different weighting vectors, W, one can obtain different aggregation operators such as:

(1)   If $w_j = 1/n$ for all j then $OWA(a_1, \ldots, a_n) = \frac{1}{n} \sum_{j=1}^{n} a_j$, the simple average

(2)   If $w_n = 1$ and $w_j = 0$ for $j \neq n$ then $OWA(a_1, \ldots, a_n) = Min_j[a]..$

(3)   If $w_1 = 1$ and $w_j = 0$ for $j \neq 1$ then $OWA(a_1, \ldots, a_n) = Max_j[a]$.

(4)   4) If $w_1 = \alpha$ and $w_n = 1 - \alpha$ and all other $w_j = 0$ then we get

$$OWA(a_1, \ldots, a_n) = \alpha Max_i[a_i] + (1 - \alpha)Min_i[a]$$

5.1   OWA Weights Using Linguistic Quantifiers

The weighting vector in OWA aggregation can be developed by using linguistic quantifiers where a linguistic quantifier is an expression corresponding to a proportion of objects such as *most, few, about 1/4*. Let Q be a linguistic expression corresponding to a quantifier such as *few*. Then following Zadeh [19] this can be represented as a fuzzy subset Q over the unit interval $I = [0, 1]$ such that for any proposition $r \in I$, Q(r) indicates the degree to which r satisfies the meaning of the quantifier Q. The fuzzy subset corresponding to a linguistic quantifier can be used to obtain an OWA weighting vector W [26]. However here we only utilize RIM, regular monotonic, quantifiers. A fuzzy subset $Q: I \to I$ represents a RIM linguistic quantifier if (1) $Q(0) = 0$ (2) $Q(1) = 1$ and (3) if $r_1 > r_2$ then $Q(r_1) \geq Q(r_2)$, it is monotonic. RIM quantifiers capture the class of quantifiers where an increase in t proportion r results in an increase in the compatibility with the linguistic quantifier being modeling. Examples of these RIM quantifiers are *at least one, all, at least α%, most, some*.

Let Q be a fuzzy subset corresponding a RIM quantifier. Then for Q we have an OWA weighting vector W such that for $j = 1$ to n we have $w_j = Q(\frac{j}{n}) - Q(\frac{j-1}{n})$. These w_j satisfy the required conditions of the OWA weights, each $w_j \geq 0$ and $\sum_{j=1}^{n} w_j = 1$.

5.2 OWA Concept Modules

Using this we can introduce a basic OWA concept module CM = <CM$_1$, CM$_2$, ..., CM$_n$:Q>. This concept module consists of a collection of constituent concepts {Con$_1$, ..., Con$_n$} used for the definition of Con, and a linguistic quantifier Q indicating the proposition of these constituent concepts required to be satisfied.

So for a given object y in the database having satisfaction Con$_i$(y) to the *i*th constituent concept in the formulation of Con(y) we get Con(y) = $\sum_{j=1}^{n} w_j$ Con$_{\rho(j)}$(y)) where $\rho(j)$ is the index of the *j*th most satisfied constituent concept in the definition of Con(y), the *j*th largest of the Con$_i$(y) and w_j are the OWA weights associated with the quantifier Q.

Equal treatment of constituent concepts in formulation of an OWA type concept is implied. For some cases different importances may be related to the constituent concepts, the Con$_i$. A procedure for including these differing importances in the OWA type calculation of Con(y) will now be given. Let $\alpha_i \in [0, 1]$ show importance associated with the Con$_i$ in formulation of Con. An assumption is that the larger α_i the more important Con$_i$ and $\alpha_i = 0$ indicates no importance for Con$_i$.

In this situation we consider a more general OWA concept object

$$Con = < Con_1, Con_2, ..., Con_n : M, Q >$$

M is an n vector whose components $m_i = \alpha_i$ are the importance weights associated with Con$_i$. Now we consider evaluation of the importance weighted OWA aggregation. [27]

Let Con$_i$(y) indicate satisfaction of database element y to the *i*th constituent concept. So we need to obtain

$$\text{Con}(y) = \text{OWAQ/M}(\text{Con}_1(y), \ldots, \text{Con}_i(y), \ldots, \text{Con}_n(y))$$

where OWA$_{Q/M}$ indicates OWA aggregation of the Con$_i$(y) based on quantifier Q and importance M.

The associated weighting vector, W(y), is obtained based on Q and M. After obtaining the vector of weights we have

$$\text{Con}(y) = \sum_{j=1}^{n} w_j(y) Con_{\rho_y(j)}(y)$$

Here $\rho_y(j)$ is the index of the jth largest of the Con$_i$(y). It is clear that the weighting vector will be influenced by ordering of the Con$_i$(y).

The associated weights are as follows

(1) Let $T = \sum_{i=1}^{n} \alpha_j$, the sum of the importance weights
(2) Let $S_j(y) = \sum_{k=1}^{j} \alpha_{\rho_y(k)}$, it is the sum of importance weights of the j most satisfied concepts by object y

$$w_j(y) = Q\left(\frac{S_j(y)}{T}\right) - Q\left(\frac{S_{j-1}(y)}{T}\right)$$

In summary the OWA operator provides a formulation used to compose concepts from other constituent concepts. A more general framework for constructing concepts from other concepts can be done using Choquet integrals [28].

Acknowledgements Petry would like to thank the Naval Research Laboratory's Base Program for sponsoring this research.

References

1. Yager, R. R. (1989). On linguistic summaries of data. In *Proceedings of IJCAI workshop on knowledge discovery in databases*, Detroit (pp. 378–389).
2. Yager, R. R. (1991). On linguistic summaries of data. In G. Piatetsky-Shapiro & B. Frawley (Eds.), *Knowledge discovery in databases* (pp. 347–363). MIT Press.
3. Kacprzyk, J., Yager, R. R., & Zadrozny, S. (2000). A fuzzy logic based approach to linguistic summaries in databases. *International Journal of Applied Mathematical Computer Science, 10*, 813–834.
4. Kacprzyk, J., & Yager, R. R. (2001). Linguistic summaries of data using fuzzy logic. *International Journal of General Systems, 30*, 133–154.
5. Kacprzyk, J., Yager R. R., & Zadrozny S. (2001). Fuzzy linguistic summaries of databases for efficient business data analysis and decision support. In W. Abramowicz & J. Zaruda (Eds.), *Knowledge discovery for business information systems* (pp. 129–152). Kluwer Academic Publishers.
6. Kacprzyk, J., & Zadrozny, S. (2010). Computing with words is an implementable paradigm: Fuzzy queries, linguistic data summaries, and natural-language generation. *IEEE Transactions on Fuzzy Systems, 18*, 461–472.

7. Bouchon-Meunier, B., & Moyse, G. (2012). Fuzzy linguistic summaries: Where are we, where can we go? In *Proceedings of the 2012 IEEE conference on computational intelligence for financial engineering and economics*, CIFEr 2012, New York, NY, USA (pp. 1–8).
8. Almeida, R., Lesot, M.-J., Bouchon-Meunier, B., Kaymak, U., & Moyse, G. (2013). Linguistic summaries of categorical time series for septic shock patient data. In *Fuzz-IEEE 2013—IEEE international conference on fuzzy systems*, Hyderabad, India (pp. 1–8).
9. Moyse, G., Lesot, M.-J., & Bouchon-Meunier, B. (2013). Linguistic summaries for periodicity detection based on mathematical morphology. In *IEEE SYmposium series on computational intelligence*, Singapore (pp. 106–113).
10. Moyse, G., Lesot, M.-J., & Bouchon-Meunier, B. (2013). *Mathematical morphology tools to evaluate periodic linguistic summaries, Flexible query answering systems* (Vol. 8132, pp. 257–268). Lecture Notes in Computer Science.
11. Moyse, G., Lesot, M.-J., & Bouchon-Meunier, B. (2015). Oppositions in fuzzy linguistic summaries. In *FUZZ-IEEE'15—IEEE International conference on fuzzy systems*, Istanbul, Turkey (pp. 1–8).
12. Lesot, M.-J., Moyse, G., & Bouchon-Meunier, B. (2016). Interpretability of fuzzy linguistic summaries. *Fuzzy Sets and Systems, 292*, 307–317.
13. Moyse, G., & Lesot, M.-J. (2016). Linguistic summaries of locally periodic time series. *Fuzzy Sets and Systems, 285*, 94–117.
14. Boran, F. E., Akay, D., & Yager, R. R. (2014). A probabilistic framework for interval type-2 fuzzy linguistic summarization. *IEEE Transactions on Fuzzy Systems, 22*, 1640–1653.
15. Boran, F. E., Akay, D., & Yager, R. R. (2016). An overview of methods for linguistic summarization with fuzzy sets. *Expert Systems with Applications, 61*, 356–377.
16. Wilbik, A., & Keller, J. M. (2013). A fuzzy measure similarity between sets of linguistic summaries. *IEEE Transactions on Fuzzy Systems, 21*, 183–189.
17. Yager, R. R. (2012). On Z-valuations using Zadeh's Z-numbers. *International Journal of Intelligent Systems, 27*, 259–278.
18. Zadeh, L. A. (2011). A note on Z-numbers. *Information Sciences, 181*, 2923–2932.
19. Zadeh, L. A. (1983). A computational approach to fuzzy quantifiers in natural languages. *Computing and Mathematics with Applications, 9*, 149–184.
20. Yager, R. R. (1985). Reasoning with fuzzy quantified statements: Part I. *Kybernetes, 14*, 233–240.
21. Yager, R. R. (1992). On the specificity of a possibility distribution. *Fuzzy Sets and Systems, 50*, 279–292.
22. Yager, R. R. (1992). Default knowledge and measures of specificity. *Information Sciences, 61*, 1–44.
23. Beliakov, G., Pradera, A., & Calvo, T. (2007). *Aggregation functions: A guide for practitioners.* Springer.
24. Yager, R. R. (1988). On ordered weighted averaging aggregation operators in multi-criteria decision making. *IEEE Transactions on Systems, Man and Cybernetics, 18*, 183–190.
25. Yager, R. R., Kacprzyk, J., & Beliakov, G. (2011). *Recent developments in the ordered weighted averaging operators: Theory and practice.* Springer.
26. Yager, R. R. (1996). Quantifier guided aggregation using OWA operators. *International Journal of Intelligent Systems, 11*, 49–73.
27. Yager, R. R. (1997). On the inclusion of importances in OWA aggregations. In R.R. Yager & J. Kacprzyk (Eds.), *The ordered weighted averaging operators: theOry and applications* (pp. 41–59). Kluwer Academic Publishers.
28. Wang, Z., Yang, R., & Leung, K.-S. (2010). *Nonlinear integrals and their applications in data mining.* World Scientific.

Fuzzy or Neural, Type-1 or Type-2—When Each Is Better: First-Approximation Analysis

Vladik Kreinovich and Olga Kosheleva

Abstract In many practical situations, we need to determine the dependence between different quantities based on the empirical data. Several methods exist for solving this problem, including neural techniques and different versions of fuzzy techniques: type-1, type-2, etc. In some cases, some of these techniques work better, in other cases, other methods work better. Usually, practitioners try several techniques and select the one that works best for their problem. This trying often requires a lot of efforts. It would be more efficient if we could have a priori recommendations about which technique is better. In this paper, we use the first-approximation model of this situation to provide such a recommendation.

1 Formulation of the Problem

Need for data processing. In many practical situations, we are interested in a quantity y which is difficult—or even impossible—to measure directly. For example, we may be interested in the distance to a faraway star, in the amount of oil in a given well, in tomorrow's temperature, etc. Since we cannot directly measure this quantity, a natural idea is to measure it indirectly—i.e., to measure the values of easier-to-measure related quantities $x_1, \ldots, x_n$, and then provide the best estimate for y based on the results $\tilde{x}_i$ of these measurements. For this purpose, we need to know the dependence $y = f(x_1, \ldots, x_n)$ between the desired quantity y and the auxiliary easier-to-measure quantities x_i.

The resulting computation of the estimate $\tilde{y} = f(\tilde{x}_1, \ldots, \tilde{x}_n)$ based on the measurement results $\tilde{x}_i$ is what is usually called *data processing*.

V. Kreinovich (✉) · O. Kosheleva
University of Texas at El Paso, 500 W University Ave, El Paso, TX 79968, USA
e-mail: vladik@utep.edu

O. Kosheleva
e-mail: olgak@utep.edu

© The Author(s), under exclusive license to Springer Nature Switzerland AG 2023
Y. P. Kondratenko et al. (eds.), *Artificial Intelligence in Control and Decision-making Systems*, Studies in Computational Intelligence 1087,
https://doi.org/10.1007/978-3-031-25759-9_4

Sometimes, the dependence is known, but not always. In some practical situations. we know the desired dependence $y = f(x_1, \ldots, x_n)$—at least approximately, with reasonable accuracy.

For example, to estimate the distance to a faraway star, we can measure the direction to this star in two different seasons, when the Earth is at two opposite sides of the Sun. In this case, the distance can be obtained by using known trigonometric formulas. Similarly, we know the equations that describe the atmosphere's dynamics.

However, in many other situation, we do not know the dependence—or we know the approximate dependence, but this approximate dependence is very accurate. For example, no accurate formula is known for predicting future changes in the country's economy.

Need for—generally understood—machine learning. In situations when we do not know the desired dependence, we need to determine this dependence from the experimental data. In other words, we collect situations k in which we know the values $y^{(k)}$ and $x_i^{(k)}$ of both y and x_i, and we want to find the dependence $y = f(x_1, \ldots, x_n)$ that fits all these data, i.e., for which, for all k, we have $y^{(k)} \approx f\left(x_1^{(k)}, \ldots, x_n^{(k)}\right)$. The determination of the dependence from the available data is known as *machine learning*; see, e.g., [3].

Often, in addition to the values $y^{(k)}$ and $x_i^{(k)}$, we have some information about the dependence $y = f(x_1, \ldots, x_n)$. For example, often we know that the desired dependence is linear or quadratic—or belongs to a known few-parametric class of dependencies. Often, we know some imprecise ("fuzzy") rules that describe this dependence, e.g., that if x_1 is low and x_2 is high, then y is medium.

Many techniques are used to solve this problem. Many different techniques are used to solve this "machine learning" problem of determining the dependence from the experimental data.

In many situations, neural network techniques work the best—especially techniques of deep learning; see, e.g., [5]. In other situations, better results are obtained if we use fuzzy techniques (see, e.g., [2, 6, 7, 9, 10, 13])—i.e., when we first use fuzzy techniques to translate the imprecise of-then rules into a precise first-approximation dependence, and then adjust the parameters of this dependence so as to fit the available data. There are many version of fuzzy techniques: we can use the traditional fuzzy techniques, we can use intuitionistic fuzzy techniques (see, e.g., [1]), we can use type-2 fuzzy approach [7], etc. There are many other machine learning techniques; see, e.g., [3].

For each of these techniques:

- sometimes, this technique works well—and its results are better than others, while
- in other situations, this techniques is not working as well.

Remaining challenge. At present, it is not clear a priori which of the techniques will work better. Practitioners usually try different techniques—and then use the technique that works the best for a given situation. This trying of several different techniques takes a lot of efforts.

It is therefore desirable to be able to decide when each of the available techniques is better—and thus, in effect, come up with a new combined methodology that would enable us to utilize the best features of each technique without the need to try them all.

What we do in this paper. In this paper, on the example of fuzzy-neural dichotomy, we show how such a combination can be attained.

2 Analysis of the Problem

Neural and fuzzy machine learning: what is the main difference? What is the main difference between neural and fuzzy approaches? In both cases, we adjust the parameters of the model to fit all the data:

- in a neural network, we adjust the parameters of all the neurons,
- in the fuzzy case, we adjust the parameters of the membership functions (and, sometimes, the parameters of the corresponding "and"- and "or"-operations, also known as t-norms and t-conorms).

We often even use the same techniques for this adjustment: e.g., gradient descent. So what is the main different between these two techniques?
  The main difference is that:

- In the neural network, a change in each parameter of each neuron, in principle, affects all possible values of the resulting function $f(x_1, \ldots, x_n)$. In this sense, all these parameters are *global*.
- In contrast, in fuzzy techniques, a change in a membership function—e.g., in a membership function describing when x_1 is small—only affects the values $f(x_1, \ldots, x_n)$ corresponding to small x_1. In this sense, all these parameters are *local*.

What is the main difference between type-1, type-2, etc.? From this general viewpoint, what is the main difference between different versions of fuzzy techniques? For example, if we use triangular membership functions—and not necessarily symmetric ones—then we need 3 parameters to describe each function:

- the value v_{0-} at which the membership function starts increasing,
- the value v_1 at which the membership function reaches its largest value 1, and
- the value v_{0+} at which the membership function reaches the value 0 and stops decreasing.

If we use symmetric membership functions, we only need two parameters to describe each membership function, since in this case,

$$v_1 = \frac{v_{0-} + v_{0+}}{2}.$$

If we use trapezoid membership functions, in the general case, we need four parameters:

- the value v_{0-} at which the membership function starts increasing,
- the value v_{1-} at which the membership function first reaches its largest value 1,
- the value v_{1+} at which the membership function starts decreasing, and
- the value v_{0+} at which the membership function reaches the value 0 and stops decreasing.

For symmetric trapezoid functions, we only need three parameters.

In the case of interval-values membership functions $\left[\underline{\mu}(x), \overline{\mu}(x) \right]$, we need, in effect, to describe two different membership functions $\underline{\mu}(x)$ are $\overline{\mu}(x)$ and thus, we need twice as many parameters. For example, if we use symmetric trapezoid membership functions, we need $2 \cdot 3 = 6$ parameters to describe each membership function.

From this viewpoint, the main difference between different fuzzy representations is in how many parameters we use to describe the local behavior of the desired function in the corresponding region.

Now, we are ready to reformulate the above challenge in precise mathematical terms. In view of the above analysis, in order to select a proper technique, we need to decide:

- first, whether it is better to describe the dependence globally (as, e.g., in the neural network approach) or to divide it into regions (as in fuzzy approach) and have a separate description in each region;
- second, if it is more efficient to divide into regions, what is the best size of a region, and how many parameters should we use to describe the dependence in each region.

Let us describe thus reformulated problem in precise terms.

3   First-Approximation Model and the Resulting Recommendation

Simplifying assumption. For simplicity, let us consider a 1-D version of the problem, when we want to find a good approximation to an unknown function of one variable $y = f(x)$.

How many parameters can we use? Of course, the more parameters we use, the more accurately we can represent a function. However, the more parameters we use, the more time it will take to process the data, the more space will be needed to store all these parameters. So, in practice, there is a limit N to how many parameters we can use.

From this viewpoint, the question is: what is the best approximation that we can attain for the given number of parameters?

Gauging approximation accuracy. Most real-life dependencies are smooth, even analytical; see, e.g., [4, 12]. Such functions can be expanded in Taylor series

$$f(x) = a_0 + a_1 \cdot (x - x_0) + a_2 \cdot (x - x_0)^2 + \cdots,$$

and a natural way to approximate such a function is to use the first few terms in its Taylor series, i.e., to use the approximation

$$f(x) \approx a_0 + a_1 \cdot (x - x_0) + a_2 \cdot (x - x_0)^2 + \cdots + a_{p-1} \cdot (x - x_0)^{p-1}. \quad (1)$$

This is a standard way to approximate real-life dependencies in physics (see, e.g., [4, 12]), this is how computers compute functions like $\exp(x)$, $\sin(x)$, etc.

This approximation means that we ignore the following terms in the Taylor series. The largest of these terms is the term proportional to $(x - x_0)^p$. If we divide the whole interval of possible values of x—whose size we will denote by I—into regions of smaller size s, then this ignored term has the size s^p.

The expression s^p thus describes the size of the approximation error.

Resulting optimization problem. Once we select the degree of the approximating polynomial—i.e., the value p—we will know that to describe the dependence on each subregion, we need p parameters.

Overall, we have N parameters. Thus, we can have N/p subregions.

The size s of each subregion can be obtained if we divide the size I of the whole interval by the number N/p of the subregions. Thus, this size is equal to

$$s = \frac{I}{N/p} = \frac{p \cdot I}{N}.$$

So, the resulting approximation error is equal to

$$s^p = \left(\frac{p \cdot I}{N} \right)^p. \quad (2)$$

The values I and N are given, the only parameter that we can control is p. Thus, we must select the value p for which the approximation error (2) attains the smallest possible value.

Solving the resulting optimization problem. The function $\ln(x)$ is strictly increasing. Thus, minimizing the expression (2) is equivalent to minimizing its logarithm

$$\ln(s^p) = p \cdot \left(\ln(p) + \ln \left(\frac{I}{N} \right) \right). \quad (3)$$

Differentiating this expression by p and equating the derivative to 0, we conclude that

$$\ln(p) + \ln\left(\frac{I}{N}\right) + 1 = 0.$$

By applying $\exp(x)$ to both sides of this equality, we conclude that

$$p \cdot \frac{I}{N} \cdot e = 1,$$

hence

$$p = \frac{N}{I \cdot e}. \tag{4}$$

In this case, the optimal size of the subregion is equal to

$$s = \frac{p \cdot I}{N} = \frac{1}{e}. \tag{5}$$

So, we can make the following conclusions.

First conclusion: local or global? The optimal size of the region does not depend on the desired accuracy:

- if this size is sufficiently small—smaller than $1/e$ in some natural units—then it is better to use global techniques like neural networks;
- if this size is larger, then it is better to use local techniques like fuzzy.

In other words, in qualitative terms:

- if the inputs vary a lot, then local (e.g., fuzzy) techniques are better;
- on the other hand, if the inputs do not deviate much, global (e.g., neural) techniques are better.

Second conclusion: how to reach better accuracy? Suppose that we have a local model—e.g., a fuzzy model with a fixed number of triangular membership functions. This model provides some accuracy. What should we do if we want higher accuracy?

Of course, for this purpose, we will need to use more parameters. In general, these parameters can be allocated differently:

- we can divide the original interval into a larger number of subintervals—i.e., use a larger number of membership functions,
- or we can keep the same number of subintervals, but provide a more sophisticated description of the function on each subinterval, description that requires more parameters—e.g., use type-2 fuzzy, or use a more general class of membership functions.

Our analysis shows that the second idea leads to better results. If we had 3 or 5 membership functions before, we should continue to use the same number of membership functions—but make these functions more sophisticated.

Comment. Our result may explain why we humans always divide all objects into 7 ± 2 classes—i.e., in effect, use a subdivision into a fixed number of subregions; see, e.g., [8, 11].

Thus, we provide a natural explanation of this feature of human data processing.

4 Future Work

What we have analyzed is the first approximation, an approximation in which:

- we did not take into account specific features of each application area, and
- we only took into account number of regions and number of parameters.

Additional analysis is needed:

- Additional analysis is needed to decide between different local techniques with the same number of parameters: e.g., should we go to type-2 or should we use more sophisticated membership functions.
- Additional analysis is also needed to decide which of "global" machine learning techniques would work better in a given situation.

We hope that this simple paper will inspire researchers to extend our results and to solve remaining challenges.

Acknowledgements This work was supported in part by the National Science Foundation grants 1623190 (A Model of Change for Preparing a New Generation for Professional Practice in Computer Science), and HRD-1834620 and HRD-2034030 (CAHSI Includes), and by the AT&T Fellowship in Information Technology.

It was also supported by the program of the development of the Scientific-Educational Mathematical Center of Volga Federal District No. 075-02-2020-1478, and by a grant from the Hungarian National Research, Development and Innovation Office (NRDI).

The authors are greatly thankful to Yuriy P. Kondratenko for his encouragement, and, of course, to Janusz Kacprzyk for his numerous ideas and discussions that motivated this work.

References

1. Atanassov, K. (1999). *Intuitionistic fuzzy sets: Theory and applications.* Springer.
2. Belohlavek, R., Dauben, J. W., & Klir, G. J. (2017). *fuzzy logic and mathematics: A historical perspective.* Oxford University Press.
3. Bishop, C. M. (2006). *Pattern recognition and machine learning.* Springer.
4. Feynman, R., Leighton, R., & Sands, M. (2005). *The Feynman lectures on physics.* Addison Wesley.
5. Goodfellow, I., Bengio, Y., & Courville, A. (2016). *Deep leaning.* MIT Press.
6. Klir, G., & Yuan, B. (1995). *Fuzzy sets and fuzzy logic.* Prentice Hall.
7. Mendel, J. M. (2017). *Uncertain rule-based fuzzy systems: Introduction and new directions.* Springer.

8. Miller, G. A. (1956). The magical number seven plus or minus two: Some limits on our capacity for processing information. *Psychological Review, 63*(2), 81–97.
9. Nguyen, H. T., Walker, C. L., & Walker, E. A. (2019). *A first course in fuzzy logic*. Chapman and Hall/CRC.
10. Novák, V., Perfilieva, I., & Močkoř, J. (1999). *Mathematical principles of fuzzy logic*. Kluwer.
11. Reed, S. K. (2010). *Cognition: Theories and application*. Wadsworth Cengage Learning.
12. Thorne, K. S., & Blandford, R. D. (2017). *Modern classical physics: Optics, fluids, plasmas, elasticity, relativity, and statistical physics*. Princeton University Press.
13. Zadeh, L. A. (1965). Fuzzy sets. *Information and control, 8*, 338–353.

Matrix Resolving Functions in Game Dynamic Problems

A. A. Chikrii and G. Ts. Chikrii

Abstract We consider a non-stationary problem of bringing the trajectory of a conflict-controlled process to a given terminal set that varies over time. The basic method for research is the method of resolving functions. While the main scheme of the method uses scalar resolving functions, this paper uses matrix functions of a diagonal form with different elements on the diagonal. This circumstance makes it possible to cover more general game situations. A number of schemes for solving the problem are proposed. Sufficient conditions for the termination of the game in the finite time in the class of quasi and stroboscopic strategies are obtained. The way is indicated for expanding the possibilities and effective application of the method of resolving functions.

Keywords Method of resolving functions · Conflict-controlled process · The Pontryagin condition · Set-valued mapping · Superposition measurability · Measurable choice

1 Introduction

In the theory of dynamic pursuit-evasion games, along with methods focused on making optimal decisions, several methods provide a guaranteed result in a conflict confrontation. Among the former are, first of all, the Isaacs method associated with the basic equation of the theory of differential games [1], Pontryagin's method of alternating integral [2] and the method of Pshenichnyi T_ε-operators [3]. The guaranteed result is provided by Pontryagin's first direct method [2], Krasovskii extreme targeting rule [4] and the method of resolving functions [5]. The second group

A. A. Chikrii (✉) · G. Ts. Chikrii
Glushkov Institute of Cybernetics of NAS of Ukraine, 40 Glushkov Avenue, Kyiv 03187, Ukraine
e-mail: g.chikrii@gmail.com

A. A. Chikrii
Yurij Fedkovich Chernivtsi National University, 2 Kotsuibynskyi Street, Chernivtsi 58012, Ukraine

of methods, without focusing the research on the problem of optimality, justifies well-known from practice methods.

Problems of the mathematical theory of control under conditions of conflict and uncertainty are an important part of artificial intelligence. They make it possible to develop and make effective decisions in the face of confrontation between the parties and implement them in control systems.

This work is devoted to the development of the already mentioned method of resolving functions. If we can say about the rule of extreme targeting that it is based on the apparatus of support functions, then the method of resolving functions is based on the inverse Minkowski functionals [5]. With their help, scalar functions are constructed that characterize the payoff of the pursuer at a given moment. When the total win reaches a predetermined value, the game ends. In the process of confrontation, the pursuer builds his control on the control of the evader based on measurable choice theorems [6], using the property of superposition measurability.

The attractive side of the method of resolving functions is the fact that it gives a complete justification of the pursuit along the Euler line, the rule of parallel pursuit and beam pursuit (motion camouflage) [7, 8], which are well known to designers of rocket and space technology. It also allows effective use of the modern apparatus of set-valued mappings and their selections [9] in the justification of game structures and obtaining meaningful results based on them.

The problems of evasion [10], in particular the problem of escape of one controlled object from a group of pursuers [11], made for the emergence of the method of resolving functions. With the help of the special *minmaxmin* function [12], the situation of being surrounded by a group of one was formalized [13], which laid the foundation for research in this direction.

Further development of the method led to the extension of the ideology to functional-differential systems, to conflict-controlled processes for equations with partial and fractional derivatives [14], to descriptor processes, and systems of Ito stochastic equations [15]. Of special interest is fuzzy control [16, 17] under conflict conditions, based on the method of resolving functions.

In this paper, matrix functions are introduced instead of scalar resolving functions. If scalar functions attract the shifted solid part of the terminal set to the intersection with some set-valued mapping only in the cane spanned by this set, then in the matrix case, not only attraction but also rotation is realized. The latter circumstance expands the possibilities of the pursuer and provides additional opportunities in terminating the game.

Sufficient conditions for the termination of the game in a finite time are established with the help of appropriate quasi-strategies. Under an additional assumption, the same result is realized in the class of Hajek's stroboscopic strategies [18], which prescribe Krasovskii counter-control. A modified scheme of the method is given, which makes it possible to terminate the game with the help of counter-controls. Each of these schemes contains rather rigid assumptions on the terminal set.

In order to get rid of them, a scheme for bringing the trajectory of a conflict-controlled process to a terminal affine manifold, defined with the help of the selection of the solid part of the terminal set, is explored. The studies [19, 20] developing this approach are also devoted to the investigation of game problems based on the method of resolving functions [5]. Matrix resolving functions were introduced in [21], and the technique was further developed in [22].

2 Problem Statement. Strategies of the Player Behavior

Let the dynamics of the conflict-controlled process in a finite-dimensional Euclidean space E^n be given by a system of non-stationary quasi-linear equations

$$\dot{z} = C(t)z + \varphi(t, u, v), \quad z(t_0) = z_0, \quad t \geq t_0 \geq 0 \tag{1}$$

where $C(t)$ is a matrix function, whose values at each t are square matrices of order n. The elements of the matrix function are measurable functions, which are locally summable. The control parameters of the players represent measurable selections of the control domains $U(t)$ and $V(t)$, which are measurable compact-valued mappings with direct images in. E^n for $t \in [t_0, +\infty)$. Vector-function $\phi(t, u, v)$—the control unit, satisfies the conditions of Caratheodori, i.e. is measurable in t and jointly continuous in (u, v). In addition, it is assumed that

$$\|\varphi(t, u, v)\| \leq \phi_0(t), \quad u \in U(t), \quad v \in V(t), \quad t \in [t_0, +\infty), \tag{2}$$

$\Phi_0(t)$ is some locally summable function.

Also, a cylindrical terminal set is given:

$$M^*(t) = M_0 + M(t), \quad t \in [t_0, +\infty), \tag{3}$$

where M_0 is a linear subspace in E^n, and $M(t)$ is a measurable compact-valued mapping whose direct images belong to the orthogonal complement L to M_0 in E^n.

Let us determine the awareness of the players during the game, taking the side of the first player (u). The second player (v) chooses arbitrary measurable selections of the set-valued mapping $V(t)$, which exist by virtue of the theorem on measurable choice [6]. Let us denote its totality Ω_v.

If the first player at the moment t, $t \geq t_0$, has information about the initial state of the process (t_0, z_0) and the control history of the second player

$$v_t(\cdot) = \{v(s) : v(s) \in V(s), \ s \in [t_0, t]\},$$

i.e. $u(t) = u(t_0, z_0, t, v_t(\cdot))$, then they say that its control is prescribed by a quasi-strategy [4]. In this case the control $u(t)$ must be a measurable selection of a mapping $U(t)$.

In the case when the first player makes a decision at the moment t only on the basis of information only about the initial state and the instantaneous value of the control of the second player, i.e. $u(t) = u(t_0, z_0, t, v(t))$, then one speaks of the counter-control according to Krasovskii [4], which is prescribed by the stroboscopic strategy of Hajek [18]. It goes without saying that in this case $u(t)$ also must be a measurable selection of the set-valued mapping $U(t)$.

The goal of the first player is to bring the trajectory of the process (1) to the terminal set (3), the second player prevents this.

3   The Pontryagin Condition. Method Scheme

Denote π the orthoprojector acting from E^n to L, and introduce a set-valued mapping

$$\varphi(t, U(t), v) = \{\phi(t, u, v) : u \in U(t)\}, \quad v \in V(t), \quad t \geq t_0.$$

By virtue of assumptions about the parameters of the conflict-controlled process (1)–(3) and the theorem on direct image [6], this mapping is measurable in t and continuous in v by the Hausdorff technique [9].

Let us set

$$W(t, \tau, v) = \pi \Phi(t, \tau)\phi(\tau, U(\tau), v),$$

$$W(t, \tau) = \bigcap_{v \in V(\tau)} W(t, \tau, v), \quad t \geq \tau \geq t_0,$$

where $\Phi(t, \tau)$ is the transition matrix of the homogeneous system (1)—the Cauchy matrix [23, 24]. The set-valued mapping $W(t, \tau, v)$ is measurable in τ and continuous in v, while the mapping $W(t, \tau)$ is measurable in τ [25, 26].

Let us denote

$$\Delta(t_0) = \{(t, \tau) : t_0 \leq \tau \leq t < +\infty\}$$

Pontryagin's condition. The set-valued mapping $W(t, \tau)$ has non-empty images in the set $\Delta(t_0)$ and is closed-valued.

Taking into account the Pontryagin condition, from the theorem on the measurable choice it follows that for any t, $t \geq t_0$, there exists at least one measurable-in-τ selection $\gamma(t, \tau)$ of the set-valued mapping $W(t, \tau)$, $(t, \tau) \in \Delta(t_0)$.

Let us denote its totality

$$\Gamma_t = \{\gamma(t, \tau) : \gamma(t, \tau) \in W(t, \tau), \ \tau \in [t_0, t]\}.$$

We choose some measurable in τ selection $\gamma(t, \tau)$ from Γ_t and introduce a function

$$\xi(t) = \xi(t, t_0, z_0, \gamma(t, \cdot)) = \pi \Phi(t, t_0)z_0 + \int\limits_{t_0}^{t} \gamma(t, \tau)d\tau.$$

By virtue of initial assumptions the selection $\gamma(t, \tau)$ is summable in τ function, $\tau \in [t_0, t]$, for any $t, t > t_0$.

Denote $k = \dim L$, and consider a square diagonal matrix R of order k with the non-negative elements on its diagonal

$$R = \begin{pmatrix} \rho_1 & & & \\ & \rho_2 & & 0 \\ & & \cdot & \\ & & \cdot & \\ 0 & & & \cdot \\ & & & & \rho_k \end{pmatrix} = diag\{\rho_1, ...\rho_k\}, \quad \rho_i \geq 0.$$

Consider a set-valued mapping

$$R_k(t, \tau, v) = \{(\rho_1, ...\rho_k), \rho_i \geq 0 : [W(t, \tau, v) - \gamma(t, \tau)] \cap$$

$$\cap \begin{pmatrix} \rho_1 & & & \\ & \rho_2 & & 0 \\ & & \cdot & \\ & & & \cdot \\ 0 & & & \cdot \\ & & & & \rho_k \end{pmatrix} [M(t) - \xi(t)] \neq \emptyset\}. \tag{4}$$

Among the direct images of the mapping $R_k(t, \tau, v)$, there is the v-dimensional zero since the Pontryagin condition is satisfied and, therefore, $0 \in W(t, \tau, v) - \gamma(t, \tau)$.

Each element $\rho(t, \tau, v)$ of set-valued mapping $R_k(t, \tau, v)$, namely its selection, presents a set of k dimensional scalar functions $(\rho_1(t, \tau, v), ..., \rho_k(t, \tau, v))$. If at a moment $t, t > t_0, \xi(t) \in M(t)$, then $R_k(t, \tau, v) = \underbrace{[0, +\infty) \times ... \times [0, +\infty)}_{k}$ for

any $\tau, v \in V(\tau), \tau \in [t_0, t]$.

Let us form from the selections of the set-valued mapping $R_k(t, \tau, v)$ for $\xi(t) \overline{\in} M(t)$ a maxmin scalar function

$$\tilde{\rho}(t, \tau, v) = \sup_{\rho(t,\tau,v) \in R_k(t,\tau,v)} \min_{i=1,...,k} \rho_i(t, \tau, v). \tag{5}$$

Let us assume that the exact upper bound in expression (5) is achieved and denote

$$R^*(t, \tau, v) = \{\rho(t, \tau, v) \in R_k(t, \tau, v) : \tilde{\rho}(t, \tau, v) = \min_{i=1,...,k} \rho_i(t, \tau, v)\}. \tag{6}$$

It is a marginal set of the selections of mapping $R_k(t, \tau, v)$. We see that in the case $\rho_1(t, \tau, v) = ... = \rho_k(t, \tau, v)$ function $\tilde{\rho}(t, \tau, v)$ coincides with the scalar resolving function [5]. The selections of the mapping $R^*(t, \tau, v)$ will be called the extreme selections.

Condition 1. The maxmin function $\tilde{\rho}(t, \tau, v)$ and set-valued mapping $R^*(t, \tau, v)$, consisting of extreme vectors, are closed-valued and jointly $L \times B$—measurable in (τ, v), $v \in V(\tau)$, $\tau \in [t_0, t]$ for any $t > t_0$.

Therefore, under Condition1, by the theorem on measurable choice [6] among the extreme selections of the mapping $R^*(t, \tau, v)$ there is at least one, which is jointly $L \times B$-measurable in (τ, v), $v \in V(\tau)$, $\tau \in [t_0, t]$, for $t \geq t_0$. Let us fix it for later, denoting $\rho^*(t, \tau, v)$.

Consider the set

$$T(t_0, z_0, \gamma(\cdot, \cdot)) = \{t \geq t_0 : \inf_{v(\cdot) \in \Omega_v} \int_{t_0}^{t} \tilde{\rho}(t, \tau, v(\tau))d\tau \geq 1\}. \tag{7}$$

Since the function $\tilde{\rho}(t, \tau, v)$ is jointly $L \times B$ measurable in (τ, v), then it is superposition measurable [26]. Therefore, the integral in the relationship (7) has sense.

From previous, it follows that, if at some $t, t > t_0, \xi(t) \in M(t)$, then $\tilde{\rho}(t, \tau, v) = +\infty$ for $v \in V(\tau)$, $\tau \in [t_0, t]$, and it is natural to set the value of the integral in (7) equal to $+\infty$, and corresponding inequality, in this case, is readily fulfilled. If the inequality in expression (7) is not fulfilled for all $t, t > t_0$, then we set $T(t_0, z_0, \gamma(\cdot, \cdot)) = \emptyset$.

Let us formulate some assumptions; which must be satisfied by the extreme selections, used in what follows.

Condition 2. If $T \in T(t_0, z_0, \gamma(\cdot, \cdot)) \neq \emptyset$, then for the extreme selection $\rho^*(T, \tau, v)$ its truncated selections $\sigma(T, \tau, v) = (\sigma_1(T, \tau, v), ..., \sigma_k(T, \tau, v))$,

$$\sigma_i(T, \tau, v) = \begin{cases} \rho_i^*(T, \tau, v), & \tau \in [t_0, t_i), \\ 0, & \tau \in [t_i, T], \end{cases}$$

represent selections of the set-valued mapping $R_k(T, \tau, v)$, for any $t_i, t_i \in [t_0, T]$, $i = 1, ..., k$.

Note that for $t_i = t_0$ or $t_i = T$, $i = 1, ..., k$, Condition 2 is readily fulfilled by virtue of the Pontryagin condition and given previous constructions. Consequently, it concerns only inner points of the interval $[t_0, T]$, which are potential points of switching from one control law to another in the method of resolving function. We assume that the solid part of the terminal set satisfies the following condition.

***Condition* 3.** If the game (1)–(3) runs on the interval $[t_0, T]$, then for any diagonal matrix function $A(\tau)$ of order k with measurable non-negative elements $a_i(\tau)$, $i = 1, ..., k$, $\tau \in [t_0, T]$, which have the property $\int_{t_0}^{T} A(\tau)d\tau = E$, ($E$- the unit matrix of order k), the inclusion takes place

$$\int_{t_0}^{T} A(\tau)M(T)d\tau \subset M(T), \tag{8}$$

where the left-hand side of the inclusion contains the Aumann integral of the set-valued mapping [27].

Drawing a parallel with the method of resolving functions [5], it can be noted that, if all diagonal elements in the matrix function $A(\tau)$ coincide: $a_1(\tau) = ... = a_k(\tau) = a(\tau)$, i.e. $A(\tau) = a(\tau)E$, $\int_{t_0}^{T} a(\tau)d\tau = 1$ and the mapping $M(t)$ is convex-valued, then equality has been placed in (8). In the case of exact capture in the pursuit problem, i.e. $M(t) = \{0\}$, $t \geq t_0$, the inclusion (8) also becomes equality. If the set $M(T)$ is a solid polyhedron in subspace L with faces parallel to the coordinate axes, then the inclusion (8) is satisfied in the assumptions, formulated in Condition 3.

4   Sufficient Conditions for the Game Termination

The following statement is true.

Theorem 1 *Assume that for a conflict-controlled process (1)–(3) the Pontryagin condition holds. In addition, suppose that for a given initial state (t_0, z_0) of the process (1) there is a measurable in τ the selection $\gamma(t, \tau)$,*

$$\gamma(t, \tau) \in W(t, \tau), \ t_0 \leq \tau \leq t < +\infty,$$

such that $T(t_0, z_0, \gamma(\cdot, \cdot)) = \emptyset$, and Conditions 1,2, and 3 are fulfilled.

Then at the moment T a trajectory of the process (1) can be brought to the terminal set (3) from the initial state (t_0, z_0) using an appropriate quasi-strategy.

Proof Let $v(\tau)$ be an arbitrary measurable selection of the set-valued mapping $V(\tau)$, $\tau \in [t_0, T]$.

Consider the case $\xi(T) \overline{\in} M(T)$. For scalar functions—components of the $L \times B$—extreme selection $\rho^*(T, \tau, v), v \in V(\tau), \tau \in [t_0, T]$, the following evident equalities have place

$$\rho_i^*(T, \tau, v) \geq \min_{i=1,...,k} \rho_i^*(T, \tau, v), \ i = 1, ..., k,$$

Taking into account the connection of the maxmin function $\tilde{\rho}(T, \tau, v)$ with the extreme selections of the set-valued mapping $R^*(T, \tau, v)$, from the inequality in relation (7) we have

$$\int_{t_0}^{T} \rho_i^*(T, \tau, v(\tau))d\tau \geq 1, \ i = 1, ..., k. \tag{9}$$

Let us introduce a set of test functions

$$f_i(t) = 1 - \int_{t_0}^{t} \rho_i^*(T, \tau, v(\tau))d\tau, \ i = 1, ..., k.$$

Functions $f_i(t), i = 1, ..., k$, are absolutely continuous [28], they do not increase, since functions $\rho_i^*(T, \tau, v)$ are non-negative, and $f_i(t_0) = 1$. But since $f_i(T) \leq 0$, $i = 1, ..., k$ then, under the inequality (9), from the known theorem of analysis it follows that there exist moments $t_i^*, \ t_i^* \in [t_0, T], i = 1, ..., k$, such that $f_i(t_i^*) = 0$, in addition $t_i^* = t_i^*(v(\cdot))$.

Analogously to the scheme of the method of resolving functions [5], the segments of time $[t_0, t_i^*)$ and $[t_i^*, T]$ will be called the active and passive intervals.

Let us specify the control method of the first player. We introduce the truncated selections

$$\sigma_i(T, \tau, v) = \begin{cases} \rho_i^*(T, \tau, v), & \tau \in [t_0, t_i^*), \\ 0, & \tau \in [t_i^*, T] \end{cases}$$

for the extreme selection $\rho^*(T, \tau, v)$ of the set-valued mapping $R^*(T, \tau, v)$, where t_i^* are the zeroes of the test functions $f_i(t), i = 1, ..., k$.

Next, we consider the set-valued mapping

$$U(\tau, v) = \{u \in U(\tau) : \pi\Phi(T, \tau)\phi(\tau, u, v) - \gamma(T, \tau) \in$$

$$\in A(T, \tau, v)[M(T) - \xi(T)]\}, \ v \in V(\tau), \ \tau \in [t_0, T]. \tag{10}$$

where the matrix function $A(T, \tau, v)$ has the form

$$A(T, \tau, v) = \begin{pmatrix} \sigma_1(T, \tau, v) & & & \\ & \sigma_2(T, \tau, v) & & 0 \\ & & \cdot & \\ & & & \cdot & \\ 0 & & & \cdot & \\ & & & & \sigma_k(T, \tau, v) \end{pmatrix},$$

in addition, its diagonal elements have the property $\int_{t_0}^{T} \sigma_i(T, \tau, v(\tau))d\tau = 1, i = 1, ..., k$.

In view of Condition2 the set-valued mapping (10) has non-empty images, it is closed-valued and $L \times B$—measurable [6]. Therefore, there exists at least one $L \times B$—measurable selection $u(\tau, v)$ of the mapping $U(\tau, v)$, which is a superposition measurable function [26]. Let us put control of the first player on the interval $[t_0, T]$ equal to $u(\tau) = u(\tau, v(\tau))$, i.e. a counter-control.

In the case when $\xi(T) \in M(T)$, we choose control of the first player analogously, setting in expression (10) the matrix function $A(T, \tau, v)$ equals to the zero matrix.

Using the Cauchy formula for the solution of the system (1), we obtain a relation for its trajectory projection on the subspace L at the moment T

$$\pi z(T) = \pi \Phi(T, t_0)z_0 + \int_{t_0}^{T} \pi \Phi(T, \tau)\phi(\tau, u(\tau), v(\tau))d\tau$$

Taking into account the control law of the first player, in their number the inclusion in relation (10) for the case $\xi(T)\overline{\in}M(T)$, we obtain

$$\pi z(T) \in (E - \int_{t_0}^{T} A(T, \tau, v(\tau))d\tau\xi(T) + \int_{t_0}^{T} A(T, \tau, v(\tau))M(T)d\tau. \qquad (11)$$

From here, due to Condition 3, taking into account equality $\int_{t_0}^{T} A(T, \tau, v(\tau))d\tau = E$, it immediately follows $\pi z(T) \in M(T)$.

In case $\xi(T) \in M(T)$, if $A(T, \tau, v(\tau))$ is the zero matrix, then from the inclusion (11) we have $\pi z(T) = \xi(T) \in M(T)$.

5 Counter-Controls in the Game Problem of Approach

In the previous section, to terminate the game and approach the terminal set in some guaranteed time, the first player, when making a solution on the choice of control at the moment t, used information about the opponent's control history $v_t(\cdot)$. This is necessary to find the switching moments $t_i^*, i = 1, ..., k$—zeroes of corresponding test functions that are prescribed by the quasi-strategy. These moments separate the active and passive segments, interrupting the accumulation of the components $\rho_i^*(T, \tau, v), i = 1, ..., k$, of the extreme selection $\rho^*(T, \tau, v)$. At the same time, in each of the mentioned sections, the first player uses counter-controls.

Next, we determine the conditions on the parameters of the conflict-controlled process, under which there are no switching times, and the approach process can be implemented using counter-controls.

Keeping the notations, we formulate the following requirements.

Condition 4. Let T be a moment, such that $T \in T(t_0, z_0, \gamma(\cdot, \cdot)) \neq \emptyset$, and $\xi(T) \overline{\in} M(T)$. Then, if $\rho^*(T, \tau, v)$ is an extreme selection of the mapping $R^*(T, \tau, v)$, then an arbitrary $L \times B$ -measurable function $\rho(T, \tau, v)$, whose elements satisfy the conditions

$$0 \leq \rho_i(T, \tau, v) \leq \rho_i^*(T, \tau, v), \ \ v \in V(\tau), \ \ \tau \in [t_0, T], \ \ i = 1, ..., k,$$

represents a test function of the set-valued mapping $R_k(T, \tau, v)$.

Condition 4 is a more stringent assumption than Condition 2 of the previous section. It is an analog of the convexity condition for the mapping $R_k(T, \tau, v)$ for the case $\rho_1 = ... = \rho_k$ in the main scheme of the method of resolving functions [5].

Condition 5. Let the moment T, $T \in T(t_0, z_0, \gamma(\cdot, \cdot)) \neq \emptyset$, and $\xi(T) \overline{\in} M(T)$, be given. Then the functions $\inf_{v \in V(\tau)} \rho_i^*(T, \tau, v)$, $i = 1, ..., k$, are measurable in τ, $\tau \in [t_0, T]$, and the following equality is satisfied:

$$\inf_{v(\cdot) \in \Omega_v} \int_{t_0}^{T} \rho_i^*(T, \tau, v(\tau))d\tau = \int_{t_0}^{T} \inf_{v \in V(\tau)} \rho_i^*(T, \tau, v)d\tau.$$

For more details on the fulfillment of this equality, see [29].

The following assertion is true.

Theorem 2 *Let for the game problem (1)–(3) the Pontryagin condition be fulfilled. Suppose that for the initial state (t_0, z_0) of the process (1) there exists a measurable in τ the selection $\gamma(t, \tau)$, $\gamma(t, \tau) \in \Gamma_t$, $t_0 \leq \tau \leq t < +\infty$, and the moment T, $T \in T(t_0, z_0, \gamma(\cdot, \cdot)) \neq \emptyset$, such that Conditions 3, 4, and 5 are satisfied. Then a trajectory of the process (1) can be brought to the terminal set at the moment T with the help of some counter-control for any counter-action of the opponent.*

Proof Consider the case when $\xi(T) \overline{\in} M(T)$. From the inequality in relation (7) for the set of guaranteed times, it follows the inequality (9) that holds for arbitrary $v(\cdot) \in \Omega_v$. Taking into account the equality in Condition4, we introduce functions

$$\overline{\rho}_i(T, \tau) = \frac{\inf_{v \in V(\tau)} \rho_i^*(T, \tau, v)}{\int_{t_0}^{T} \inf_{v \in V(\tau)} \rho_i^*(T, \tau, v)d\tau}, \ \ i = 1, ..., k, \ \ v \in V(\tau).$$

Since $\int_{t_0}^{T} \inf_{v \in V(\tau)} \rho_i^*(T, \tau, v)d\tau \geq 1$, then

$$\overline{\rho}_i(T, \tau) \leq \inf_{v \in V(\tau)} \rho_i^*(T, \tau, v) \leq \rho_i^*(T, \tau, v), \ \ i = 1, ..., k,$$

for any $v \in V(\tau)$, and, therefore, the vector-function

$$\overline{\rho}(T, \tau) = (\overline{\rho}_1(T, \tau), ..., \overline{\rho}_k(T, \tau))$$

represents a measurable selection of the set-valued $R_k(T, \tau, v)$.

Let us introduce a set-valued mapping

$$U^*(\tau, v) = \{u \in U(\tau) : \pi \Phi(T, \tau)\phi(\tau, u, v) - \gamma(T, \tau) \in$$

$$\in \overline{R}(T, \tau)[M(T) - \xi(T)]\}, \qquad (12)$$

where

$$\overline{R}(T, \tau) = \begin{pmatrix} \overline{\rho}_1(T, \tau) & & & \\ & \overline{\rho}_2(T, \tau) & & 0 \\ & & \cdot & \\ & & & \cdot \\ & 0 & & \cdot \\ & & & \overline{\rho}_k(T, \tau) \end{pmatrix},$$

$v \in V(\tau), \tau \in [t_0, T]$, in addition, the diagonal elements have the property

$$\int_{t_0}^{T} \overline{\rho}_i(T, \tau)d\tau = 1, \ i = 1, ..., k.$$

Since $\overline{\rho}(T, \tau) \in R_v(T, \tau, v)$ for $v \in V(\tau), \tau \in [t_0, T]$, then the mapping $U^*(\tau, v)$ has non-empty direct images and, by virtue of assumptions on the parameters of the conflict-controlled process, it is closed-valued and $L \times B$—measurable. By the theorem on measurable choice, there exists a $L \times B$—measurable selection $u^*(\tau, v)$, which is a superposition of measurable function [26].

If $\xi(T)\overline{\in}M(T)$, then we put control of the first player over the entire interval $[t_0, T]$ equal to $u^*(\tau) = u^*(\tau, v(\tau))$. If, otherwise, $\xi(T) \in M(T)$, then we choose control of the first player similarly, with the only difference that in expression (12) we set matrix function $\overline{R}(T, \tau)$ equal to the zero-matrix.

In conclusion, if $\xi(T)\overline{\in}M(T)$, then from the Cauchy formula and the control choice law (12) it follows the inclusion

$$\pi z(T) \in (E - \int_{t_0}^{T} \overline{R}(T, \tau)d\tau)\xi(T) + \int_{t_0}^{T} \overline{R}(T, \tau)M(T)d\tau. \qquad (13)$$

Taking into account the equality $\int_{t_0}^{T} \overline{R}(T, \tau)d\tau = E$ and Condition 3, we infer

that $\pi z(T) \in M(T)$. An analogous result follows from the inclusion (13) in the case $\overline{R}(T, \tau) = \{0\}$. The theorem is proved.

In the framework of the original statement of the game problem, now we consider another scheme of the method o resolving functions, which provides sufficient conditions for termination of the game in the class of stroboscopic strategies. The scheme itself and the sufficient conditions somewhat differ from the structures proposed earlier, so we call them modified.

Keeping the previous notations, under the Pontryagin condition and fixed selection, $\gamma(t, \tau), \gamma(t, \tau) \in W(t, \tau)$, $t_0 \leq \tau \leq t < +\infty$, we consider a set-valued mapping

$$R_k(t, \tau) = \bigcap_{v \in V(\tau)} R_k(t, \tau, v), \quad t_0 \leq \tau \leq t < +\infty. \tag{14}$$

The mapping $R_k(t, \tau)$ has no-empty direct images in its domain of definition since at least the zero k-dimensional vector belongs to the set-valued mapping $R_k(t, \tau, v)$ for all values of arguments from its domain of definition.

If at some $t > t_0, \xi(t) \in M(t)$, then from formulas (4), (14) it follows that

$$R_k(t, \tau) = \underbrace{[0, +\infty) \times \ldots \times [0, +\infty)}_{k}, \quad \tau \in [t_0, t]. \tag{15}$$

If $\xi(t) \overline{\in} M(t)$ then the mapping $R_k(t, \tau)$ is closed-valued and measurable in τ. The selections of the set-valued mapping $R_k(t, \tau)$ represent vector functions $\rho(t, \tau) = (\rho_1(t, \tau), \ldots, \rho_k(t, \tau))$, where $\rho_i(t, \tau)$ are scalar functions Let us introduce a scalar maxmin function

$$\tilde{\rho}(t, \tau) = \sup_{\rho(t, \tau) \in R_k(t, \tau)} \min_{i=1,\ldots,k} \rho_i(t, \tau). \tag{16}$$

We assume that its internal minimum is achieved. Let us denote

$$R^*(t, \tau) = \{\rho(t, \tau) \in R_k(t, \tau) : \tilde{\rho}(t, \tau) = \min_{i=1,\ldots,k} \rho_i(t, \tau)\}.$$

Condition 6. Function $\tilde{\rho}(t, \tau)$ and set-valued mapping $R^*(t, \tau)$ are measurable in τ for any $t > t_0$.

The set-valued mapping $R^*(t, \tau)$ is closed-valued and, as follows from Condition6, it is measurable in τ. Therefore, among its extreme selections, there is at least one measurable in τ selection. We denote it $\rho^*(t, \tau)$, $t_0 \leq \tau \leq t$.

Consider a set

$$\Theta(t_0, z_0, \gamma(\cdot, \cdot)) = \{t \geq t_0 : \int_{t_0}^{t} \tilde{\rho}(t, \tau)d\tau \geq 1\}. \tag{17}$$

Since the function $\tilde{\rho}(t, \tau)$ is measurable in τ, then the integral has sense. Also, note that generally speaking, $\int_{t_0}^{t} \tilde{\rho}(t, \tau)d\tau$ is no continuous function in t. This explains the inequality in expression (17).

If for some t, $t > t_0, \xi(t) \in M(t)$, then, by (15), (16), $\tilde{\rho}(t, \tau) = +\infty$ for all $\tau \in [t_0, t]$ and the inequality in formula (17) is readily satisfied.

Condition 7. If $\Theta \in \Theta(t_0, z_0, \gamma(\cdot, \cdot)) \neq \emptyset$, then for the extreme selection $\rho^*(\Theta, \tau)$ of the set-valued mapping $R^*(\Theta, \tau)$ its truncated selections $\sigma(\Theta, \tau) = (\sigma_1(\Theta, \tau), ..., \sigma_k(\Theta, \tau))$,

$$\sigma_i(\Theta, \tau) = \begin{cases} \rho_i^*(\Theta, \tau), & \tau \in [t_0, t_i), \\ 0, & \tau \in [t_i, \Theta], \end{cases}$$

represent selections of the set-valued mapping $R_k(\Theta, \tau)$ for all t_i, $t_i \in [t_0, \Theta]$, $i = 1, ..., k$.

Theorem 3 *Let for the conflict-controlled process (1)–(3) the Pontryagin condition holds and suppose for the initial state (t_0, z_0) there is a measurable in τ the selection $\gamma(t, \tau), \gamma(t, \tau) \in \Gamma_t$, $t_0 \leq \tau \leq t < +\infty$, and a time Θ, $\Theta \in \Theta(t_0, z_0, \gamma(\cdot, \cdot)) \neq \emptyset$, such that Conditions 3, 6, and 7 are fulfilled.*

Then a trajectory of the system (1) can be brought to the set (3) with help of a counter-control.

Proof Let $\xi(\Theta) \overline{\in} M(\Theta)$. Consider an extreme selection $\rho^*(\Theta, \tau)$ of the set-valued mapping $R^*(\Theta, \tau)$. Its components satisfy the inequality

$$\rho_i^*(\Theta, \tau) \geq \min_{i=1,...,k} \rho_i^*(\Theta, \tau), \quad \tau \in [t_0, \Theta], \quad i = 1, ..., k.$$

From here, taking into account the inequality in relation (17), we infer

$$\int_{t_0}^{\Theta} \rho_i^*(\Theta, \tau)d\tau \geq 1, \quad i = 1, ..., k. \tag{18}$$

Let us introduce test functions

$$f_i^*(t) = 1 - \int_{t_0}^{t} \rho_i^*(\Theta, \tau)d\tau, \quad i = 1, ..., k.$$

They are absolutely continuous, do not increase, and $f_i^*(t_0) = 1$.

But, since from (18) it follows that $f_i^*(\Theta) \leq 0$, then there exist moments $\bar{t}_i$, $\bar{t}_i \in$ $[t_0, \Theta]$, $i = 1, ..., k$, such that $f_i^*(\bar{t}_i) = 0$.

We call the intervals of time $[t_0, \bar{t}_i)$, $[\bar{t}_i, \Theta]$, $i = 1, ..., k$, active and passive, respectively. It should be emphasized that here the moments of switching $\bar{t}_i$ do not depend on the control history of the second player.

Let us introduce truncated selections of the extreme selection $\rho^*(\Theta, \tau)$

$$\sigma_i(\Theta, \tau) = \begin{cases} \rho_i^*(\Theta, \tau), & \tau \in [t_0, \bar{t}_i), \\ 0, & \tau \in [\bar{t}_i, \Theta] \end{cases}.$$

Here the moments of switching $\bar{t}_i$, $i = 1, ..., k$, are the zeroes of the test functions $f_i^*(t)$. To define controls of the first player, we consider the set-valued mapping

$$U_*(\tau, v) = \{u \in U(\tau) : \pi \Phi(\Theta, \tau)\phi(\tau, u, v) - \gamma(\Theta, \tau) \in$$

$$\in A(\Theta, \tau)[M(\Theta) - \xi(\Theta)]\}, \quad \tau \in [t_0, \Theta]. \tag{19}$$

where the matrix function $A(\Theta, \tau)$ has the form

$$A(\Theta, \tau) = \begin{pmatrix} \sigma_1(\Theta, \tau) & & & \\ & \sigma_2(\Theta, \tau) & & 0 \\ & & \ddots & \\ & 0 & & \ddots \\ & & & & \sigma_k(\Theta, \tau) \end{pmatrix}$$

and its diagonal elements satisfy the condition

$$\int_{t_0}^{\Theta} \sigma_i(\Theta, \tau)d\tau = 1, \quad i = 1, ..., k.$$

In view of Condition 7, the set-valued mapping $U_*(\tau, v)$ is closed-valued and $L \times B$-measurable [6]. Therefore, there exists at least one $L \times B$-measurable selection $u_*(\tau, v)$ of the mapping $U_*(\tau, v)$, which is a superposition measurable function [26].

If $\xi(\Theta) \bar{\in} M(\Theta)$ we set control of the first player equal to $u_*(\tau) = u_*(\tau, v(\tau))$, $\tau \in [t_0, \Theta]$.

If otherwise, $\xi(\Theta) \in M(\Theta)$, then we choose control of the first player analogously to the previous case, wherein in the relation (19) the matrix function $A(\Theta, \tau)$ represents the zero-matrix.

Therefore, if $\xi(\Theta) \bar{\in} M(\Theta)$, then from the Cauchy formula, taking into account the control law (19), we obtain an inclusion

$$\pi z(\Theta) \in (E - \int_{t_0}^{\Theta} A(\Theta, \tau)d\tau)\xi(\Theta) + \int_{t_0}^{\Theta} A(\Theta, \tau)M(\Theta)d\tau. \qquad (20)$$

Since $\int_{t_0}^{\Theta} A(\Theta, \tau)d\tau = E$, then, given Condition 3, the inclusion (20) implies $\pi z(\Theta) \in M(\Theta)$.

If $\xi(\Theta) \in M(\Theta)$, then from the inclusion (20) for the zero-matrix $A(\Theta, \tau)$ it readily follows that $\pi z(\Theta) \in M(\Theta)$.

6 Method Scheme with Fixed Selections of the Solid Part of the Terminal Set

In the main scheme of the method of resolving functions [5], in the formula for the set-valued mapping

$$R(t, \tau, v) = \{\rho \geq 0 : [W(t, \tau, v) - \gamma(t, \tau)] \cap \rho[M(t) - \xi(t)] \neq \emptyset\},$$

the attraction of the set-valued mapping $M(t) - \xi(t)$ to its intersection with the mapping $W(t, \tau, v) - \gamma(t, \tau)$ is implemented with the help of a scalar resolving function

$$\rho(t, \tau, v) = \sup\{\rho : \rho \in R(t, \tau, v)\}.$$

In this work, the possibilities of method are expanded. This procedure is realized by a matrix resolving function with different diagonal elements, thereby including not only shortening—lengthening but also rotation.

While in the case of scalar resolving functions, the sufficient approach condition, in addition to the Pontryagin condition and the finiteness of the total value of the accumulated resolving function, is the convexity of the mapping of the solid part of the terminal set, in the matrix case, this is a more stringent Condition3.

To get rid of these restrictive conditions on the set (3) and effectively use matrix resolving functions we consider a method scheme focused on on approaching some a priori fixed selection of the set-valued mapping $M(t)$, the choice of which is in the power of the first player.

Let us assume that in the game problem (1)–(3) the Pontryagin condition is fulfilled and $\gamma(t, \tau), \gamma(t, \tau) \in W(t, \tau), (t, \tau) \in \Delta(t_0)$, is a measurable selection of the set-valued mapping $W(t, \tau)$.

Let $m(t)$ be a measurable selection of the set-valued mapping $M(t), t \geq t_0$. We put $\eta(t, m) = m - \xi(t), m \in M(t)$, and introduce a set-valued mapping

$$R_k(t, \tau, v, m) = \{(\rho_1, ..., \rho_k), \ \rho_i \geq 0 :$$

$$\begin{pmatrix} \rho_1 & & \\ & \rho_2 & 0 \\ & & \cdot \\ & & \quad \cdot \\ 0 & & \quad \cdot \\ & & \qquad \rho_k \end{pmatrix} \eta(t, m) \in W(t, \tau, v) - \gamma(t, \tau)\},$$

$$v \in V(\tau), \quad m \in M(t), \quad (t, \tau) \in \Delta(t_0).$$

The direct images of the set-valued mapping $R_k(t, \tau, v, m)$ contain the k- dimensional zero for all values of arguments from the domain of definition, under the Pontryagin condition.

Each element $\rho(t, \tau, v, m) = (\rho_1(t, \tau, v, m), ..., \rho_k(t, \tau, v, m))$, belonging to the set-valued mapping $R_k(t, \tau, v, m)$, represents its selection and is a set of k scalar functions.

If at some moment $t, t > t_0, \eta(t, m) = 0$ for $m \in M(t)$, then $R_k(t, \tau, v, m) = \underbrace{[0, +\infty) \times ... \times [0, +\infty)}_{k}$ for all $v \in V(\tau), \tau \in [t_0, t]$.

With the help of selections of the set-valued mapping $R_k(t, \tau, v, m)$, in the case $\eta(t, m) \neq 0, m \in M(t)$, we build a maxmin scalar function

$$\tilde{\rho}(t, \tau, v, m) = \sup_{\rho(t, \tau, v, m) \in R_k(t, \tau, v, m)} \min_{i=1, ..., k} \rho_i(t, \tau, v, m),$$

$$v \in V(\tau), \quad m \in M(t), \quad (t, \tau) \in \Delta(t_0). \tag{21}$$

Assuming that the external minimum in (19) is achieved; consider the set-valued mapping

$$R^*(t, \tau, v, m) = \{\rho(t, \tau, v, m) \in R_k(t, \tau, v, m) :$$

$$\tilde{\rho}(t, \tau, v, m) = \min_{i=1, ..., k} \rho_i(t, \tau, v, m)\}. \tag{22}$$

By analogy with the previous constructions, the selections of the set-valued mapping $R^*(t, \tau, v, m)$ will be called extreme.

Condition 8. The maxmin function $\tilde{\rho}(t, \tau, v, m)$ and the set-valued mapping $R^*(t, \tau, v, m)$ are jointly $L \times B$-measurable in (τ, v), $v \in V(\tau), \tau \in [t_0, t]$, for any $m, m \in M(t), t > t_0$.

If in the game problem (1)–(3) Condition 8 is satisfied, then, by the theorem on measurable choice [6], among extreme selections of the mapping $R^*(t, \tau, v, m)$ there is at least one, which is jointly $L \times B$-measurable in (τ, v), $v \in V(\tau), \tau \in [t_0, t]$, for any $m \in M(t), t > t_0$. Let us fix it and denote $\rho^*(t, \tau, v, m)$.

Consider a set

$$T(t_0, z_0, \gamma(\cdot, \cdot), m(\cdot)) = \{t \geq t_0 : \inf_{v(\cdot) \in \Omega_v} \int_{t_0}^{t} \tilde{\rho}(t, \tau, v(\tau), m(t)) d\tau \geq 1\},$$

$$m(t) \in M(t). \tag{23}$$

Under Condition 8 the integral is well defined.

If for some t, $t > t_0$, and $m \in M(t)$, $\eta(t, m) = 0$, then $\tilde{\rho}(t, \tau, v, m) = +\infty$ for $v \in V(\tau), \tau \in [t_0, t]$, and we set the value of the integral in (23) equal to $+\infty$. This leads to the fulfillment of the inequality in relation (23). If the inequality in relation (23) is not satisfied for all $t > t_0$ for the given selection $m(t)$, $m(t) \in M(t)$, then we set $T(t_0, z_0, \gamma(\cdot, \cdot), m(\cdot)) = \emptyset$.

Condition 9. Let $T_m \in T(t_0, z_0, \gamma(\cdot, \cdot), m(\cdot))$. Then for the extreme selection $\rho^*(T_m, \tau, v, m), v \in V(\tau)$, $m \in M(t)$, $\tau \in [t_0, T_m]$, its truncated selections $\sigma(T_m, \tau, v, m) = (\sigma_1(T_m, \tau, v, m), ..., \sigma_k(T_m, \tau, v, m))$,

$$\sigma_i(T_m, \tau, v, m) = \begin{cases} \rho_i^*(T_m, \tau, v, m), & \tau \in [t_0, t_i), \\ 0, & \tau \in [t_i, T_m], \end{cases} \tag{24}$$

are the selections of the set-valued mapping $R_k(T_m, \tau, v, m)$ for all t_i, $t_i \in [t_0, T_m]$.

Theorem 4 *Let for the game problem (1)–(3) the Pontryagin condition is satisfied. Suppose that for the initial state (t_0, z_0) of the process (1) there exists a measurable in τ the selection $\gamma(t, \tau), \gamma(t, \tau) \in W(t, \tau)$, $t_0 \leq \tau \leq t < +\infty$, and a selection $m(t)$ of the set-valued mapping $M(t)$, such that $T(t_0, z_0, \gamma(\cdot, \cdot), m(\cdot)) \neq \emptyset$, $T_m \in T(t_0, z_0, \gamma(\cdot, \cdot), m(\cdot))$, and Conditions 3, 8, and 9 are fulfilled.*

Then a trajectory of the process (1) can be brought to the affine manifold $M_0 + m(T_m)$ at the moment T_m, using the control history of the second player.

Proof Let the game run on the interval of time $[t_0, T_m]$. We put $m = m(T_m) \in M(T_m)$.

Consider the case $\eta(t, m) \neq 0$. For the components of the jointly $L \times B$—measurable in (τ, v), $v \in V(\tau), \tau \in [t_0, T_m]$, and the extreme selection $\rho^*(T_m, \tau, v, m)$ the following inequality is valid

$$\rho_i^*(T_m, \tau, v, m) \geq \min_{i=1,...,k} \rho_i^*(T_m, \tau, v, m), \quad i = 1, ..., k.$$

From relation (23) and inclusion $T_m \in T(t_0, z_0, \gamma(\cdot, \cdot), m)$ it follows

$$\int_{t_0}^{T_m} \rho_i^*(T_m, \tau, v(\tau), m) d\tau \geq 1, \quad i = 1, ..., k. \tag{25}$$

where $v(\tau)$ is an arbitrary admissible selection of the mapping $V(\tau), \tau \in [t_0, T_m]$.

Consider the test functions

$$f_i(t, m) = 1 - \int_{t_0}^{t} \rho_i^*(T_m, \tau, v(\tau), m) d\tau, \quad i = 1, ..., k.$$

They are absolutely continuous, do not increase, and $f_i(t_0, m) = 1$. But, since $f_i(T_m, m) \leq 0, i = 1, ..., k$, by the inequality (25), then there exist moments t_i, $t_i \in [t_0, T_m]$, $i = 1, ..., k$, such that $f_i(t_i, m) = 0$. Similar to the scheme of the method of resolving functions, the intervals of time $[t_0, t_i)$, $[t_i, T_m]$ will be called active and passive. For the extreme selection $\rho^*(T_m, \tau, v, m)$, we introduce a truncated selection with the components (24), where t_i, $i = 1, ..., k$, are the zeroes of the test functions $f_i(t, m)$ under Condition 9.

Consider a set-valued mapping

$$U_m(\tau, v) = \{u \in U(\tau) : \pi \Phi(T_m, \tau) \phi(\tau, u, v) -$$
$$- \gamma(T_m, \tau) \subset A(T_m, \tau, v, m) \eta(T_m, m)\}, \quad v \in V(\tau), \quad \tau \in [t_0, T_m] \tag{26}$$

where

$$A(T_m, \tau, v, m) = \begin{pmatrix} \sigma_1(T_m, \tau, v, m) & & & \\ & \sigma_2(T_m, \tau, v, m) & & 0 \\ & & \cdot & \\ & & \cdot & \\ & 0 & \cdot & \\ & & & \sigma_k(T_m, \tau, v, m) \end{pmatrix},$$

in addition, $\int_{t_0}^{T_m} \sigma_i(T_m, \tau, v(\tau), m) d\tau = 1$.

Under Condition 8, the matrix function $A(T_m, \tau, v, m)$ is jointly $L \times B$-measurable in (τ, v) the selection of the mapping $R_k(T_m, \tau, v, m)$. That is why [30], the set-valued mapping $U_m(\tau, v)$ is also jointly $L \times B$-measurable in the mentioned variables and, therefore, by the theorem on measurable choice, it has a $L \times B$-measurable selection $u_m(\tau, v)$ [6], which is superposition measurable function. We set control of the first player equal to

$$u_m(\tau) = u_m(\tau, v(\tau)), \quad \tau \in [t_0, T_m].$$

If $\eta(T_m, m)\} = 0$, then, in the relation (26) we set $A(T_m, \tau, v, m)$ equal to the zero-matrix and choose control of the first player similarly.

From the Cauchy formula, taking into account the control law of the first player in the case $\eta(T_m, m) \neq 0$, we obtain

$$\pi z(\mathrm{T}_m) = (E - \int_{t_0}^{\mathrm{T}_m} A(\mathrm{T}_m, \tau, \upsilon(\tau), m)d\tau)\eta(\mathrm{T}_m, m) + m. \qquad (27)$$

Therefore, $\pi z(\mathrm{T}_m) = m$, since $\int_{t_0}^{\mathrm{T}_m} A(\mathrm{T}_m, \tau, \upsilon(\tau), m)d\tau = E$.

In the case when $\eta(\mathrm{T}_m, m) = 0$ the equality $\pi z(\mathrm{T}_m) = m$ follows from formula (27).

7 Conclusion

The work is devoted to the development of the method of resolving functions in a case when diagonal matrix functions are used instead of scalar functions. This generalization makes it possible the possibilities of the method.

Sufficient conditions for the termination of the game in a finite time in the classes of quasi-strategies and counter-controls are obtained using the measurable choice theorem. In so doing, the construction of controls essentially relies on the properties of special set-valued mappings and the superposition measurability of their selections. To implement the technique, additional conditions are imposed on the terminal set.

To avoid these restrictions, we consider the case with a fixed selection of the solid part of the terminal set, which turns it into an affine manifold and eliminates additional assumptions on the structure of the terminal set.

A number of results on motion control is contained in boors [31–33] and also in papers [34, 35].

Acknowledgements This work was partially supported by the National Research Foundation of Ukraine. Grant 2020.02/0121 "Analytic methods and machine learning in control theory and decision making in conditions of conflict and uncertainty".

References

1. Isaacs, R. (1965). *Differential games*. John Wiley.
2. Pontryagin, L. S. (1988). *Selected scientific papers* (Vol. 2). Nauka.
3. Pschenichnyi, B. N., & Ostapenko, V. V. (1992). *Differential games*. Naukova Dumka.
4. Krasovskii, N. N. (1970). *Game problems on the encounter of motions*. Nauka.
5. Chikrii, A. A. (2013). *Conflict-controlled processes*. Springer Science and Business Media.
6. Aubin, J., & Frankowska, H. (1990). *Set-valued analysis*. Birkhauser.
7. Locke, A. S. (1955). *Guidance*. D. Van Nostrand Company.
8. Siouris, G. M. (2004). *Missile guidance and control systems*. Springer.
9. Shouchuan, H. S., & Papageorgion, N. S. (1997). *Handbook of multi-valued analysis theory* (vol. 1). Springer.
10. Chikrii, A. A. (2020). Conflict situations involving controlled object Groups. Part.1. Collision avoidance. *Journal of Automation and Information Sciences, 52*(7), 19–37.

11. Pontryagin, L. S. (1971). A linear differential evasion game. *Trudy MIAN, 112*, 30–63. (in Russian).
12. Chikrii, A. A. (1976). Linear problem of avoiding several pursuers. *Engineering Cybernetics, 14*(4), 38–42.
13. Pschenichnyi, B. N. (1976). Simple pursuit by several objects. *Kibernetika, 3*, 115–116. (in Russian).
14. Chikrii, A. A., Matychyn, I., Gromaszek, K., & Smolarc, L. (2011). *Control of fractional-order dynamic systems under uncertainty, modelling and optimization* (pp. 3–56). Publ. Lublin, Univ. of Technology.
15. Vlasenko, L. A., Rutkas, A. G., & Chikrii, A. A. (2020). On a differential Game in a stochastic system. *Proceedings of the Steklov Institute of Mathematics, 309*(1), 185–198.
16. Kacprzyk, J. (1997). *Multistage fuzzy control: A prescriptive approach*. John Wiley & Sons Inc.
17. Lodwick, W. A., & Kacprzyk, J. (Eds.) (2010). *Fuzzy optimization*, STUDFUZ 254. Springer.
18. Hajek, O. (1975). *Pursuit games* (p. 12). Academic Press.
19. Grigorenko, N. L. (1990). *Mathematical methods of control of several dynamic processes*. Izdat. Mosc. Gos. Univ.
20. Blagodatskikh, A. I., & Petrov, N. N. (2009). *Conflict interaction of groups of controlled objects*. Udmurtskiy Universitet.
21. Chikrii, A. A., & Chikrii, G. T. (2015). Matrix resolving functions in game problems of dynamics. *Proceedings of the Steklov Institute of Mathematics, 291*, 56–65.
22. Petrov, N. N. (2021). Matrix resolving functions in a linear problem of group pursuit with multiple captures. *Trudy Instituta Matematiki i Mekhaniki Ur O RAN, 27*(2), 185–196.
23. Gantmakher, F. R. (1967). *Theory of matrices*. Nauka.
24. Gaishun, I. V. (1999). *Introduction into theory of linear non-stationary systems*. Institute of Mathematics of NAS of Belarus.
25. Nikolskii, M. S. (1984). *First direct method of L.S. Pontryagin in differential games*. Izdat. Mosc. Gos. University.
26. Nikolskii, M. S. (1992). Stroboscopic strategies and first direct method of L.S. Pontryagin in quasi-linear differential games of pursuit-evasion. *Problems of Control and Information Theory, 11*(5), 373–377.
27. Aumann, R. J. (1965). Integrals of set-valued functions. *Journal of Mathematical Analysis and Applications, 12*, 1–12.
28. Natanson, I. P. (1974). *Theory of functions of real variable*. Nauka.
29. Joffe, A. D., & Tikhomirov, V. M. (1974). *Theory of extreme problems*. Nauka.
30. Chikrii, A. A. (2010). An analytical method in dynamic pursuit games. *Proceedings of the Steklov Institute of Mathematics, 271*, 69–85.
31. Kuntsevich, V. M., et al. (Eds.) (2018). Control systems: Theory and applications. In *Automation, control and robotics*. River Publishers.
32. Kondratenko, Y. P., et al. (Eds.) (2021). Advanced control systems: Theory and applications. In *Automation, control and robotics*. River Publishers.
33. Kondratenko, Y. P., et al. (Eds.) (2022). Recent developments in control systems. In *Automation, control and robotics*. River Publishers.
34. Chikrii, G. Ts. (2021). Principle of time stretching for motion control in condition of conflict. Advanced control systems: Theory and applications. In *Automation, control and robotics* (pp. 53–82). River Publishers.
35. Chikrii, G. T. (2009). On one problem of approach for damped oscillations. *Journal of Automation and Information Sciences, 41*(10), 1–9.

Evolving Stacking Neuro-Fuzzy Probabilistic Networks and Their Combined Learning in Online Pattern Recognition Tasks

Ye. Bodyanskiy🔘 and O. Chala🔘

Abstract The patterns (images) classification-recognition problem under conditions of overlapping classes (fuzzy) and a small training dataset is considered. The proposed method is based on a modification of the classical probabilistic neural network by D. Specht, improved with combined learning, which includes classical supervised learning, "lazy learning" based on the concept of "Neurons at data points", self-learning based on the rule "Winner takes all" and evolving learning system architecture. During the learning process, simultaneously both activation functions and the architecture itself are tuned. Fuzzy modifications of the probabilistic neural network are introduced, where one-dimensional membership functions are used instead of multidimensional activation functions, which reduce the number of adjustable parameters.

Stack architectures are proposed, where a probabilistic neural network is used as an autoencoder, and the actual recognition process is implemented using a modified radial-basis function network operating in recognition mode. A distinctive feature of the proposed stack systems is that information can be processed both in the traditional vector form and in the matrix form inherent in images. The proposed systems are characterized by a high learning speed, which allows them to solve problems that arise within the general problem of Data Stream Mining.

Keywords Image classification · Small training dataset · Overlapping classes · Probabilistic neural networks · Supervised learning · Lazy learning

Ye. Bodyanskiy
Control Systems Research Laboratory, Kharkiv National University of Radio Electronics, Kharkiv, Ukraine

O. Chala (✉)
Artifical Intelligence Departement, Kharkiv National University of Radio Electronics, Kharkiv, Ukraine
e-mail: olha.chala@nure.ua

Y. P. Kondratenko et al. (eds.), *Artificial Intelligence in Control and Decision-making Systems*, Studies in Computational Intelligence 1087,
https://doi.org/10.1007/978-3-031-25759-9_6

1　Introduction

Probabilistic Neural Networks (PNNs) that were proposed by Specht [30], were designed to solve classification problems—pattern recognition, and are the "closest relatives" of radial basis function and generalized regression neural networks [2–6]. These constructions are based on Bayesian inference, Parzen windows [24], and Nadaraya-Watson estimates [21] using kernel activation functions. Most often, as such functions Gaussians are used, although in principle all radial-basis structures having a bell shape can be used.

Although probabilistic networks are not competitive in case of accuracy comparing to deep neural networks' (DNN) classification, the extremely high speed of their tuning based on lazy learning ("Neurons at data points", "Just-in-time-models" [22, 33]) and information processing in a number of cases favours these networks, especially in situations when data are fed to the system in online real time mode.

It should be noted that most neural networks (both DNN and traditional shallow ones) solve the problems of crisp classification, that is, images belonging to different classes that do not overlap in a feature space. If this is not the case, then neuro-fuzzy systems should be used instead of neural networks, which, in addition to their own pattern recognition, estimate the level of fuzzy membership of each observation to each of the possible classes. Thus, fuzzy probabilistic networks have been proposed in [11–13], which have proven their effectiveness in solving fuzzy classification problems, but their computational "cumbersome" limits their ability to process information online.

Thereby it is appropriate to introduce into consideration the evolving stacking neuro-fuzzy probabilistic networks, which allow solving the task of fuzzy classification (overlapping classes) at the high speed under conditions when data are fed sequentially in online mode for processing. In doing so for the learning of such system it is appropriate to use the idea of evolving systems for the architecture tuning, competitive self-learning for the centroids tuning, lazy learning for its initialization, and standard supervised learning for spread parameters of kernel activation functions.

2　Neuro-Fuzzy-Probabilistic Network

2.1　Architecture of Neuro-Fuzzy-Probabilistic Network

The proposed modified fuzzy-probabilistic neural network (NFPN) architecture has a four-layer architecture is shown in Fig. 1.

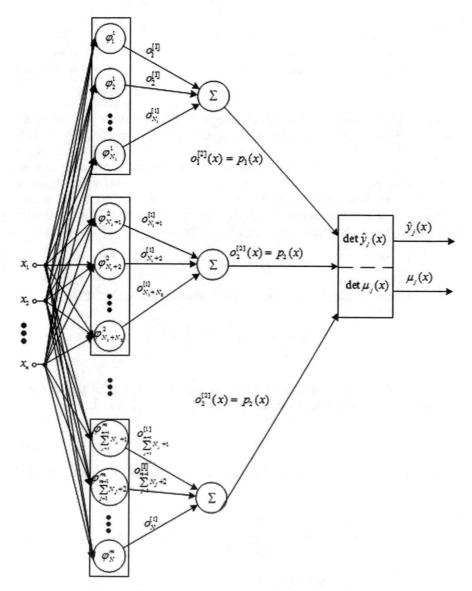

Fig. 1 The architecture of the proposed adaptive probabilistic neural network

Initial information for system synthesis is a training dataset of images formed with batch of N-dimensional vectors $X = \{x(1),\ x(2),\ \ldots, x(k),\ \ldots, x(N)\}$, $x(k) = (x_1(k),\ \ldots, x_i(k),\ \ldots, x_n(k))^T \in R^N$, where N_1 observations belong to the first class Cl_1, N_2—to the second class Cl_2, …, N_m—to Cl_m, that is:

$$k = \underbrace{1, 2, \ldots, N_1}_{\substack{Cl_1 \text{ contains } N_1 \\ observations}} , \underbrace{N_1 + 1, \ldots, N_1 + N_2}_{\substack{Cl_2 \text{ contains } N_2 \\ observations}}, \ldots, N_1 + N_2 + N_3, \underbrace{\ldots, N}_{\substack{Cl_m \text{ contains } N_m \\ observations}}$$

(1)

The number of neurons in the first hidden layer (pattern layer) is equal N, the activation function of each neuron is denoted as $\varphi_{\tau_j}(x, c_{\tau_j})$, where $\tau_j = N_1 + N_2 + \ldots + +N_{j-1} + 1, \ldots, N_1 + N_2 + \ldots N_j$, $j = 1, 2, \ldots, m$—the class number, $c_{\tau_j} = (c_{\tau_j 1}, \ldots, c_{\tau_j i}, \ldots, c_{\tau_j n})^T$ the centroid of the bell-shaped activation function. Tuning of the NFPN takes place at the pattern layer with the help of lazy learning (the concept of "Neurons in data points") so that $c_{\tau_j} = x(k) = x(\tau_j)$, if $x(k) \in Cl_j$.

When vector x with unknown classification is fed to the system to the first hidden layer, the signals at the outputs of the first layer of neurons appear in the form

$$o_j^{[1]}(x) = \varphi_{\tau_j}^j(x, c_{\tau_j})$$

(2)

which can be written in the case of Gaussian activation function in the form

$$\varphi_{\tau_j}^j(x, c_{\tau_j}) = \exp\left(-\frac{\|x - c_{\tau_j}\|^2}{2\sigma^2}\right)$$

(3)

(here σ^2 is the width parameter of the activation function), and if the output is pre-centroided and normalized so that $\|x(k)\| = \|c_{\tau_j}\| = \|x\| = 1$, then

$$o_j^{[1]}(x) = \exp(\sigma^{-2}(x^T c_{\tau_j} - 1))$$

(4)

The second hidden layer contains m adders, the outputs of which are generated signals

$$o_j^{[2]}(x) = \sum_{\tau_j = N_1 + N_2 + \ldots + N_{j-1} + 1}^{N_1 + N_2 + \ldots + N_j} o_j^{[1]}(x) = p_j(x)$$

(5)

which are Parzen estimates of unknown probabilities distributions $p_j(x)$.

The third hidden layer takes into account a priori probabilities

$$P_j(x) = \frac{N_j}{N}$$

(6)

and the cost of classification errors S_j so that

$$o_j^{[3]}(x) = o_j^{[2]}(x) N_j N^{-1} S_j$$

(7)

It is easy to see that the main advantage of NFPN is the simplicity of its synthesis and learning, and the main disadvantage is the significant increase in the number of neurons in the first hidden layer with the increase in the number of observations N in the training sample.

2.2 Evolving of NFPN in the Layer of Patterns

It is possible to limit the number of neurons in the first hidden layer using the ideas of the theory of evolving systems of computational intelligence [18]. Let's suppose that the total number of neurons in this layer cannot exceed a certain number $M < N$ and at the same time each class Cl_j contains $\frac{M}{m}$ neurons. Let's take a look at the process of formation-evolution layer for a class $j = 1, 2, …, m$, in this case, we will use the index $\tau_j = 1, 2, …, \frac{M}{m}$ to simplify the designation of a particular neuron. The process of forming the layer can be written in the form of a sequence of steps—stages [9]:

Step 0: the maximal number of neurons in this layer is M, and the threshold of indistinguishability between two neighbouring centroids of activation functions c_{τ_j} and c_{τ_j+1} is Δ.

Step 1: when the observation $x(1)$ that belongs to class j enters to the network, the first activation function φ_1^j of the class is formed with centroid $c_1^j = x(1)$ and width parameter σ^2, which is chosen rather arbitrarily. However, if the output is centered and normalized so that $-1 \leq x_i(k) \leq 1$, it is convenient to use the value $\sigma \approx 0,33$.

Step 2: when the second observation $x(2)$ from the same class enters to the network, the condition is checked:

$$\left\| x(2) - c_1^j \right\| \leq \Delta \tag{8}$$

and if it is satisfied, then $x(2)$ does not form a new class; after that the condition is checked

$$0 < \left\| x(2) - c_1^j \right\| \leq 2\Delta \tag{9}$$

and if it satisfied, the centroid c_1^j is adjusted according to Kohonen's self-learning rule [14] "Winner takes all":

$$c_1^j(2) = c_1^j(1) + \eta(2)(x(2) - c_1^j(1)) \tag{10}$$

that is, earlier formed centroid $c^j(1)$ "pulls up" to a new observation $x(2)$ (here $0 < \eta(2) < 1$- learning rate parameter). Finally, if the following inequality holds

$$2\Delta < \left\| x(2) - c_1^j \right\|, \tag{11}$$

a new activation function φ_2^j is being formed with centroid $c_2^j = x(2)$:.

Step h: when observation $x(h)$ $(h < m^{-1}M < N)$ is fed to the network, first of all seeking the neuron-winner with centroid c_{h-1}^{j*} which is the closest in adopted metric to this observation. The conditions are then checked

$$\left\| x(h) - c_{h-1}^{j*} \right\| \leq \Delta \tag{12}$$

$$0 < \left\| x(h) - c_{h-1}^{j*} \right\| \leq 2\Delta, \tag{13}$$

$$2\Delta < \left\| x(h) - c_{h-1}^{j*} \right\|, \tag{14}$$

after that, either this observation is ignored, or the centroid c_{h-1}^{j*} is drawn to the observation $x(h)$, or the new activation function with the centroid $c_h^j = x(h)$ is formed.

The process of constructing the first layer continues until each layer is formed by $m^{-1}M$ neurons. After that, this layer is considered to be formed, and new observations coming into the network can only change the location of the centroids according to Kohonen's self-learning rule.

The process of adjusting the spread parameter of activation function σ_{τ_j} can be implemented too using supervised learning (controlled learning) defined by the adopted goal function. Since in PNN usually Gaussians are used, it is appropriate to introduce the method of spread parameter setup for Gaussian functions, this parameter is variance.

Let's introduce into consideration goal-function—learning criterion

$$E(x, \sigma^2) = \sum_{j=1}^{m} E_j(x, \sigma^2) = \frac{1}{2} \left\| e(x, \sigma^2) \right\|^2 \tag{15}$$

where

$$E(x, \sigma^2) = \frac{1}{2} e_j^2(x, \sigma^2) = \frac{1}{2}(y_j(x) - \hat{y}_j(x))^2, \tag{16}$$

$$e(x, \sigma^2) = (e_1(x, \sigma_1^2), \ldots, e_j(x, \sigma_j^2), \ldots, e_m(x, \sigma_m^2))^T \tag{17}$$

$$y_j(x) = \begin{cases} 1, \ if \ x(k) \in Cl_j, \\ 0 \ otherwise \end{cases} \tag{18}$$

—learning reference signal formed by zeroes and ones, so-called "one-hot coding".

Introducing further derivatives

$$\frac{\partial E(x, \sigma^2)}{\partial \sigma_l^{-2}} = -e_j(x, \sigma_j)\frac{1 - \hat{y}_j(x)}{o^{[2]}(x)} \sum_{\tau_j} \frac{\partial o_j^{[1]}(x)}{\partial \sigma_l^{-2}}, \tag{19}$$

$$\frac{\partial E(x, \sigma^2)}{\partial \sigma_l^{-2}} = e_l(x, \sigma_l)\frac{1 - \hat{y}_l(x)}{o^{[2]}(x)} \sum_{\tau_l} \frac{\partial o_l^{[1]}(x)}{\partial \sigma_l^{-2}}, \, l \neq j \tag{20}$$

or equivalently

$$\frac{\partial E(x, \sigma^2)}{\partial \sigma_j^{-2}} = e_j(x, \sigma_j)\frac{\hat{y}_j(x) - \delta_{jl}}{o^{[2]}(x)} \sum_{\tau_j} \frac{\partial o_j^{[1]}(x)}{\partial \sigma_j^{-2}} \tag{21}$$

where

$$\delta_{jl} = \begin{cases} 1, \, if \, j = l, \\ 0 \, otherwise, \end{cases} \tag{22}$$

$$\frac{\partial o_j^{[1]}(x)}{\partial \sigma_j^{-2}} = -\frac{1}{2}\|x - c_{\tau_j}\|^2 \exp\|x - c_{\tau_j}\|^2 = (x^T c_{\tau_j} - 1)exp\left(\sigma_j^{-2}(x^T c_{\tau_j} - 1)\right). \tag{23}$$

Finally, the algorithm of learning the spread parameter can be written in the form

$$\sigma_j^{-2}(k + 1) = \sigma_j^{-2}(k) - \eta(k)\frac{\partial E_j(x(k), \sigma_j^2)}{\partial \sigma_j^{-2}} =$$

$$= \sigma_j^{-2}(k) + \eta(k)\sum_{j=1}^m e_j(x(k), \sigma_j^2)\frac{\hat{y}_j(x(k)) - \delta_{jl}}{o^{[2]}(x(k))} \sum_{\tau_l} \frac{\partial o_j^{[1]}(x)}{\partial \sigma_j^{-2}} =$$

$$= \sigma_j^{-2}(k) + \eta(k)\sum_{j=1}^m e_j(x(k), \sigma_j^2(k))\frac{\hat{y}_j(x(k)) - \delta_{jl}}{o^{[2]}(x(k))} \sum_{\tau_j} \exp\left(-\frac{\|x(k) - c_{\tau_j}\|^2}{2\sigma^2(k)}\right)$$

$$\|x(k) - c_{\tau_j}\|^2 \tag{24}$$

where $0 < \eta(k) < 1$—learning rate parameter of gradient search.

Thus, the introduced evolving fuzzy-probabilistic neural network implements the learning principles according to Specht [30], evolving systems of [17–19], self-learning of Kohonen [14] and classical supervised learning.

2.3   Results of Experiment

A number of classification experiments with time estimation were performed. Nine datasets with different amounts of samples were formed randomly from the original one. The classification was processed on each on them eight times and eventually, time of each one was averaged. The results of the experiments are represented in Fig. 2.

The horizontal line indicates the number of samples in the dataset, the vertical one—the growth of classification time with increasing data amount. From the result it can be seen while the dataset is growing by 60%, from 1000 to 1600 samples, classification time increased by less than 30%.

The comparative analysis of classification accuracy of the proposed neural network and popular machine learning classification methods such as k-nearest neighbours (KNN) and support vector machine (SVM) were held. The results of the experiment represented in Table 1 and Fig. 3.

In the first row of the table, the provided results of the KNN classification method are shown, in the second one—SVM, in the third one—using the proposed neural network. It can be seen from Table 1 that the considered evolving fuzzy-probabilistic neural network provides higher accuracy in the tasks of recognition low-resolution images.

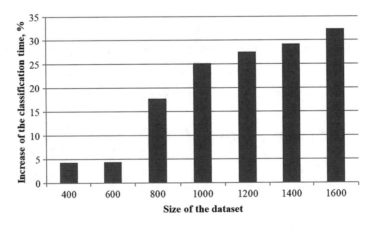

Fig. 2 The dependency between size of the dataset and time consumption

Table 1 The algorithms accuracy comparison in relation to the size of the dataset

Algorithms for comparison	Classification accuracies, %					
	400	600	800	1000	1200	1600
KNN	62	71	74.7	75	73.5	75.7
SVM	64.3	78. 9	81.2	84.1	87.6	89.8
ENFPN	88.7	91.2	91.2	96.4	96.4	96.4

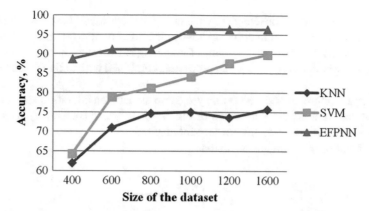

Fig. 3 The dependency between size of the dataset and accuracy

The dependence of accuracy on the sample size that presented in Fig. 3 shows that the proposed ENFPN algorithm achieves maximum accuracy with the dataset containing 1000 elements or 62% of the initial one. At the same time, KNN and SVM algorithms achieve maximum accuracy only with a full-size dataset. Therefore, using the proposed network, it is possible to decrease the size of the training sample and thereby significantly reduce the training time. This advantage is key when recognizing images online mode.

The dependence of accuracy on the sample size that is presented in Fig. 3 shows that the proposed ENFPN algorithm achieves maximum accuracy with the dataset containing 1000 elements or 62% of the initial one. At the same time, KNN and SVM algorithms achieve maximum accuracy only with a full-size dataset. Therefore, using the proposed network, it is possible to decrease the size of the training sample and thereby significantly reduce the training time. This advantage is key when recognizing images in an online mode.

The evolving fuzzy probabilistic neural network is proposed to classify vector observations that are processed in a sequential online mode with a limited number of neurons, characterized by a high rate of learning, which distinguishes it from both deep and traditional shallow multilayer networks and allows for a large amount of data to be processed within the overall problem of Data Stream Mining.

3 Probabilistic Neuro-Fuzzy System with One Dimensional Membership Function

The main disadvantages of the neuro-fuzzy network as discussed above is some bulkiness, because a standard PNN can be a bit tricky: its' first hidden pattern layer in the general case contains a number of R-neurons, which is equal to the number of observations in the training dataset.

To overcome this inconvenience, we can use the approach based on the hybrid systems of computational intelligence [2–4], specifically neuro-fuzzy systems [29, 33]. The last ones have a vast amount of advantages over classic neural networks, keeping multipurpose approximating properties combined with online mode learning abilities.

The architecture of the neuro-fuzzy system under consideration that has four layers of data processing represented in Fig. 4. On the zero (receptive) layer of the system, vectors of observations are sequentially fed.

Here input data are organized so that:

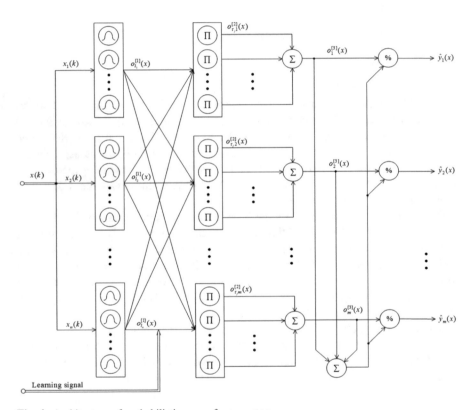

Fig. 4 Architecture of probabilistic neuro-fuzzy system

$$\begin{cases} if \ x(k) \in Cl_1 \ then \ k = \tau_1 = 1, 2, \ \ldots, N_1, \\ if \ x(k) \in Cl_2 \ then \ k = \tau_2 = N_1 + 1, \ \ldots, N_1 + N_2, \\ \vdots \\ if \ x(k) \in Cl_j \ then \ k = \tau_j = N_1 + N_2 + \cdots N_{j-1}, \ \ldots, \\ N_1 + N_2 + \cdots N_j, \\ \vdots \\ if \ x(k) \in Cl_m \ then \ k = \tau_m = N_1 + N_2 + \cdots N_{m-1} + 1, \ \ldots, \\ N_1 + N_2 + \cdots N_m = N. \end{cases} \tag{25}$$

These observations are fed to the first hidden fuzzification layer formed with fuzzy membership functions $\varphi_{l_i}(x_i, c_{l_i})$, $l_i = 1, 2, \ldots, h_i$, $i = 1, 2, \ldots, n$ as which the usual one-dimensional Gaussians are used:

$$\varphi_{l_i}(x_i, c_{l_i}) = \left(\frac{\left(x_i - c_{l_i}\right)^2}{2\sigma^2} \right) \tag{26}$$

where c_{l_i} is the center of corresponding membership function, and σ is the spread parameter.

The number of these membership functions at each input x_i can be different: from two in the case of a binary input variable that takes only two values "yes" or "no" to N as in a standard probabilistic neural network PNN, i.e.

$$2 \le h_i \le N \tag{27}$$

for each input or for the system in general

$$2n \le h = \sum_{i=1}^{n} h_i \le N_n \tag{28}$$

The output signals of the first layer

$$o_{l_i}^{[1]} = \varphi_{l_i}(x_i, c_{l_i}) \tag{29}$$

are fed to inputs on second hidden aggregation layer formed with N standard multi-plication blocks that are forming bell shaped activation functions in multidimectional Gaussians form

$$\varphi_{\tau_j}^{j}(x_i, c_{\tau_j}) = \prod_{i=1}^{n} \exp \left(-\frac{\left(x_i - c_{l_i}\right)^2}{2\sigma^2} \right) = \exp \left(-\frac{\|x - c_{\tau_j}\|^2}{2\sigma^2} \right) \tag{30}$$

where vector-center $c_{\tau_j} = \left(c_{l_1}, \ldots, c_{l_i}, \ldots, c_{l_n} \right)^T$ is formed basing on one-dimensional centers for each class j. Here the number of multiplication blocks for each class is N_j. Thus, in the second layer, the N multidimensional kernel activation functions are formed, and on its output, the signals appeared

$$o_{\tau_j i}^{[2]} = \varphi_{\tau_j}^j(x(\tau_j), \, c_{\tau_j}). \tag{31}$$

It is easy to see that the first two layers of the proposed system coincide with input layers of the popular neuro-fuzzy system of Takagi–Sugeno-Kang or ANFIS [29, 33] and implement the same function as the PNN's pattern layer. The advantages of the system popped up in case when some components of input vectors-images of the training dataset coincide. This situation appears when the input signals' individual components are either binary, or rank, or nominal variables, which is common in real tasks. Here, the number of membership functions in the corresponding input h_i is smaller than the amount of training data N.

A third hidden layer is formed by $m + 1$ adders, where from the first m Parzen estimates of the data distribution density in each class are formed

$$o_j^{[3]} = \sum_{\tau_j = N_1 + N_2 + \cdots N_{j-1} + 1}^{N_1 + N_2 + \cdots + N_j} o_{\tau_j j}^{[2]} = p_j(x) \tag{32}$$

and $(m + 1)$st adder adds all output signals of this layer.

Finally, the output layer of the system computes the probability of membership of each observation x—which is not from the training dataset—to each class

$$\hat{y}_j(x) = \frac{o_j^{[3]}(x)}{o^{[3]}(x)} = \frac{o_j^{[3]}(x)}{\sum_{j=1}^m o_j^{[3]}(x)}. \tag{33}$$

It is easy to see that the output layer performs the operation similar to the defuzzification in neuro-fuzzy systems. Still, in our case, it has purely probabilistic meaning.

We propose to use a modified lazy learning procedure based on the concept "Neurons at data points" [22, 29] for the tuning of the Probabilistic neuro-fuzzy system (PNFS). Using this concept, classic PNN tunes simultaneously with feeding image $x(\tau_j)$ from training set to the system. As a result the multidimensional activation function $\varphi_{\tau_j}^j(x, \, c_{\tau_j})$ is formed with current observation as its center $c_{\tau_j} = x(\tau_j)$. This means that learning is almost instantaneous, but the number of activation functions is determined by the amount of the training dataset N.

Using a neuro-fuzzy approach allows to narrow down the number of membership functions, based on which multidimensional activation function are formed in the second hidden layer. In the simplest case, the components of the observation $x(\tau_j)$ form N membership functions c_{l_j} where individual coordinates determine centers

$x(\tau_j)$, and we do not get any gain in the number of functions. Although with significant N value, these one-dimensional centers $c_{\tau_j i}$ can be placed very close, and almost coincide on the axis x_i. Note also that within neuro-fuzzy approach, all input signals' coordinates are preprocessed so that

$$0 \le x_i(\tau_j) \le 1. \tag{34}$$

The gain in the number of functions appeared when within two different training observations $x(\tau_j)$ and $x_i(\delta_j)(\tau_j \neq \delta_j)$ some coordinates coincide $x_i(\tau_j) = x_i(\delta_j). x_i(\tau_j) = x_i(\delta_j)$.

Such situations are common when, for instance, one of the coordinates is binary value 0 or 1. Then it is necessary to form only two membership functions with centers in 0 and 1. The profit in the number of functions appears with nominal and rank variables as well. In this case, it is needed to create membership functions where their number is equal to the number of ranks.

Also, to reduce the number of membership functions, we specify the maximum possible number h_i for each coordinate x_i, then the distance between two neighboring functions is defined by the value

$$\Delta_i = \frac{1}{h_i - 1}. \tag{35}$$

Clearly for the binary variables $\Delta_i = 1$, in this case, centers are placed in points 0 and 1. Centers are placed in points 0, Δ, 2Δ, ..., 1 for the continuous numeric variables. If any value from the training dataset $x_i(\tau_j)$ appeared to be in the interval between two centers $l_i\Delta$, $(l_i + 1)\Delta$, it assigned to the closest one. Further, based on these one-dimensional functions, multidimensional activation functions are formed in the second hidden layer. The number of multiplication blocks in this layer describer by the N value, and it is assumed that there are no identical training observations.

Hence, PNFS learning is ended up to the tuning of one-dimensional membership functions centers and multiplication blocks in the aggregation layer. Clearly, such process happens almost instantly.

The two datasets were chosen to evaluate the proposed system. First is "ML handwritten digits" from the UCI repository as a dataset where tags are rank values, and the second one is "Fashion MNIST," which has nominal tags values. The first dataset has 1797 images that are matrices 8×8. The fact that they are low resolution gives the ability to show how well the proposed network works in a fuzzy case. The second dataset has 60 000 instances in the training and 6000 in the test dataset. Each sample of these datasets is a matrix 24×24.

As we can see, the resolution is higher than the first dataset, but due to the size of dataset, we can see how the proposed network performs in online mode. Both of these datasets contain overlapping classes. The samples from the both datasets represented in the Fig. 5. The fuzzy case that appears in the "ML hand-written digits" dataset is represented in Fig. 6 that shows only part of the full visualization.

Fig. 5 Example of the
observation from: **a** dataset
"Fashion MNIST", **b** dataset
"ML hand-written digits"

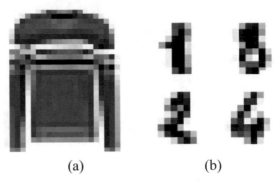

(a)   (b)

Fig. 6 Overlapping classes
in the "ML hand-written
digits" dataset

Table 2 The accuracy and time consumption of algorithms chosen for the comparison and the proposed one	Algorithms for comparison	Classification accuracies	Time, sec
	KNN	82	0.2
	EFPN	96.4	7
	PNFS	93.1	5.5

For the comparison analysis, the popular machine learning method K-Nearest neighbours (KNN), Evolving NFPN described above, and the probabilistic neuro-fuzzy system (PNFS) were taken. During the first part of the experiment, "ML hand-written digits" were applied, and the spent time, along with the accuracy, were tracked. The obtained results are represented in Table 2.

As we can see, the probabilistic neuro-fuzzy system works faster than the Evolving fuzzy-probabilistic neural network; however, it yields to EFPNN in classification accuracy. It has to take into account that the K-Nearest neighbours algorithm was performed on GPU and the probabilistic neuro-fuzzy system on CPU, which led to the gap in the computation time. Nevertheless, if the probabilistic neuro-fuzzy system would have performed on the CPU, the consumed time would be comparable, because the speed of the modern CPU is higher than the recent GPU from five to ten times.

For the second part of the experiment, the "Fashion MNIST" dataset was used. Out of the original full-size dataset, we randomly picked nine subsets of different size. Each of these datasets was formed randomly. With the initial "Fashion MNIST" dataset, the biggest formed dataset has 15 000 samples. Also, the wide-spread parameter was chosen empirically for the initial dataset to obtain the best results. For the large dataset, the wide-spread parameter is 0.74. The results are represented in Fig. 7.

As seen from the graphic, the KNN algorithm provides the smallest increase in computation time, depending on the dataset size. However, this algorithm has a lower accuracy compared to algorithms based on probabilistic neural networks.

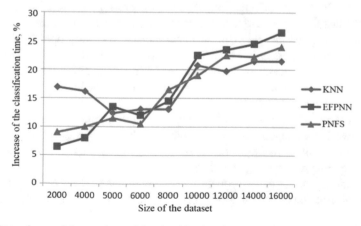

Fig. 7 Dependency of dataset size and the classification time

The proposed PNFS algorithm requires lower computational burden compared to the EFPNN algorithm with similar classification accuracy. On datasets with a larger size, more than 8000 elements, this algorithm shows a nearly linear increase in computational costs with an increase in the sample size.

The increase in computational costs when going from small datasets to the long are probably related to features of software implementation. The small dataset fills evenly in RAM, and therefore as it grew, the additional processing costs increased insignificantly. The long dataset requires more resources for swapping pages by the memory operator, which significantly increases RAM's cost.

4 Matrix Fuzzy-Probabilistic Neural Network in Image Recognition Task

The problem of pattern recognition is quite common within the general problem of Data Mining, and there exist now a lot of approaches, methods, and algorithms for its solution. Among them, the most effective today are the ones based on computational intelligence approaches [2–4]. Currently, DNNs [21–23] are the most used to solve this problem. They provide the high quality of the received solution but are characterized by low learning speed, so their tuning can be implemented only in batch mode on a plurality of epochs. Thus, in situations where the size of the dataset which to be processed is unlimited, and the same data are fed to the recognition system in a sequential online mode, DNNs are ineffective. In this case, when the speed of the recognition system comes to the fore, DNNs (like the traditional CNNs of the three-layer perceptron type) can hardly be used.

It should be noted that although the accuracy of classification PNN is yielded to the modern DNN, their speed in some cases gives them a significant advantage.

It should also be noted that the input information of most known neural networks, including the PNN, comes in the form of vectors, which means that image recognition tasks must first be vectorized, which means, matrix signals must be converted into vectors. In principle, this can be accomplished by using operations such as convolution and pooling as it is done in a convolutional DNN. However, the use of these operations significantly reduces the speed of the recognition system in general. Therefore, it is expedient not to vectorize the input images, but to feed them to the receptive layer in a matrix form.

That is why we propose to introduce into consideration fuzzy probabilistic neural network for image recognition, which provides a high speed of information processing that comes in the matrix form and can restore the observation classes arbitrarily overlapped in the space of features.

The proposed matrix fuzzy-probabilistic neural network (MFPNN) contains four layers of information processing: the first hidden layer of patterns, the second hidden layer, formed by elements of summation, the third hidden layer of probability distributions correction, and, finally, the fourth output layer, which determines the levels of membership of the classified images to specific classes.

Information comes to the zero (receptive) layer of the system is given by a sample of images in the form of $(n_1 \times n_2)$—matrices $X = \{x(1), x(2), \ldots, x(k), \ldots, x(N)\}, x(k) = \{x_{i_1 i_2}(k)\} \in R^{n_1 \times n_2}$, where N_1 observations belong to the first class Cl_1, N_2—to the second class Cl_2, $\ldots$, N_m to Cl_m according to (1).

The number of radial-basis neurons [8] is defined by the amount of observations N, with the activation functions of this layer in the future we will denote as $\varphi(x, c_{\tau_j}, \sigma_{\tau_j}^2)$ where τ_j varies in the range $N_1 + N_2 + \ldots + N_j + 1$, $N_1 + N_2 + \ldots N_j + 2$, $N_1 + N_2 + \ldots N_j$, $c_{\tau_j} \in R^{n_1 \times n_2}$—matrix center of activation function, that is $\{c_{\tau_j i_1 i_2}^j\}$—$n_1 \times n_2$ matrix, $\sigma_{\tau_j}^2$—a parameter that specifies the "width" of the corresponding bell-shaped activation function. As an activation function can be used either matrix Gaussian

$$\varphi(x, \ c_{\tau_j}, \sigma_{\tau_j}^2) = \exp\left(-\frac{Tr(x - c_{\tau_j})(x - c_{\tau_j})^T}{2\sigma_{\tau_j}^2}\right) \tag{36}$$

or any other bell-shaped function, such as matrix Cauchian

$$\varphi(x, \ c_{\tau_j}, \sigma_{\tau_j}^2) = \frac{1}{1 + \sigma_{\tau_j}^{-2} Tr(x - c_{\tau_j})(x - c_{\tau_j})^T} \tag{37}$$

where

$$Tr(x - c_{\tau_j})(x - c_{\tau_j})^T = d_F^2(x, c_{\tau_j}) \tag{38}$$

—matrix Frobenius metrics, which specifies the distance between the matrix centers c_{τ_j} and the input image x.

Network tuning usually occurs only at the level of pattern layer based on the concept of "Neurons at data points" [7], that is, according to the rule

$$c_{\tau_j} = x(k), \ if \ x(k) \in Cl_j, \ \tau_j = N_1 + N_2 + \ldots + N_{j-1} + 1, \ldots, N_1 + N_2 + \ldots N_j \tag{39}$$

When all of the radial-basis neurons of the first hidden layer are applied to an image x with an unknown classification, on the outputs of current layer appear N signals $o_k^{[1]} = \varphi(x, \ c_{\tau_j}, \sigma_{\tau_j}^2) \ \forall \ j = 1, 2, \ldots, m; \ k = 1, 2, \ldots, N$. The second hidden layer consists of m elementary adders (one for each class), and on its output signals are formed

$$o_j^{[2]}(x) = \sum_{\tau_j = N_1 + N_2 + \ldots + N_{j-1} + 1}^{N_1 + N_2 + \ldots + N_j} o_{\tau_j}^{[1]}(x) \tag{40}$$

which are the Parzen estimates of the probability distribution density, i.e.

$$p_j(x) = \sum_{\tau_j=N_1+N_2+\ldots+N_{j-1}+1}^{N_1+N_2+\ldots+N_j} o_{\tau_j}^{[1]}(x). \tag{41}$$

In the third hidden layer, these estimates are corrected to take into account empirical a priori probabilities as it was presented with equitation (6) so the classification error can be described with (7). inally, in the output layer, the class m to which the image x belongs is determined, that is the "class-winner":

$$j^* = \arg \max_j o_j^{[3]}(x). \tag{42}$$

In principle, the described procedure does not differ from the work of the standard PNN except for the fact that the input of the system obtains not traditional vectors-images but matrices-images.

In real-world problems, the situation is often made more complicated by the fact that an image subject to classification can simultaneously belong to several classes at the same time with different levels of membership. This situation occurs when the classes formed by the training dataset X are mutually overlapping, in other words, a fuzzy situation arises.

The centroids for each class are taken into consideration in the form

$$c_j = \frac{1}{N_j} \sum_{\tau_j=N_1+N_2+\ldots+N_{j-1}+1}^{N_1+N_2+\ldots+N_j} c_{\tau_j} \tag{43}$$

and the radii of these classes—

$$r_j = \max d_F(x(\tau_j), c_j). \tag{44}$$

Then, if there is a situation for two classes Cl_j and Cl_l Cl_l

$$d_F(c_j, c_l) < r_j + r_l, \tag{45}$$

then these classes overlap.

To estimate the level of the membership of the observation x to class Cl_j we can use as an estimate that arises in the popular method of fuzzy C-means clustering (FCM) in the form [12, 15]:

$$\begin{aligned}
\mu_j(x) &= \frac{d_F^{-2}(x, c_j)}{\sum_{l=1}^m d_F^{-2}(x, c_l)} \\
&= \frac{\left(Tr(x - c_{\tau_j})(x - c_{\tau_j})^T\right)^{-1}}{\sum_{l=1}^m \left(Tr(x - c_{\tau_j})(x - c_{\tau_j})^T\right)^{-1}}
\end{aligned}$$

$$
= \frac{1}{1 + d_F^2(x, c_j) \sum_l = 1l \neq j^m d_F^{-2}(x, c_l)}
$$

$$
= \frac{1}{1 + \sigma_j^{-2} d_F^2(x, c_j)} = \frac{1}{1 + \sigma_j^{-2} Tr(x - c_j)(x - c_j)^T} \tag{46}
$$

where

$$
\sigma_j^2 = \left(\sum_l = 1l \neq j^m d_F^{-2}(x, c_l) \right)^{-1}. \tag{47}
$$

It's easy to see those fuzzy memberships are determined with Cauchian (37), which can also be used as activation functions of the second hidden layer (40). It should be noted that the width parameters σ_j^2 during evaluating fuzzy memberships are determined automatically using the relation (47).

Thus, in a fuzzy case in the output layer of the network what is determined are both the winner class and the membership level of each image x to each class Cl_j

In Data Stream Mining tasks, the situation when the training dataset is given not in the form of a fixed data batch, but as the sequence $x(1), \ldots, x(k), \ldots, x(N), x(N+1), \ldots$ that increases in size over time. This automatically increases the number of neurons in the pattern layer, which means the network in the training process "builds up" its architecture. Therefore, the configuration of such a network should occur in a sequential mode with the help of those or other recurrent learning algorithms. Assume that the training dataset received an observation $x(N + 1)$ that belongs to the class Cl_j. At the same time in the pattern layer, one extra neuron is added to the group that "corresponds" exactly for this class. The procedure for clarifying the centroid of this class can be represented as

$$
c_j(N_j + 1) = \frac{1}{N_j + 1} \sum_{\tau_j} c_{\tau_j} \tag{48}
$$

or in a recurrent form

$$
c_j(N_j + 1) = \begin{cases} c_j(N_j) + (N_j + 1)^{-1}(x(N + 1) - c_j(N_j)) \ if \ x(N + 1) \in Cl_j, \\ c_j(N_j) \ if \ x(N + 1) \notin Cl_j \end{cases}
$$

$$
\tag{49}
$$

It's easy to see that (50) coincides with the T. Kohonen's self-organizing maps (SOM) training rule [14], "winner takes all" (WTA).

The increasing number of neurons in the pattern layer may be unwanted because the network may become too cumbersome, especially if the information comes in the form of a stream of images. In this case, it is advisable to fix the number of neurons for each class by some number N_j (basically, this number may be the same for all classes), and the training procedure can be implemented on a "sliding window" in the form

$$c_j(N_j + 1) = \begin{cases} c_j(N_j) + \overline{N_j}^{-1}(x(N + 1) - x(N - \overline{N_j} + 1)) \, if \, x(N_j + 1) \in Cl_j, \\ c_j(N_j) \, if \, x(N + 1) \notin Cl_j \, . \end{cases}$$

$$(50)$$

In fuzzy cases, when classes can overlap, a situation when the observation $x(N + 1)$ belongs to the Cl_l is placed closer to the centroid $Cl_j(N)$ than to the "native" $Cl_l(N)$, that is $d_F(x(N + 1), c_j(N)) < d_F(x(N + 1), c_l(N))$ can arise. However if $x(N + 1) \in Cl_l$ and if in the procedure (16), the specified centroid $Cl_l(N + 1)$ is "tightened" to $x(N + 1)$, in this case, $c_j(N + 1)$ should "push" from $x(N + 1)$.

In this situation, the tuning procedure for centroid c_j can be written in the form

$$c(N_j + 1) = \begin{cases} c_j(N) + \eta(N + 1)(x(N + 1) - c_j(N_j)) \, if \\ x(N + 1) \in Cl_j, \; d_F(x(N + 1), c_j(N)) < d_F(x(N + 1), c_l(N)), \\ c_j(N) - \eta(N + 1)(x(N + 1) - c_j(N_j)) \, if \\ x(N + 1) \in Cl_l, \; d_F(x(N + 1), c_j(N)) < d_F(x(N + 1), c_l(N)), \\ c_j(N) \, if \, x(N + 1) \notin Cl_j, Cl_l \end{cases}$$

$$(51)$$

where $0 < \eta(k + 1) < 1$—the parameter of the learning rate, which is selected according to A. Dvoretzky's conditions [10].

It's easy to see that procedure (51) is an algorithm for learning vector quantization (LVQ) [4, 28], which is widely used in problems of image classification-recognition.

Thus, the introduced matrix probabilistic neural network is a hybrid of the standard PNN, the Kohonen's SOM, and LVQ, as well as the J. Bezdek's FCM, which is meant to solve the problems of images classification under conditions of classes that arbitrarily overlap in the space of features.

For the implementation of the proposed MFPNN the two well-known datasets such as UCI dataset "ML hand-written digits datasets" (Fig. 8) and "Fashion MNIST".

The second dataset has 60 000 observations in the training dataset and 6000 in the test dataset which allows evaluating the effectiveness in online mode.

Both of them were processed using the K-Nearest Neighbours (KNN), Convolutional Neural Network (CNN) and MFPNN. The results of the experiment are represented in the Tables 3 and 4. The first table contains the results of the accuracy and time consumption evaluation of all mentioned approaches with ML hand-written digits dataset.

The accuracy of the matrix fuzzy-probabilistic neural network is higher compared to the KNN but yield to CNN. KNN and CNN algorithm were performed on the GPU and the proposed approach on the CPU. Due to hardware support, the speed in the first and second cases increases by up to 10 times. Thus, the classification time in accordance with the proposed approach with the same hardware implementation

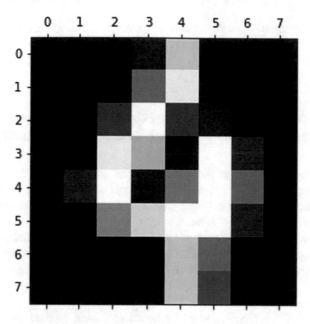

Fig. 8 The sample of the "ML hand-written digits datasets" dataset

Table 3 The accuracy and time consumption of algorithms with ml hand-written digits dataset

Algorithms for comparison	Classification accuracies	Time, sec
KNN	82	0.2
CNN	87.6	6.4
MFPNN	83.3	6

Table 4 The accuracy of algorithms chosen for comparison and the proposed one in relation to the size of the dataset

Algorithms for comparison	Classification accuracies, %					
	500	1000	2000	6000	12,000	60,000
KNN	51	71.3	76.8	81	82.7	83.3
CNN	62.8	73.1	77	83.7	86	91
MFPNN	67.3	75.9	80.1	81.5	82.3	84.7

would be comparable with the classification using the KNN and several times shorter than the classification time using the CNN.

The Table 4 describes the accuracy of algorithms for a different number of elements of the Fashion MNIST dataset.

From the presented data we can conclude that the accuracy of the MFPNN is growing faster comparing to the others mentioned approaches as the number of elements is increased. Therefore, the proposed algorithm allows achieving higher accuracy when training on a smaller number of elements, which reduces the training time in online mode.

5 Fast Image Recognition Using Double Hyper Basis Function Neural Network and Its Combined Learning

Currently, for the solution of this problem most effective ones are convolutional neural networks (CNNs). They are one of the most advanced areas in the intensively developing deep neural networks [21–23]. These systems provide high recognition accuracy, however, their usage is faced with several problems. First, learning of these networks requires a huge amount of training data, which are not always available during solving various practical tasks. Meanwhile, the usage of transfer learning does not always allow solving the problem. Secondly, usually, CNN has a tremendous number of tuning synaptic weights, and its learning is too time-consuming to use them in online mode.

Coming back to CNN, it should be noted that such type of network consists of two sections: autoencoder—formed with a sequence of convolution and pooling layers, and multilayer perceptron (MLP), which solves recognition-approximation task. Autoencoder transforms the input image, which is represented in matrix form, to the vector which is then fed to inputs of multilayer perceptron. Exactly learning of multilayer perceptrons requires heaps of time. Here, it is important to mention that the main requirement for MLP is universal approximating properties. Not only MLP but radial-basis function networks (RBFNs) [26–30] have such properties. Such systems are "the closest relatives" to PNN which implement kernel approximation [16] and do not yield to MLP in accuracy. The CNN which uses RBFN with Epanechnikov's kernels activation function [5] instead of MLP is proposed in [1]. This network showed good result in the pattern recognition task, nevertheless has some disadvantages. First, RBFN suffers from the "curse of dimensionality", when the number of tuning synaptic weight grows exponentially with the increasing dimension of the input vector. This leads to the forming of a small-sized vector on the output of the last pooling layer of the autoencoder, which carries on to the accuracy lost. Secondly, the usage of Epanechnikov kernels can lead to "gaps" in the feature space of RBFN. Thirdly, in such a network, the time is mainly spent on convolution-pooling, instead of approximation task.

To overcome the disadvantages of CNN and RBFN is possible through the usage of modified one instead of standard PNN autoencoder, in order to the input signal to be formed as an image-matrix [5], but not as a vector. The "curse of dimensionality" effect can be solved by using the so-called hyper basis function network (HBFN) instead of standard RBFN. The HBFN is characterized by the usage of hyperellipsoids with arbitrary orientation of the axes in place of receptive fields—hyperspheres of RBFN activation functions. The introduction of additional learning stages for the parameters of these hyperellipsoids helps to improve the approximating properties of the network, decreases the number of tuning synaptic weights (the protection of "curse of dimensionality"), and defend against gaps in the feature space.

Like CNN, the proposed neural network for solving the image recognition problem also consists of two sections: an autoencoder—a matrix probabilistic neural network and an approximator—a hyper basis neural network with tunable receptive fields. At the same time, the architectures of both network sections are quite close and are based on the use of multidimensional kernel activation functions [16].

The autoencoder is based on a matrix probabilistic neural network [5] and contains three layers of information processing: a pattern layer, a second hidden layer formed by m adders (here m is the number of classes in a dataset), and a third (output) layer for correcting the probability distribution.

The training sample is an array of N images, each of which is ($n_1 \times n_2$) matrix-image $x(k) = \{x_{i_1 i_2}(k)\}$, where $k = 1, 2, \ldots, N$—image number in the training dataset. Also, it is supposed that N_1 the images in the initial array belong to the first class Cl_1, N_2—to the second Cl_2 and, finally, N_m—to the mth class Cl_m:

$$\sum_{j=1}^{m} N_j = N \qquad (52)$$

The number of kernel activation functions (R-neurons) in PNN is determined by the size of the dataset N, herewith, the activation functions of the autoencoder's pattern layer will be denoted as $\varphi^A(x, c_{\tau_j}, \sigma_{\tau_j}^2)$ where variable—observation number τ_j varies in the interval $N_1 + N_2 + \cdots + N_{j-1}+1 \leq \tau_j \leq N_i + N_2 + \cdots + N_j, c_{\tau_j} \in R^{n_1 \times n_2}$—matrix-center of the activation function, defined in the learning process, and $\sigma_{\tau_j}^2$ is the parameter of the spherical receptive field of the corresponding bell-shaped activation function. Usually, in PNN, the traditional Gaussian is used as activation functions, and in this case, its matrix modification

$$\varphi^A(x, c_{\tau_j}, \sigma_{\tau_j}^2) = \exp\left(-\frac{Tr(x - c_{\tau_j})(x - c_{\tau_j})^T}{\sigma_{\tau_j}^2}\right). \qquad (53)$$

where original data were previously transformed as

$$\|x\| = \|c_{\tau_j}\| = 1 \qquad (54)$$

where $Tr(\circ)$ is the matrix trace symbol, $Tr(x - c_{\tau_j})(x - c_{\tau_j})^T = d_F^2(x, c_{\tau_j})$ is the Frobenius matrix metric specifying the distance between the center c_{τ_j} and input image x.

When the image x is fed in the autoencoder, on the input layer, at the outputs of all R-neurons of the first hidden layer appear N signals

$$o_k^{[1]} = \varphi^A(x, c_{\tau_j}, \sigma_{\tau_j}^2) \,\forall\, j = 1, 2, \ldots m, k = 1, 2, \ldots N. \qquad (55)$$

The second hidden layer of the autoencoder is formed by m adders, each of which belongs to a specific class Cl_j. Parzen estimates of the probability distribution densities are calculated at the outputs of these adders:

$$p_j(x) = o_j^{[2]}(x) = \sum_{\tau_j = N_1 + N_2 + \cdots N_{j-1} + 1}^{N_1 + N_2 + \cdots N_j} o_{\tau_j}^{[1]}(x). \tag{56}$$

In the third layer, these estimates are specified (if necessary) taking into account the empirical a priori probabilities according to (6) and the probabilities of the presented image x belongs to the j-th class appear at the outputs of the autoencoder:

$$o_j^A(x) = o_j^{[2]}(x) N_j N^{-1}. \tag{57}$$

Thus, an m-dimentional signal will appear at the output of the autoencoder $o^A(x) = (o_1^A(x), \ldots, o_j^A(x), \ldots, o_m^A(x))^T$, which is essentially a preliminary assessment of the classification results.

The neural network approximator is based on the hyper basis neural network (HBFN), which is a modification of the popular radial-basis function networks (RBFN) with receptive hyperellipsoidal fields with arbitrary orientation of the axes. It is assumed that in the process of HBFN learning, the parameters of these hyperellipsoids can be adjusted simultaneously with synaptic weights.

The sequence as a training dataset feds on the inputs of $o^A(x(k)) = o^A(k) = (o_1^A(x), \ldots, o_j^A(x), \ldots, o_m^A(x))^T$, $k = 1, 2, \ldots, N$, which is transmitted to h hyper basis R-neurons, at the outputs of which signals are generated

$$\varphi_l^H(o^A(k), c_l, \Sigma_l^{-1}) = \exp\left(-(o^A(k) - c_l)^T \Sigma_l^{-1}(o^A(k) - c_l)\right) =$$
$$= \exp\left(-\|o^A(k) - c_l\|_{\Sigma_l^{-1}}^2\right), l = 1, 2, \ldots, h, \ h >> m \tag{58}$$

where $Cl \in R^m$ is the vector center of an activation function $\varphi_l^H(\circ)$, Σ_l^{-1} is the covariance matrix that determines the shape, size and orientation of the receptive field axes of the corresponding activation function.

The output signals of R-neurons are fed to the HBFN output layer, which is formed not by sigmoidal adaptive linear elements (adalines), but by elementary Rosenblatt perceptrons with softmax activation functions. Thus, signals are generated at the outputs of HBFN and the recognition system as a whole.

$$\hat{y}_j(k) = w_{j0} + \sum_{l=1}^h w_{jl}\varphi_l^H\left(\|o^A(k) - c_l\|_{\Sigma_l^{-1}}^2\right) = \sum_{l=0}^h w_{jl}\varphi_l^H\left(\|o^A(k) - c_l\|_{\Sigma_l^{-1}}^2\right),$$

$$\varphi_0^H(\circ) \equiv 1, \ y_j^*(k) = \text{softmax}\, \hat{y}_j(k) = \frac{\exp \hat{y}_j(k)}{\sum_{p=1}^m \exp \hat{y}_p(k)}. \tag{59}$$

It is interesting to note that if the signal at the PNN output determines the level of probability that the image presented to the network belongs to a specific class, then the signal at the output of the introduced HBFN sets the levels of fuzzy membership of this image to the same class.

Hence, the recognition system contains only five layers of information processing, while the first and fourth layers are formed by R-neurons with hyperbasis and radial-basic activation functions.

The learning of the system under consideration occurs separately for the autoencoder and approximator based on different principles.

So, autoencoder tuning is implemented using lazy learning on the principle of "Neurons at data points" and almost instantly:

$$c_{\tau_j} = x(k), \ if \ x(k) \in Cl_j,$$
$$\tau_j = N_1 + N_2 + \cdots N_{j-1} + 1, \ldots N_1 + N_1 + \cdots N_{j-1} + N_j \qquad (60)$$

This means that the center of the kernel activation function is established at the point with the coordinates of the input image with a known classification.

Such approach provides sufficiently high classification accuracy and a high learning rate [22].

Tuning a hyper basis neural network is implemented based on the traditional controlled learning method with one-hot coding of reference signal, i.e. external learning signal elements $y_j(k)$ can only take two values: 1, if $x(k)$ belongs to a specific class and 0 otherwise.

Standard crossentropy is used as the HBFN learning rate:

$$E = -\sum_{k=1}^{N} \sum_{j=1}^{m} y_j(k) \ln y_j^*(k) = -\sum_{k=1}^{N} \sum_{j=1}^{m} y_j(k) \ln \frac{\exp \hat{y}_j(k)}{\sum_{p=1}^{m} \exp \hat{y}_p(k)} \qquad (61)$$

Further, to shorten the notation, we introduce additional notation:

$$w_j = \left(w_{j0}, w_{j1}, \ldots, w_{jl}, \ldots, w_{jh} \right)^T, \qquad (62)$$

$$\varphi^H(o^A(k), \ c_l, \ \Sigma_l^{-1}) = \left(1, \varphi_1^H(o^A(k), \ c_1, \ \Sigma_l^{-1}), \ldots, \varphi_l^H(o^A(k), \ c_l, \ \Sigma_l^{-1}), \ldots, \right.$$
$$\left. \varphi_h^H(o^A(k), \ c_h, \ \Sigma_l^{-1}) \right)^T$$

$$(63)$$

and write the signals at the adders output of in a more compact form

$$\hat{y}_j(k) = w_j^T \varphi^H \left(\left\| o^A(k) - c_l \right\|_{\Sigma_l^{-1}}^2 \right), \qquad (64)$$

$$y_j^*(k) = \frac{\exp w_j^T \varphi^H \left(\left\| o^A(k) - c_l \right\|_{\Sigma_l^{-1}}^2 \right)}{\sum_{p=1}^m \exp w_p^T \varphi^H \left(\left\| o^A(k) - c_l \right\|_{\Sigma_l^{-1}}^2 \right)} \tag{65}$$

where w_j and $\varphi^H \left(\left\| o^A(k) - c_l \right\|_{\Sigma_l^{-1}}^2 \right)$—$(h+1) \times 1$ vectors of synaptic weights at the j-th output of the system and signals at the outputs of R-neurons of HBFN, respectively.

Introducing further $(m \times 1)$—vectors $\hat{y}(k) = \left(\hat{y}_1(k), \ldots, \hat{y}_j(k), \ldots, \hat{y}_m(k) \right)^T$, $y^*(k) = \left(y_1^*(k), \ldots, y_j^*(k), \ldots, y_m^*(k) \right)^T$, $y_m^*(k)$ and $(m(h+1))$—synaptic weight matrix $w = \left(w_1, w_1, \ldots, w_j, \ldots, w_m \right)^T$ the output signal can also be written as compactly as possible:

$$\hat{y}(k) = w \varphi^H \left(\left\| o^A(k), \; c_l, \; \Sigma_l^{-1} \right\| \right) = w \varphi^H(k). \tag{66}$$

In order to tune synaptic weights, centers, and matrices of the receptive fields, the gradient learning procedure can be used:

$$w_{jl}(k+1) = w_{jl}(k) + \eta_w(k+1)$$
$$\left(y_j(k+1) - w_j^T(k) \, \varphi^H \left(\left\| o^A(k+1) - c(k) \right\|_{\Sigma^{-1}(k)}^2 \right) \right)$$
$$\varphi_l^H \left(\left\| o^A(k+1) - c_l(k) \right\|_{\Sigma_l^{-1}(k)}^2 \right), \tag{67}$$

$$c_l(k+1) = c_l(k) - \eta_c(k+1)$$
$$\left(y_j(k+1) - w_j^T(k+1) \, \varphi^H \left(\left\| o^A(k+1) - c(k) \right\|_{\Sigma^{-1}(k)}^2 \right) \right)$$
$$w_{jl}(k+1) \left(\varphi_l^H \left(\left\| o^A(k+1) - c_l(k) \right\|_{\Sigma_l^{-1}(k)}^2 \right) \right)$$
$$\Sigma_l^{-1}(k) \left(o^A(k+1) - c(k) \right), \tag{68}$$

$$c_l(k+1) = c_l(k) - \eta_c(k+1)$$
$$\left(y_j(k+1) - w_j^T(k+1) \, \varphi^H \left(\left\| o^A(k+1) - c(k) \right\|_{\Sigma^{-1}(k)}^2 \right) \right)$$
$$w_{jl}(k+1) \left(\varphi_l^H \left(\left\| o^A(k+1) - c_l(k) \right\|_{\Sigma_l^{-1}(k)}^2 \right) \right) \Sigma_l^{-1}(k) \left(o^A(k+1) - c(k) \right), \tag{69}$$

where $\eta_w(k+1), \eta_c(k+1), \eta_\Sigma(k+1)$—learning rate parameters for appropriate tuning variables.

The expression for adjusting of synaptic weights can be rewritten in vector and matrix forms:

$$w_{jl}(k+1) = w_j(k) + \eta_w(k+1)$$

$$\left(y_j(k+1) - w_j^T(k)\varphi^H\left(\left\|o^A(k+1-c(k)\right\|_{\Sigma^{-1}(k)}^2\right)\right)$$
$$\varphi_l^H\left(\left\|o^A(k+1) - c_l(k)\right\|_{\Sigma_l^{-1}(k)}^2\right), \tag{70}$$

and

$$w(k+1) = w(k) + \eta_w(k+1)\left(y(k+1) - w(k)\varphi^H\right.$$
$$\left(\left\|o^A(k+1) - c(k)\right\|_{\Sigma^{-1}(k)}^2\right)\right)\varphi^H\left(\left\|o^A(k+1) - c(k)\right\|_{\Sigma^{-1}(k)}^2\right). \tag{71}$$

Thus, in contrast to traditional CNNs, where only synaptic weights of neurons are tuned in the proposed system, the parameters of activation functions are simultaneously clarified, which provides additional flexibility and speed.

The computational experiment was performed using the "Amazigh Handwritten Character" dataset from the Kaggle repository as well as system under consideration. The dataset consists of 780 scanned images in each of 33 classes, hence 25 740 total.

The system under consideration processed "Amazigh Handwritten Character" in both cases: first—the spread parameter was fixed and equal to 0.33, and the second—the parameter was tuned by the system. The obtained accuracy in case with the fixed value was 8% lower than in the second, where value of the parameter was equal to 0.42, which allowed archive the highest accuracy.

In the second experiment, the dataset was processed with system under consideration without autoencoder and with one, considering the spread parameter, which was deduced previously. In the first case the obtained accuracy is 2% lower than in the second case, where the accuracy reached 89%.

The computational experiment described the comparison analysis of the proposed system and CNN proved that the speed of the proposed system us approximately one time faster that the CNN. Thereat, the accuracy of the HBFN is slightly (approximately 5%) lower, thus, the approach we develop allows solving classification task under strict restrictions on processing time of information.

6 Conclusion

Fuzzy matrix modifications of a probabilistic neural network are introduced to solve the problem of classification-recognition of images in the stream of observations that are successively received for processing, and these images can be presented in vector or matrix forms, i.e. directly in the form of images. The introduced systems have a stack architecture, and the learning algorithms combine the advantages of traditional probabilistic neural networks of D. Specht, self-organizing maps and vector quantization training according to T. Kohonen, evolving systems of N. Kasabov-P. Angelov. The introduced systems are able to learn under conditions of both small and long training samples and to operate under conditions of classes of any form presented in features space. The proposed classification systems are simple in numerical implementation and have increased speed compared to known image recognition systems and, above all, convolutional neural networks.

References

1. Amirian, M., & Schwenker, F. (2020). Radial basis function networks for convolutional neural networks to learn similarity distance metric and improve interpretability. *IEEE Access, 8*, 123087–123097.
2. Angelov, P. P. (2002). Evolving rule-based models: A tool for design of flexible adaptive systems. Physica-Verlag HD: Imprint : Physica, Heidelberg.
3. Angelov, P. P., & Kasabov, N. K. (2005). Evolving computational intelligence systems. In *Proceedings of the 1st International Workshop on Genetic Fuzzy Systems, Granada* (pp. 76–82)
4. Bengio, Y., Cun, Y. L., & Hinton, G. (2015). Deep learning. *Nature, 521*, 436–444.
5. Bodyanskiy, Y., Deineko, A., & Pliss, I. et al. (2020). Matrix fuzzy-probabilistic neural network in image recognition task. In *2020 IEEE Third International Conference on Data Stream Mining & Processing (DSMP). IEEE, Lviv, Ukraine* (pp. 33–36).
6. Bodyanskiy, Y., Gorshkov, Y., & Kolodyazhniy, V. (2003). Resource-allocating probabilistic neuro-fuzzy network. In: *Proceedings of 2nd Conferences of European Union Sosciety for Fuzzy Logic and Technology (EUSFLAT) 2003. Zittau, Germany* (pp. 392–395).
7. Bodyanskiy, Y., Gorshkov, Y., Kolodyazhniy, V., & Wernstedt, J. (2003). Probabilistic neuro-fuzzy network with non-conventional activation functions. *Lecture Notes in Artificial Intelligence, 2773*, 973–979.
8. Bodyanskiy, Y., Gorshkov, Y., Wernstedt, J., & Kolodyazhniy, V. (2003). A learning of probabilistic neural network with fuzzy inference. In *Proceedings of Sixth International Conference on Artificial Neural Nets and Generic Algorithms "ICANNGA 2003."* Springer, Wien (pp. 13–17)
9. Bodyanskiy, Y., Pliss, I., Chala, O., & Deineko, A. (2020). Evolving fuzzy-probabilistic neural network and its online learning. In *10th International Conference on Advanced Computer Information Technologies. Deggendorf, Germany.*
10. Dvoretzky, A. (1956). On stochastic approximation. In *Berkeley Symposium on Mathematical Statistics and Probability.*
11. Goodfellow, I., Bengio, Y., & Courville, A. (2016). *Deep learning.* MIT Press.
12. Kacprzyk, J., & Pedrycz, W. (2015). *Springer handbook of computational intelligence.* Springer.
13. Kasabov, N. K. (2007). *Evolving connectionist systems: The knowledge engineering approach* (2nd ed.). Springer.
14. Kohonen, T. (2001). *Self-organizing maps.* Springer.
15. Kruse, R., Borgelt, C., Klawonn, F., et al. (2013). *Computational intelligence.* Springer.
16. Kung, S. Y. (2014). *Kernel methods and machine learning.* Cambridge University Press.
17. Leonard, J. A., Kramer, M. A., & Ungar, L. H. (1992). Using radial basis functions to approximate a function and its error bounds. *IEEE Transactions on Neural Networks, 3*, 624–627.
18. Lughofer, E. (2011). *Evolving fuzzy systems - methodologies.* Springer.
19. Moody, J., & Darken, C.J. (1989). Fast learning in networks of locally tuned processing units. *Neural Computation, 1*, 281–294.
20. Mumford, C. L., & Jain, L. C. (2009). *Computational intelligence.* Springer.
21. Nadaraya, E. A. (1964). On estimating regression. *Theory of Probability & Its Applications, 9*, 141–142. https://doi.org/10.1137/1109020.
22. Nelles, O. (2001). *Nonlinear systems identification.* Springer.

23. Park, J., & Sandberg, I. W. (1991). Universal approximation using radial-basis-function networks. *Neural Computation, 3*, 246–257.
24. Parzen, E. (1962). On estimation of a probability density function and mode. *The Annals of Mathematical Statistics, 33*, 1065–1076. https://doi.org/10.1214/aoms/1177704472
25. Poggio, T., & Girosi, F. (1990). Networks for approximation and learning. *Proceedings of the IEEE, 78*, 1481–1497.
26. Rutkowski, L. (2004). Adaptive probabilistic neural networks for pattern classification in time-varying environment. *IEEE Transactions on Neural Networks, 15*, 811–827.
27. Schalkoff, R. J. (1997). *Artificial neural networks*. McGraw-Hill.
28. Schmidhuber, J. (2015). Deep learning in Neural networks: An overview. *Neural Networks, 61*, 85–117.
29. Souza, P. V. C. (2020). Fuzzy neural networks and neuro-fuzzy networks: A review the main techniques and applications used in the literature. *Applied Soft Computing, 92*.
30. Specht, D. F. (1990). Probabilistic neural networks. Neural. *Network, 3*, 109–118.
31. Specht, D. F. (1990). Probabilistic neural networks and polynomial ADALINE as complementary techniques to classification. *IEEE Transactions on Neural Networks, 1*, 111–121.
32. Tkachenko, R., Tkachenko, P., Izonin, I., et al. (2019). Committee of the combined RBF-SGTM neural-like structures for prediction tasks. *International Conference on Mobile Web and Intelligent Information Systems* (pp. 267–277). Springer.
33. Zahirniak, D. R., Chapman, R., Rogers, S. K. et al. (1990). Pattern recognition using radial basis function network. *Aerospace Application of Artificial Intelligence* 249–260

Artificial Intelligence Techniques in Modelling and Optimization

Intelligent Information Technology for Structural Optimization of Fuzzy Control and Decision-Making Systems

Oleksiy V. Kozlov, Yuriy P. Kondratenko, and Oleksandr S. Skakodub

Abstract This chapter is devoted to the development and research of intelligent information technology (IIT) for structural optimization of fuzzy systems (FS) based on the evolutionary search of the optimal membership functions. The proposed IIT uses a combination of different (two or more) bioinspired evolutionary algorithms and allows finding the optimal membership functions of linguistic terms at solving the compromise problems of multi-criteria structural optimization of various FSs to increase their efficiency, as well as to reduce the degree of complexity of further parametric optimization. To study the effectiveness of the considered IIT, the search of the optimal membership functions is conducted for a FS of the multi-purpose mobile robot (MR) designed to move along inclined and vertical ferromagnetic surfaces, with the implementation of three evolutionary algorithms: genetic, artificial immune systems, biogeographic. The analysis of the obtained results showed that the usage of the proposed IIT gives the opportunity to significantly increase the efficiency of the MR control, as well as to reduce the total number of parameters at further parametric optimization of linguistic terms, which confirms its high efficiency.

Keywords Fuzzy system · Linguistic terms membership functions · Structural optimization · Intelligent information technology · Bioinspired evolutionary algorithm · Fuzzy controller · Mobile robot

1 Introduction

The theory of soft computing, fuzzy sets, and fuzzy logic are now widely introduced into the practice of fundamental and applied scientific research as well as the development of different types of computer and technical systems [1–3]. As proved in

O. V. Kozlov (✉) · Y. P. Kondratenko · O. S. Skakodub
Department of Intelligent Information Systems, Petro Mohyla Black Sea National University, Mykolaiv, Ukraine
e-mail: kozlov_ov@ukr.net

Y. P. Kondratenko
Institute of Artificial Intelligence Problems of MES and NAS of Ukraine, Kyiv, Ukraine

© The Author(s), under exclusive license to Springer Nature Switzerland AG 2023
Y. P. Kondratenko et al. (eds.), *Artificial Intelligence in Control and Decision-making Systems*, Studies in Computational Intelligence 1087,
https://doi.org/10.1007/978-3-031-25759-9_7

a number of fundamental works [4–6], the mathematical apparatus of fuzzy logic makes it possible to successfully mimic the mechanisms of human thinking and decision-making, simulate various processes and plants that are difficult to formalize, as well as solve problems of high complexity in conditions of incomplete information and uncertainty. One of the most promising areas of fuzzy logic application is the development of intelligent decision-making and control systems for automation of complicated non-stationary, multi-mode and multi-connected technical plants, which include multipurpose mobile robots, unmanned vehicles and drones, planes and spaceships, ships, and floating docks, industrial lines of continuous production, alternative energy facilities, pyrolysis reactors, etc. [7–9].

Modern research in the field of creation and development of fuzzy systems for decision-making and control is carried out mainly in the direction of designing highly efficient methods and information technologies of their synthesis and structural-parametric optimization [10–12]. In connection with the intensive development and increase in the power of computer technology, bioinspired intelligent methods and information technologies have become quite promising for solving the problems of FS synthesis and optimization, which include evolutionary and multi-agent methods that simulate natural processes of natural selection and collective behavior of various social groups of animals, insects and microorganisms [13, 14]. These methods are stochastic methods of global optimization and have a number of advantages compared to classical search methods: (a) allow efficient optimization of fuzzy systems of various configurations and large dimensions; (b) do not impose additional restrictions on the objective functions; (c) make it possible to find global extrema of large non-smooth and multimodal search spaces [15, 16].

This chapter is devoted to the development and research of intelligent information technology for structural optimization of fuzzy systems, in particular, for finding the optimal membership functions (MF) of linguistic terms (LT), based on the use of a combination of several bioinspired intelligent algorithms. The chapter is organized in the following way. The statement of the problem, brief literature overview in the studied area, and the purpose of this research are presented in Sect. 2. Section 3 describes in detail the proposed IIT for structural optimization of fuzzy systems. In turn, Sect. 4, includes the description of several bioinspired global optimization algorithms, that are adapted for searching the optimal linguistic terms membership functions (LTMF) of FS. Section 5 presents the results of studying the effectiveness of the designed information technology on a specific example of the fuzzy control system for a multi-purpose caterpillar mobile robot with a detailed discussion of the results of computer simulation. Finally, the conclusions and references list are given at the end of this chapter.

2 Problem Statement and Related Works

The functional structure of the generalized fuzzy system (decision-making or control) is shown in Fig. 1, where the following abbreviations are used: SPO Mechanism is the mechanism of structural-parametric optimization; FU is the fuzzification unit; FIE is the fuzzy inference engine; AGG is the aggregation unit; ACT is the activation unit; ACC is the accumulation unit; DFU is the defuzzification unit; $x_1, x_2, …, x_i,$ $…, x_n$ are fuzzy system input variables; $y_1, y_2, …, y_j, …, y_m$ are FS output variables; $\mathbf{S}$ is the vector of structure variants of a fuzzy system; $\mathbf{P}$ is the vector of parameters of the FS; $\mathbf{Q}$ is the vector of output variables of control plant/operating environment, that are used for structural-parametric optimization of a fuzzy system.

The generalized fuzzy MIMO-system (shown in Fig. 1), implements the nonlinear dependence f_{FS} [17]

$$\mathbf{Y} = f_{FS}(\mathbf{X}), \mathbf{Y} = (y_1, y_2, …, y_j, …, y_m), \mathbf{X} = (x_1, x_2, …, x_i, …, x_n), \quad (1)$$

where $\mathbf{X}$ is the vector of n input variables $x_1, x_2, …, x_i, …, x_n$; $\mathbf{Y}$ is the vector of m output variables $y_1, y_2, …, y_j, …, y_m$ of fuzzy system.

The fuzzification unit defines the degree of membership of the numerical values of all n input variables of the vector $\mathbf{X}^*$ to the corresponding fuzzy input linguistic terms of the system [18]. The fuzzy inference engine, in turn, based on fuzzified signals and received data from the rule base (RB) sequentially performs the operations of aggregation, activation, and accumulation [19].

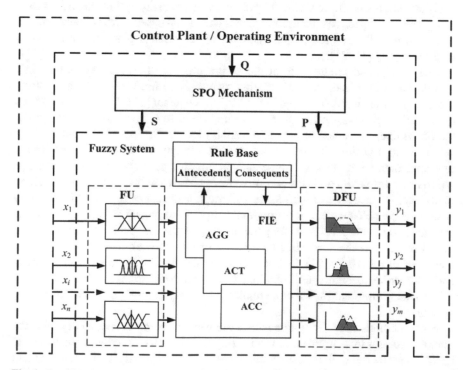

Fig. 1 Functional structure of the generalized fuzzy system

The rule base includes a set of rules made up of specific antecedents and consequents. So, for example, to implement functional dependence (1) by a fuzzy MIMO-system of Mamdani-type, using the corresponding linguistic terms, one of the RB rules for any variable x_i, $(i = 1...n)$ or y_j, $(j = 1...m)$, can be represented by expression (2) [17, 20]

$$\text{IF } "x_1 = A_1" \text{ AND } "x_2 = A_2" \text{ AND} \ldots \text{AND } "x_i = A_i" \ldots \text{AND} \ldots$$
$$\text{AND } "x_n = A_n" \text{ THEN } "y_1 = B_1" \text{ AND } "y_2 = B_2" \text{ AND} \ldots \qquad (2)$$
$$\text{AND } "y_j = B_j" \ldots \text{AND} \ldots \text{AND } "y_m = B_m",$$

where A_1, A_2, A_i, A_n, B_1, B_2, B_j, B_m are the corresponding linguistic terms of the input and output variables of the FS.

The defuzzification unit converts the consolidated fuzzy inference into a clear numerical signal for each j-th $(j = 1, 2, ..., m)$ output variable of the FS [18].

It is advisable to design the presented fuzzy system (Fig. 1) in an automated mode using the built-in mechanism of structural-parametric optimization (SPO Mechanism). At the same time, the SPO Mechanism should determine such vectors of structure variants **S** and parameters **P**, which will ensure sufficiently effective use of the FS for solving a particular problem (for example, problems of automatic control, decision making, etc.). Herewith, the efficiency of the fuzzy system can be assessed using a certain objective function J, which is calculated based on the obtained values of the vector of measured output variables of the control plant or the operating environment **Q** [11, 21].

The quality and productivity of the design process of a fuzzy system directly depend on the methods and information technologies used to optimize its structure and parameters implemented in the SPO Mechanism [17]. Moreover, structural optimization methods and IT play a very important role, since the best version of the FS structure obtained in the optimization process directly affects not only the value of the objective function J but also the computational costs in the formation/implementation of the rule base and parametric optimization of the system, as well as the complexity of its further software and hardware implementation. Further, by the optimal structure of the FS, we will mean a variant of the structure that will ensure (a) the achievement of the optimal value of the objective function J, (b) the acceptable computational costs for parametric optimization and the formation of the rule base, and (c) the acceptable complexity of the software and hardware implementation of FS [10, 13, 21]. Research aimed at developing and improving methods and information technologies for finding the optimal structure of the FS is, of course, relevant and important for the modern theory of fuzzy control and decision-making systems.

Since methods and IT for finding the optimal structure must successfully solve the complex compromise problem of minimizing the objective function, computational costs in parametric optimization, and the degree of complexity of the software and hardware implementation of the FS, it is advisable to develop them based on highly efficient stochastic global optimization techniques, which include bioinspired intelligent algorithms [22–24]. The results of cutting-edge research show that evolutionary bioinspired methods and algorithms of search make it possible to optimize fuzzy systems of various types and purposes with high efficiency [25, 26]. Thus, in a fairly large number of published papers, examples of the successful application of genetic algorithms for the structural and parametric optimization of FSs are presented [27, 28], in particular, for fuzzy automatic control systems (ACS) of thermal power facilities [29], industrial and mobile robots [30, 31], sea and air vehicles [32, 33]. In turn, the works [34] and [35] present the results of optimization of various fuzzy control and decision-making systems based on methods simulating the immune systems of living organisms. Also, for the synthesis and structural optimization of FSs, the use of differential evolution methods [36], evolutionary strategies [37], biogeographic search algorithms [38], and others [39] can be quite promising.

For a more detailed study of the influence of various components of the structure of the designed FS on its efficiency, it is advisable to solve the search problems (for example, search of optimal membership functions, optimal number of linguistic terms, the most rational procedures for aggregation, activation, accumulation, and defuzzification) separately and in turn using specially developed methods and information technologies. Herewith, one of the most important tasks of optimizing the structural components of the FS is the search and selection of the optimal membership functions for the linguistic terms of the FS input and output signals.

The purpose of this chapter is (a) to develop intelligent information technology for finding optimal membership functions of fuzzy systems based on a combination of bioinspired evolutionary algorithms of global optimization, and (b) to study the effectiveness of a real fuzzy control system using synthesized structural components.

3 Intelligent Information Technology for Finding Optimal Membership Functions of FSs Based on Bioinspired Evolutionary Algorithms

The proposed by the authors IIT for finding the optimal membership functions of the FSs based on a combination of bioinspired evolutionary algorithms of global optimization consists of the following successive stages.

Stage 1. Setting of the operating ranges for changing the input and output variables of the developed fuzzy system. At this stage, for each ith ($i = 1, 2, ..., n$) input and jth ($j = 1, 2, ..., m$) output variable of the FS, the operating ranges are set, within which this variable can change. For instance, if the input variables are fed to the FS in relative units from their maximum value, then it is advisable to set their operating ranges from -1 to 1. If some variables cannot have negative values (for example, "gas flow rate", "heating power", etc.), then their ranges of change can be represented by the interval $[0, 1]$.

Stage 2. Formation of a set of membership functions used in the optimization process. At this stage, a set of alternative LTMFs S_{MF} is created, on which the search for optimal membership functions for all linguistic terms of each ith input and jth output variable of the developed FS will be carried out. It is advisable to include in this set the most commonly used membership functions, which are: triangular *TrFN*, trapezoidal *TrpFN*, Gaussian 1st *Gs1FN* and 2nd *Gs2FN* types, bell-shaped *GbFN*, S-shaped *SFN*, Z-shaped *ZFN*, π-shaped *PiFN*, sigmoid *SgFN*, double sigmoid difference function *DsgFN*, as well as the product of two sigmoid functions *PsgFN* [40, 41]. The number of adjustable parameters k_{MF}, mathematical models, and graphic representation [8, 42] of these membership functions are given in Table 1, where a, b, c, d are adjustable parameters of LTMFs.

Stage 3. Choosing the number of linguistic terms for input and output variables of a fuzzy system. At this stage, the numbers of linguistic terms τ_i ($i = 1, 2, ..., n$) and τ_j ($j = 1, 2, ..., m$) are chosen for each i-th input and j-th output variables of the FS. In the real FSs, it is advisable to set the number of linguistic terms for input variables in the range from 2 to 7 ($\tau_i = 2...7$), and for output variables – from 3 to 9 ($\tau_j = 3...9$) [40].

Stage 4. Setting of the initial values of the parameters of membership functions of linguistic terms of the developed fuzzy system. At this stage for all LTMFs included in the set formed at *Stage 2* (Table 1), the values of their parameters are preliminarily set (a, b, c, d). In most cases, at the beginning of the search, it is advisable to choose the parameters of the membership functions in such a way that the linguistic terms for all input and output variables, depending on their number τ_i and τ_j, selected at *Stage 3*, are evenly distributed over their operating ranges, previously set at *Stage 1*.

Stage 5. Formation of a set of alternative membership functions of linguistic terms and the structure of the vector S, which determines LTMFs for the developed FS. The vector **S** of LTMF for systems with the fuzzy inference of Mamdani-type [5] or Takagi–Sugeno-type [42] can be formed, respectively, as (3) or (4) based on a set of alternative LTMFs (Table 1):

$$\mathbf{S} = \{S_{in}^i(q), S_{out}^j(k)\}, q = (1, 2, ..., \tau_i), k = (1, 2, ..., \tau_j),$$
$$i = (1, 2, ..., n), j = (1, 2, ..., m),$$

$$S_{in}^i(q) \in \{TrFN, TrpFN, GbFN, Gs1FN, Gs2FN,$$
$$PiFN, SFN, ZFN, SgFN, DsgFN, PsgFN\},$$
$$\tag{3}$$
$$S_{out}^j(k) \in \{TrFN, TrpFN, GbFN, Gs1FN, Gs2FN,$$
$$PiFN, SFN, ZFN, SgFN, DsgFN, PsgFN\},$$

$$\mathbf{S} = \{S_{in}^i(q)\}, q = (1, 2, ..., \tau_i), i = (1, 2, ..., n),$$
$$S_{in}^i(q) \in \{TrFN, TrpFN, GbFN, Gs1FN, Gs2FN,$$
$$PiFN, SFN, ZFN, SgFN, DsgFN, PsgFN\},$$
$$\tag{4}$$

Table 1 Components of the Set of Membership Functions of Linguistic Terms

S_{MF}	k_{MF}	Mathematical description	Graphic description
TrFN	3	$$\mu(x) = \begin{cases} 0, & \text{at } x \le a \text{ or } x \ge c; \\ \dfrac{x-a}{b-a}, & \text{at } a < x \le b; \\ \dfrac{c-x}{c-b}, & \text{at } b < x < c; \end{cases}$$ $a \le b \le c.$	
TrpFN	4	$$\mu(x) = \begin{cases} 0, & \text{at } x \le a \text{ or } x \ge d; \\ \dfrac{x-a}{b-a}, & \text{at } a < x \le b; \\ 1, & \text{at } b < x \le c; \\ \dfrac{d-x}{d-c}, & \text{at } c < x < d; \end{cases}$$ $a \le b \le c \le d.$	

(continued)

Table 1 (continued)

S_{MF}	k_{MF}	Mathematical description	Graphic description
GbFN	3	$\mu(x) = \dfrac{1}{1 + \left|\frac{x-c}{a}\right|^{2b}}$; $a \in (0; +\infty)$; $b \in (-\infty; +\infty)$; $c \in (-\infty; +\infty)$.	
GslFN	2	$\mu(x) = e^{-\frac{(x-b)^2}{2a^2}}$.	

(continued)

Table 1 (continued)

S_{MF}	k_{MF}	Mathematical description	Graphic description

$Gs2FN$	4	if $b < d$, then
		
		$$\mu(x) = \begin{cases} e^{-\frac{(x-b)^2}{2a^2}}, & \text{at } x < b; \\ 1, & \text{at } b \le x \le d; \\ e^{-\frac{(x-d)^2}{2c^2}}, & \text{at } x > d; \end{cases}$$
		
		if $b > d$, then
		
		$$\mu(x) = \begin{cases} e^{-\frac{(x-b)^2}{2a^2}}, & \text{at } x < d; \\ e^{-\frac{(x-b)^2}{2a^2}} \cdot e^{-\frac{(x-d)^2}{2c^2}}, & \text{at } d \le x \le b; \\ e^{-\frac{(x-d)^2}{2c^2}}, & \text{at } x > b. \end{cases}$$	

$PiFN$	4	$$\mu(x) = \min\{\mu_1(x), \mu_2(x)\};$$ $$\mu_1(x) = \begin{cases} 0, & \text{at } x \le a; \\ 2\left(\dfrac{x-a}{b-a}\right)^2, & \text{at } a < x \le \dfrac{a+b}{2}; \\ 1 - 2\left(\dfrac{b-x}{b-a}\right)^2, & \text{at } \dfrac{a+b}{2} < x \le b; \\ 1, & \text{at } x > b; \end{cases}$$ $$\mu_2(x) = \begin{cases} 1, & \text{at } x \le c; \\ 1 - 2\left(\dfrac{x-c}{d-c}\right)^2, & \text{at } c < x \le \dfrac{c+d}{2}; \\ 2\left(\dfrac{d-x}{d-c}\right)^2, & \text{at } \dfrac{c+d}{2} < x \le d; \\ 0, & \text{at } x > d. \end{cases}$$	

(continued)

Table 1 (continued)

S_{MF}	k_{MF}	Mathematical description	Graphic description
SFN	2	$$\mu(x) = \begin{cases} 0, & \text{at } x \le a; \\ 2\left(\dfrac{x-a}{b-a}\right)^2, & \text{at } a < x \le \dfrac{a+b}{2}; \\ 1 - 2\left(\dfrac{b-x}{b-a}\right)^2, & \text{at } \dfrac{a+b}{2} < x \le b; \\ 1, & \text{at } x > b. \end{cases}$$	
ZFN	2	$$\mu(x) = \begin{cases} 1, & \text{at } x \le a; \\ 1 - 2\left(\dfrac{x-a}{b-a}\right)^2, & \text{at } a < x \le \dfrac{a+b}{2}; \\ 2\left(\dfrac{b-x}{b-a}\right)^2, & \text{at } \dfrac{a+b}{2} < x \le b; \\ 0, & \text{at } x > b. \end{cases}$$	

(continued)

Table 1 (continued)

S_{MF}	k_{MF}	Mathematical description	Graphic description
SgFN	2	$\mu(x) = \frac{1}{1+e^{-a(x-c)}}.$	
DsgFN	4	$\mu(x) = \frac{1}{1+e^{-a(x-b)}} - \frac{1}{1+e^{-c(x-d)}}.$	
PsgFN	4	$\mu(x) = \frac{1}{1+e^{-a(x-b)}} \cdot \frac{1}{1+e^{-c(x-d)}}.$	

where $S_{in}^i(q)$, $S_{out}^j(k)$ are variables that determine the types of membership functions of the q-th linguistic term of the i-th input variable and the k-th linguistic term of the j-th output variable, respectively.

The vector **S** can have certain restrictions. For instance, S-shaped *SFN* and sigmoid *SgFN* membership functions (Table 1) can only be used for the rightmost linguistic terms ($q = \tau_i$, $k = \tau_j$), and, in turn, the Z-shaped *ZFN* can only be used for the leftmost terms ($q = 1$, $k = 1$) for all FS variables:

$$
\begin{aligned}
S_{in}^i(q), S_{out}^j(k) &\in \{TrFN, TrpFN, GbFN, Gs1FN, Gs2FN, \\
&\quad PiFN, ZFN, DsgFN, PsgFN\}, \text{at } q = 1, k = 1, \\
S_{in}^i(q), S_{out}^j(k) &\in \{TrFN, TrpFN, GbFN, Gs1FN, Gs2FN, \\
&\quad PiFN, DsgFN, PsgFN\}, \text{at } q = (2, ..., \tau_i{-}1), k = (2, ..., \tau_j{-}1), \\
S_{in}^i(q), S_{out}^j(k) &\in \{TrFN, TrpFN, GbFN, Gs1FN, Gs2FN, \\
&\quad PiFN, SFN, SgFN, DsgFN, PsgFN\}, \text{at } q = \tau_i, k = \tau_j.
\end{aligned}
\tag{5}
$$

*Stage 6. Choosing the initial structure (initial hypothesis) of the vector **S**, which determines the membership functions of the linguistic terms of the developed fuzzy system.* At this stage, in accordance with the number of terms τ_i ($i = 1, 2, ..., n$) and τ_j ($j = 1, 2, ..., m$) selected at *Stage 3*, the initial values of the components of the vector **S**, which determines the LTMF of the system, are chosen. These initial values can be chosen, in particular, based on an expert approach. For example, if at the beginning of the search in the Mamdani-type FS (with parameters $n = 3$, $\tau_i = \{4, 3, 2\}$, $m = 1$, $\tau_j = \{5\}$) for all linguistic terms of the input variables Gaussian 1st type membership functions *Gs1FN* are chosen, and for all terms of output variables trapezoidal membership functions *TrpFN* are selected, then the initial value of the vector **S**$_0$ will be determined by expression (6)

$$
\begin{aligned}
\mathbf{S}_0 = \{S_{in}^i(q), S_{out}^j(k)\} = \\
\{S_{in}^1(1), S_{in}^1(2), S_{in}^1(3), S_{in}^1(4), S_{in}^2(1), S_{in}^2(2), S_{in}^2(3), S_{in}^3(1), S_{in}^3(2), \\
S_{out}^1(1), S_{out}^1(2), S_{out}^1(3), S_{out}^1(4), S_{out}^1(5)\} = \\
\{Gs1FN, Gs1FN, Gs1FN, Gs1FN, Gs1FN, Gs1FN, Gs1FN, Gs1FN, Gs1FN, \\
TrpFN, TrpFN, TrpFN, TrpFN, TrpFN\}.
\end{aligned}
\tag{6}
$$

Stage 7. Formation of a complex objective function J_C for evaluating the effectiveness of the developed fuzzy system. At this stage, the type, parameters, and optimal (boundary) value of the complex objective function J_C is determined, which is used to search for optimal membership functions. Since the total number of adjustable parameters of linguistic terms depends on the types of LTMF selected at *Stage 6*, it is advisable to use (in the process of vector **S** structural optimization) the criterion J_1, which evaluate the FS performance, as well as criterion J_2, which takes into account the degree of complexity of further parametric optimization of the developed system.

Thus, the problem of finding the optimal LTMFs is reduced to the problem of multi-objective optimization [12, 43–45], for the solution of which it is necessary to find the optimal structure of the vector $\mathbf{S}$, taking into account the minimization of two criteria: J_1 and J_2. When solving this problem, it is advisable to use an a priori approach to solving problems of multi-criteria evolutionary search based on an aggregation of objective functions [13, 46, 47], according to which it is necessary to search for the optimum of a single complex objective function (global criterion) J_C, formed based on criteria J_1 and J_2 with a preliminary assessment of their significance. Following this approach, it is expedient to calculate the current value of the complex objective function J_C in the process of searching for the optimal membership functions of the FS based on the expression (7)

$$J_C = J_1 + k_{J2} J_2, \tag{7}$$

where k_{J2} is a scale factor at J_2, which determines the importance of taking into account this objective function in the process of computational search and provides scaling (normalization) of J_2 values.

For instance, when designing a one-dimensional (MISO) fuzzy automatic control system ($\mathbf{Y} = y$), the generalized integral deviation of the real transient response of this system $V_R(t, \mathbf{S})$ from the desired transient response of the reference model $V_D(t)$ can be chosen as the criterion J_1 [10, 22], which is described by the expression (8)

$$J_1(t, \mathbf{S}) = \frac{1}{t_{max}} \int_0^{t_{max}} \left[(E_V)^2 + k_{11}(\dot{E}_V)^2 + k_{12}(\ddot{E}_V)^2 \right] dt, \tag{8}$$

where E_V is the deviation of $V_R(t, \mathbf{S})$ from $V_D(t)$, $E_V = V_D(t) - V_R(t, \mathbf{S})$; t_{max} is the total time of the ACS transient process; k_{11}, k_{12} are the corresponding weighting factors for the components $(\dot{E}_V)^2$ and $(\ddot{E}_V)^2$.

In turn, the values of the J_2 criterion, which evaluates the complexity of further parametric optimization of the FS, can be calculated depending on the number of optimized parameters of linguistic terms, for example, for the Mamdani or Takagi–Sugeno FSs, based on dependencies (9) or (10), respectively:

$$J_2(\mathbf{S}) = \sum_{i=1}^{n} \sum_{q=1}^{\tau_i} k_{in}^i(q) + \sum_{j=1}^{m} \sum_{k=1}^{\tau_j} k_{out}^j(k); \tag{9}$$

$$J_2(\mathbf{S}) = \sum_{i=1}^{n} \sum_{q=1}^{\tau_i} k_{in}^i(q), \tag{10}$$

where $k_{in}^i(q)$, $k_{out}^j(k)$ are the numbers of optimized parameters of the q-th linguistic term for the i-th input variable and the k-th linguistic term for the j-th output variable, depending on their membership functions (Table 1). For example, when using the corresponding linguistic term $A = (a, b, c)$ with a triangular membership function (Table 1) [40, 41], it is necessary to perform parametric optimization of 3 parameters (a, b, c) for each linguistic term, that with 4 terms for each input and 7 terms for each output variable will require optimization in total of $(12n + 21\ m)$ parameters.

In addition, at this stage, the optimal (boundary) value of the J_{1opt} criterion is preliminarily set, and the corresponding values of J_{Copt}, J_{2opt}, and k_{J2} are selected based on the requirements and features of the FS design problem. For example, the coefficient k_{J2} can be chosen taking into account the following conditions

$$\frac{J_{Copt}}{J_{1opt}} < 2, \frac{J_{Copt}}{J_{1opt}} > 2 \text{ or} \frac{J_{Copt}}{J_{1opt}} = 2.$$

The first inequality is used when the accuracy and efficiency of the fuzzy system are more important for the designer than the complexity of the parametric optimization process and software and hardware implementation. Otherwise, the second inequality should be applied. Herewith, the exact value of the coefficient k_{J2} is selected for each specific problem.

Stage 8. Carrying out an iterative global search for the optimal structure of the vector of membership functions S_{opt} using a combination of evolutionary algorithms. To solve the problems of this class, it is advisable to use evolutionary optimization algorithms [11, 13, 16]. The authors propose to search for the global extremum of the objective function $J_C \rightarrow$ min using a combination of bioinspired evolutionary algorithms, i.e. using not one but several bioinspired evolutionary algorithms. Therefore, at this stage, to find the global optimum of the problem being solved, one by one or simultaneously using parallel processors, an iterative search procedure is carried out based on several (2 or more) different bioinspired evolutionary algorithms. To do this, pre-selected algorithms are adapted to the specifics of the problem of optimizing membership functions of the fuzzy system.

The results of preliminary studies show that different bioinspired evolutionary algorithms provide different solutions for the structural optimization of the vector **S**. A comparative analysis of such alternative solutions will make it possible to determine the optimal (in the sense of a given objective function) structure of the vector **S**. To find the optimal LTMF vector, the following bioinspired global optimization algorithms can be chosen: genetic [28], biogeography based [38], differential evolution [36], dynamic networks [48], artificial immune systems [34], as well as algorithms that simulate the spread of weeds [49], the behavior of cuckoos [50], etc.

Stage 9. Choosing the optimal structure of the S_{opt} vector based on alternative results of structural optimization of membership functions using a combination of evolutionary algorithms. At this stage, the results of the search for optimal membership functions obtained (using various bioinspired global optimization algorithms) at *Stage 8* are analyzed, and based on a comparison of the values of the complex objective function J_C, as well as its components J_1 and J_2, the best variant of the vector **S** is selected. In this case, it is also expedient to take into account the number of iterations spent by each of the algorithms to achieve the optimal value of the complex objective function J_{Copt} in the search process.

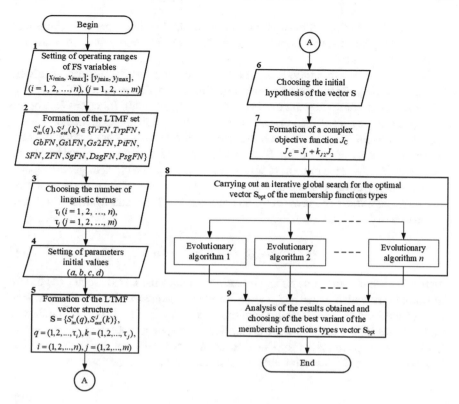

Fig. 2 Block diagram of the evolutionary-based information technology for structural optimization of fuzzy systems

Figure 2 shows a block diagram of the proposed evolutionary-based information technology for structural optimization of different types of fuzzy systems.

Further (Sect. 4), as an example, we consider the features of using such (well-tested and proven) bioinspired global optimization algorithms for searching the optimal LTMFs of FS, as genetic [28], artificial immune systems [34], and biogeography based [38].

4 Bioinspired Evolutionary Optimization Algorithms Adapted to Solve the Problem of Finding the Optimal LTMFs of the FS

Genetic Algorithm (GA). Genetic algorithms model the processes of natural selection of wildlife and use the mechanisms of selection, crossover, inheritance, and mutation to find optimal solutions in global search problems [27, 28]. Herewith, in these algorithms, the set of problem solution vectors, each of which consists of optimized variables, is represented by a population of individuals (chromosomes) consisting of a certain set of genes.

To assess the fitness of individuals (chromosomes) and the effectiveness of solving the optimization problem, the corresponding fitness function is used [29].

The procedure of finding the optimal vector of membership functions S_{opt} of the developed fuzzy system based on GA consists of the following sequence of steps.

Step 1. GA initialization. At the initialization stage, an initial population of chromosomes (variants of the vector S) is created and the following are selected: the fitness function f and the criterion for search terminating, the types of genetic operators of selection, crossover, and mutation, as well as the size of the initial population l_{max}, the probabilities of crossover P_C and mutation P_M. The created initial population N_0 consists of l_{max} chromosomes,

$$N_0 = \{S_1, S_2, ..., S_l, ..., S_{l\,max}\}, \tag{11}$$

where each l-th chromosome S_l is a certain version of the LTMF vector (3) or (4), the variables of which $S_{in}^i(q)$, $S_{out}^j(k)$ (chromosome genes) are set randomly within the generated set of membership functions (Table 1). In this case, real coding of chromosomes is used (the genotype is identical to the phenotype) [27]. Since genetic algorithms are algorithms for maximizing a function, then to minimize the complex objective function (7), it is advisable to calculate the fitness function f as

$$f = \frac{1}{J_C}. \tag{12}$$

As a condition for terminating the search, it is advisable to choose (a) achieving the optimal value of the fitness function $f \geq f_{opt}$ or (b) performing the maximum number of iterations Z_{max}. It is advisable to choose proportional selection as a selection operator, one-point or two-point crossover as a crossover operator, and a simple mutation as a mutation operator.

Step 2. Evaluation of population chromosomes and checking for search termination criteria. At this step, for each l-th version of the vector S_l ($l = \{1, 2, ..., l_{max}\}$) the value of the fitness function $f_l(Z)$ is calculated based on expression (12), and the search process termination condition is checked. When the optimal value of the fitness function $f_l(Z) \geq f_{opt}$ is reached for some vector S_l or when the maximum number of iterations $Z = Z_{max}$ is performed, the transition to *Step 7* is performed, otherwise, go to *Step 3*.

Step 3. Selection of chromosomes for crossover. When using the proportional selection mechanism [30], the average value of the population fitness function f_M is calculated based on expression (13)

$$f_M = \frac{1}{l_{max}} \sum_{l=1}^{l_{max}} f_l \tag{13}$$

and for each l-th variant of the LTMF vector S_l, the condition is checked

$$\frac{f_l}{f_M} > 1. \tag{14}$$

If condition (14) is met, then the given lth vector is allowed for a crossover at *Step 4*, and if not, then the corresponding vector is excluded from the current population.

Step 4. Chromosomes crossover. Among the chromosomes vectors allowed for crossover, parental pairs are selected based on the value of the crossover probability P_C using the random selection mechanism [28]. The selected parent pairs are crossed using a one-point or two-point crossover operator [27]. In turn, the points of discontinuity of the LTMF vector during crossover are chosen randomly.

Step 5. Chromosomes mutation. The LTMF vectors subject to mutation are randomly selected based on the mutation probability P_M value among the parent individuals of the population [30]. Next, the selected vectors are mutated using the simple mutation operator [32].

Step 6. Formation of a new generation. At this step, a new generation is formed from descendant chromosomes obtained as a result of crossover and mutation operations at *Steps 4* and *5*. For the population of the new generation to have a constant number of chromosomes (alternative variants of the optimized vector S) l_{max}, at this step, the missing vectors are added, whose membership functions (genes) are randomly generated [28]. Next, the transition to *Step 2* is carried out and the next iteration Z of the optimization process is implemented.

Step 7. End of the search procedure.

Figure 3 shows a block diagram of a genetic algorithm adapted to the specifics of the problem of finding optimal LTMFs.

Algorithm of Artificial Immune Systems (AIS). Algorithms of artificial immune systems in the process of finding optimal solutions use ideas borrowed from immunology, simulating the work of the immune systems of living organisms [34]. In these algorithms, solution vectors are represented by immune cells (antibodies), whose affinity (efficiency in combating antigens) directly depends on the values of the objective function in the problem being solved. To find optimal solutions in the process of global search, AIS algorithms operate with the mechanisms of cloning, mutation, and selection [34, 35].

The procedure for finding the optimal vector of membership functions S_{opt} for the developed fuzzy system based on the AIS algorithm consists of the following steps.

Step 1. AIS algorithm initialization. At the initialization stage, an initial population of l_{max} immune cells (variants of the vector S) is created, the affinity function f and the criterion for terminating the search are selected. Also, the parameters of the biological operators of cloning, mutation, and selection are set: the number of memory cells N_m, the number of cells in the population with the worst affinity N_w, the parameter of the cloning operator ρ or N_c, depending on the type of the cloning operator, as well as the parameter of the mutation operator r [34]. The created initial population N_0 consists of l_{max} cells corresponding to certain variants, for example, of the LTMF

Fig. 3 Block diagram of the
genetic algorithm adapted to
the specifics of the problem
of finding optimal LTMFs

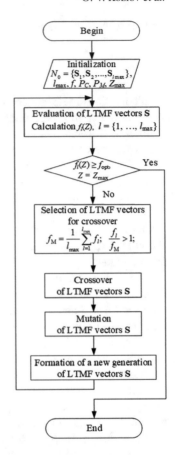

vector (3), whose variables $S_{in}^i(q)$, $S_{out}^j(k)$ (cell parameters) are set randomly within the generated set of membership functions.

The affinity function f, like the fitness function in genetic algorithms, should be calculated based on expression (12). As a condition for terminating the search, it is advisable to choose the achievement of the optimal affinity value $f \geq f_{opt}$ or the execution of the maximum number of iterations Z_{max}.

Step 2. Evaluation of cells in the population and checking the criterion for terminating the search. At this stage, for each l-th version of the vector $\mathbf{S}_l$ ($l = \{1, 2, ..., l_{max}\}$), the value of the affinity function $f_l(Z)$ is calculated, and the condition for terminating the optimization process is checked. When the optimal value of the affinity function $f_l(Z) \geq f_{opt}$ is reached for some vector $\mathbf{S}_l$, the transition to *Step 9* is performed, otherwise, go to *Step 3*.

Step 3. Cloning of memory cells. At this stage, the variants of the vector $\mathbf{S}$ (population cells) are sorted in descending order of their affinity, for example, $\mathbf{S}_1$, $\mathbf{S}_2$, ..., $\mathbf{S}_{lmax}$, where $f(\mathbf{S}_1) = f_{max}(Z)$, $f(\mathbf{S}_{lmax}) = f_{min}(Z)$ [35]. Next, from the sorted population (with the new numbering of the indexes of the variants of the $\mathbf{S}_l$ vectors in accordance with their affinity rating), N_m of the first LTMF vectors with the best affinity are selected. When using the proportional cloning operator, for each l-th vector S_l ($l = \{1, 2, ..., N_m\}$), N_l clones are generated based on expression (15) [34]

$$N_l = \frac{\rho l_{max}}{l}, l = (1, 2, ..., N_m).$$ (15)

When using the uniform clone operator, for each l-th vector (memory cell), a pre-selected number of clones $N_l = N_c$ is generated [35].

Step 4. Mutation of clones of memory cells. For all N_l clones of each l-th memory cell ($l = \{1, 2, ..., N_m\}$), the mutation procedure is performed, that is changing of r randomly selected LTMFs (cell parameters) in a random way within the set of membership functions [34].

Step 5. Evaluation of mutant clones of memory cells and checking the criterion for terminating the search. At this step, for all N_l mutant clones of each lth memory cell ($l = \{1, 2, ..., N_m\}$), the value of the affinity function $f(Z)$ is calculated, and the condition for terminating the optimization process is checked. When the optimal value of the affinity function $f(Z) \geq f_{opt}$ is reached for any mutant clone, the transition to *Step 9* is performed, otherwise, go to *Step 6*.

Step 6. Selection and updating of cells. Among all N_l mutant clones of each lth memory cell ($l = \{1, 2, ..., N_m\}$), a mutant clone with the best affinity f_{max} is selected and compared with the given memory cell. If the affinity of the selected mutant clone is higher than the affinity of its memory cell, then this mutant clone becomes a new memory cell [35].

Step 7. Formation of a new population of cells. At this step, the updated LTMF vectors are sorted in descending order of their affinity [34]. Then N_w of the last vectors with the worst affinity are removed from the sorted population and then replaced with new variants generated randomly.

Step 8. Evaluation of new cells in the population and checking the criterion for terminating the search. At this step, for all generated new variants of the vector **S**, the value of the affinity function $f(Z)$ is calculated, and the criterion for terminating the optimization process is checked. When the optimal value of the affinity function $f(Z)$ $\geq f_{opt}$ is reached for any new version of the LTMF vector or the maximum number of iterations Z_{max} is executed, the transition to *Step 9* is performed, otherwise, go to *Step 3*.

Step 9. End of the search procedure.

Figure 4 shows a block diagram of the algorithm of artificial immune systems, adapted to the specifics of the problem of finding optimal LTMFs.

Algorithm of Biogeography-Based Optimization (BBO). BBO algorithms model the patterns of the geographic distribution of animals, plants, and microorganisms in habitats in wildlife and use the mechanisms of immigration, emigration, and species mutation in the process of finding optimal solutions [23, 38]. Vectors of solutions are represented by species habitats (islands) with variables characterizing them, and the efficiency of solving the optimization problem is evaluated using the Habitat Suitability Index (HSI) [38, 39].

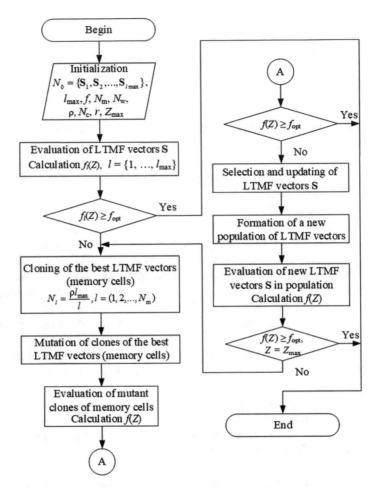

Fig. 4 Block diagram of the algorithm of artificial immune systems adapted to the specifics of the problem of finding optimal LTMFs

The procedure for finding the optimal vector of membership functions S_{opt} of the developed fuzzy system based on BBO consists of the following steps.

Step 1. BBO algorithm initialization. At the initialization stage, an ecosystem is created from l_{max} habitats (islands), the function of the habitat suitability index f and the condition for terminating the search are selected. In addition, the main parameters of the algorithm are set: the maximum level of immigration λ_{max}, the maximum level of emigration v_{max}, the coefficient of the mutation operator r, the maximum possible number of species on the island N_{max}, corresponding to the optimal value of the habitat suitability index f_{opt}, and the coefficient k_N of the relationship of the number of species N on the island from HSI [38]. Also, the dependencies of species migration on the number of species N on the islands are set: $\lambda(N)$ and $v(N)$. In turn, the created ecosystem N_0 consists of l_{max} islands corresponding to certain

Fig. 5 Dependences of
species migration on the
number of species on the
islands

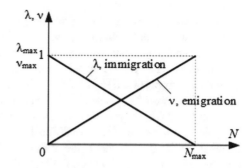

variants of the LTMF vector, the variables of which $S_{in}^i(q)$, $S_{out}^j(k)$ (characterizing the island) are set randomly within the generated set of LTMF [39]. The HSI f should be calculated based on expression (12). As a condition for terminating the search, one can choose (a) achieving the optimal value of the suitability index $f \geq f_{opt}$ or (b) performing the maximum number of iterations Z_{max}. Dependences of species migration on the number of species on islands N can be linear [38] with maximum values of immigration and emigration (Fig. 5) equal to one ($\lambda_{max} = v_{max} = 1$).

Step 2. Evaluation of ecosystem islands and checking the criterion for terminating the search. At this stage, for each l-th variant of the vector (island) S_l ($l = \{1, 2, ..., l_{max}\}$), the value of HSI $f_l(Z)$ is calculated based on expression (12), and the condition for terminating the optimization process is checked. When the optimal value of the suitability index $f_l(Z) \geq f_{opt}$ is reached for any vector of the LTMF or the maximum number of iterations Z_{max} is performed, the transition to *Step 7* is performed, otherwise, go to *Step 3*.

Step 3. Determination of Species migration levels. For each l-th variant of the vector **S**, based on the value of the suitability index f_l, the number of species N_l and the corresponding levels of immigration λ_l and emigration v_l are determined. When using linear relationships, the number of species on the island and the levels of migration are calculated based on the expressions (16)-(18) [38]

$$N_l = k_N f_l, l = (1, 2, ..., l_{max});$$ (16)

$$\lambda_l = \lambda_{max}\left(1 - \frac{N_l}{N_{max}}\right), l = (1, 2, ..., l_{max});$$ (17)

$$v_l = \frac{v_{max} N_l}{N_{max}}, l = (1, 2, ..., l_{max}).$$ (18)

Step 4. Migration of species on the islands. The migration of species from island to island is carried out as follows: a randomly chosen variable S_{MF} of the LTMF vector selected for immigration is replaced by the corresponding variable of the vector selected for emigration [39]. When choosing vectors for species immigration, all l_{max} variants (ecosystem islands) are sorted in turn, and each l-th variant of the vector **S** can be chosen with a probability proportional to λ_l. In the same way, for each vector chosen for immigration, the corresponding vector of species emigration is selected among the remaining $l_{max} - 1$ vectors of the population. Herewith, each variant of the LTMF vector can be chosen for the emigration of species with a probability proportional to v_l [38].

Step 5. Evaluation of ecosystem islands and checking the criterion for terminating the search. At this stage, for each *l*-th vector $\mathbf{S}_l$ ($l = \{1, 2, ..., l_{max}\}$), the HSI value $f_l(Z)$ is calculated, and the condition for terminating the optimization process is checked. When the optimal value of the suitability index $f_l(Z) \geq f_{opt}$ is reached for any variant of the LTMF vector, the transition to *Step 7* is performed, otherwise, go to *Sep 6*.

Step 6. Mutation of species on the islands. For each *l*-th island of the ecosystem with probability P_M, the species mutation procedure is carried out, that is changing the randomly selected variable S_{MF} in a random way within the generated set of LTMFs [42]. Herewith, the probability P_M of mutation is calculated based on the expressions (19)-(22) [38]

$$P_M = r\left(\frac{1 - P_l}{P_{l\,max}}\right);\tag{19}$$

$$P_l = \frac{\upsilon_N}{\sum\limits_{N=1}^{N_{max}+1} \upsilon_N};\tag{20}$$

$$\upsilon_N = \begin{cases} \dfrac{N_{max}!}{(N_{max} + 1 - N)!(N - 1)!}, \text{at} N \in (1, ..., N'); \\ \upsilon_{N_{max}+2-N}, \text{at} N \in (N' + 1, ..., N_{max} + 1); \end{cases}\tag{21}$$

$$N' = \text{ceil}\left(\frac{N_{max} + 1}{2}\right),\tag{22}$$

where P_l is the probability that the *l*-th island has N species; υ_N is the parameter that characterizes the relationship between the number of species N and the probability P_l; N' is the nearest integer greater than or equal to $(N_{max} + 1)/2$ (calculated using the ceil operation).

Next, go to *Step 2*.

Step 7. End of the search procedure.

Figure 6 shows a block diagram of a BBO algorithm adapted to the problem of finding optimal LTMFs.

To study the effectiveness of the proposed intelligent information technology for structural optimization of the LTMFs, Sect. 5 of this chapter presents the process of searching for optimal membership functions for a fuzzy ACS of a multi-purpose caterpillar mobile robot able to move along inclined and vertical ferromagnetic surfaces [51].

Fig. 6 Block diagram of the
BBO algorithm adapted to
the specifics of the problem
of finding optimal LTMFs

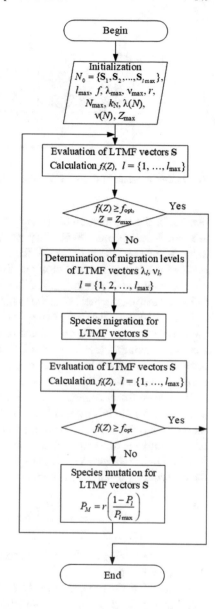

5 Search for Optimal Membership Functions for the Fuzzy ACS of the Multi-purpose Mobile Robot

Multi-purpose mobile robots with magnetic clamping devices make it possible to effectively perform various technological operations on inclined and vertical ferromagnetic surfaces in such industries as shipbuilding, ship repair, gas and oil processing, agriculture, etc. [52–57]. MR of this class, in turn, are complex technical

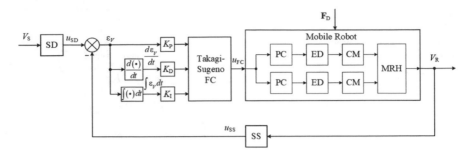

Fig. 7 The structure of fuzzy ACS of the MR speed

plants, for the automation of which intelligent control systems based on fuzzy logic are successfully used [58–60]. In paper [61], a fuzzy automatic control system of the Takagi–Sugeno type is presented for controlling and stabilizing the speed of movement of the caterpillar MR along inclined and ferromagnetic surfaces. This system has the structure shown in Fig. 7 [61], where the following designations are accepted: SD is the setting device; FC is the fuzzy controller; PC is the power converter; ED is the electric drive; CM is the caterpillar mover; MRH is the mobile robot hull; SS is the speed sensor; V_S, V_R are set and real values of the speed of MR movement; u_{SD}, u_{FC}, u_{SS} are output signals of SD, FC, and SS, respectively; ε_V is the speed control error; $\mathbf{F_D}$ is the disturbance vector; K_P, K_D, K_I are normalizing coefficients that are used to convert the FC input signals to relative units from their maximum values.

The given fuzzy ACS is designed for the caterpillar MR with the following characteristics: length is 1 m; width is 0.8 m; caterpillar drive wheel radius is 0.15 m; the total weight of the MR with equipment is 150 kg. The detailed mathematical model of this multi-purpose caterpillar mobile robot is given in papers [52, 61].

The Takagi–Sugeno type fuzzy speed controller in the developed ACS implements the control law based on the dependence (23)

$$u_{FC} = f_{FC}\left(K_P\varepsilon_V,\ K_D\frac{d\varepsilon_V}{dt},\ K_I\int \varepsilon_V dt\right). \tag{23}$$

The normalizing coefficients, in turn, have the following values: $K_P = 5$; $K_D = 0.33$; $K_I = 60$ [61].

Further, for the presented fuzzy ACS of the MR's speed, the search for optimal membership functions of the linguistic terms of the FC is performed following the main stages of the developed by the authors intelligent information technology.

At *Stage 1* of the implementation of the proposed IIT, the operating ranges for changing 3 input variables of the MR fuzzy control system are set. Herewith, for all input variables of FC $\left(K_P\varepsilon_V,\ K_D\frac{d\varepsilon_V}{dt},\ K_I\int \varepsilon_V dt\right)$ the operating ranges are set from −1 to 1 [61].

At *Stage 2*, a set of membership functions is formed that are used in the optimization process. This set includes the following LTMFs: triangular *TrFN*, trapezoidal *TrpFN*, Gaussian 1st *Gs1FN* and 2nd *Gs2FN* types, bell-shaped *GbFN*, S-shaped

SFN, Z-shaped *ZFN*, π-shaped *PiFN*, sigmoid *SgFN*, double sigmoid difference function *DsgFN*, as well as the product of two sigmoid functions *PsgFN* [62–64]. Mathematical models and graphic representations of these membership functions are given in Table 1.

At *Stage 3*, for the speed FC, the numbers of linguistic terms of the input variables are chosen. In particular, for the first input variable ($K_{P\varepsilon V}$), 5 linguistic terms are set (BN—big negative; SN—small negative; Z—zero; SP—small positive; BP—big positive). Also, for the second $\left(K_D \frac{d\varepsilon_V}{dt}\right)$ and third $\left(K_I \int \varepsilon_V dt\right)$ variables, 3 LTs are used (N—negative; Z—zero; P—positive) [61]. Thus, $\tau_1 = 5$, $\tau_2 = 3$ $\tau_3 = 3$.

Further, at *Stage 4*, the initial values of the parameters (a, b, c, d) of the entire set of LTMF (Table 1) are preset for the developed fuzzy ACS. Herewith, parameters (a, b, c, d) of the membership functions are chosen so that for all 3 input variables of the fuzzy ACS, a uniform distribution of linguistic terms over their operating ranges is ensured.

At the 5th stage of the implementation of the proposed IIT, taking into account the restrictions (5), the structure of the vector **S** is formed, which determines the set of membership functions of linguistic terms for the considered fuzzy ACS in the form of expression (24)

$$
\begin{aligned}
\mathbf{S} = \{S_{in}^i(q)\} = \{&S_{in}^1(1), S_{in}^1(2), S_{in}^1(3), S_{in}^1(4), S_{in}^1(5), \\
&S_{in}^2(1), S_{in}^2(2), S_{in}^2(3), S_{in}^3(1), S_{in}^3(2), S_{in}^3(3)\}, \\
S_{in}^i(q) \in \{&TrFN, TrpFN, GbFN, Gs1FN, Gs2FN, \\
&PiFN, ZFN, DsgFN, PsgFN\}, \text{at } q = 1, \\
S_{in}^i(q) \in \{&TrFN, TrpFN, GbFN, Gs1FN, Gs2FN, \\
&PiFN, DsgFN, PsgFN\}, \text{at } q = (2, ..., \tau_i-1), \\
S_{in}^i(q), \in \{&TrFN, TrpFN, GbFN, Gs1FN, Gs2FN, \\
PiFN, SFN, SgFN, &DsgFN, PsgFN\}, \text{at } q = \tau_i, i \in \{1, 2, 3\}.
\end{aligned}
\tag{24}
$$

In turn, for all linguistic terms of the vector (24), at *Stage 6*, triangular membership functions are initially chosen. In this case, the initial value of the vector $\mathbf{S}_0$ is determined by the expression (25)

$$
\begin{aligned}
\mathbf{S}_0 = \{S_{in}^i(q)\} = \{&TrFN, TrFN, TrFN, TrFN, TrFN, \\
&TrFN, TrFN, TrFN, TrFN, TrFN, TrFN\}.
\end{aligned}
\tag{25}
$$

The appearance of the linguistic terms of the FC with membership functions corresponding to the vector (25) with the set parameters (a, b, c) is shown in Fig. 8. The total number of adjustable parameters (a, b, c) for all linguistic terms (25) of the fuzzy controller for mobile robot's ACS is 33 parameters.

The total number of rules of the rule base of this fuzzy controller is $\prod_{i=1}^{3} \tau_i = \tau_1 \cdot \tau_2 \cdot \tau_3 = 45$, each of which is represented by the expression (26) [61]

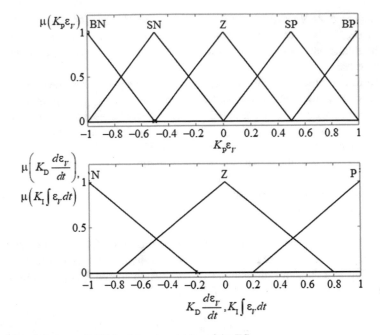

Fig. 8 Linguistic terms *TrFN* for 3 input variables of the FC

$$\text{IF } "K_P\varepsilon_V = A_{1r}" \text{ AND } "K_D\frac{d\varepsilon_V}{dt} = A_{2r}" \text{ AND } "K_I \int \varepsilon_V dt = A_{3r}"$$

$$\text{THEN } "u_{FC} = k_{1r}(K_P\varepsilon_V) + k_{2r}(K_D\frac{d\varepsilon_V}{dt}) + k_{3r}\left(K_I \int \varepsilon_V dt\right)", \qquad (26)$$

where r is the rule number in the RB; A_{1r}, A_{2r}, A_{3r} are the corresponding linguistic terms of the FC input variables; k_{1r}, k_{2r}, k_{3r} are weight coefficients of consequents of the RB rules.

The optimal values of the weight coefficients of consequents of the RB rules for this controller are found using a hybrid multi-agent PSO method of parametric optimization with an elite strategy [61]. The fragment of the rule base of this fuzzy controller with the found optimal coefficients of consequents is presented in Table 2.

As a complex objective function J_C at the 7th stage of the implementation of the information technology, in this case, the expression (7) is chosen.

In turn, the objective function J_1, which evaluates the efficiency of MR speed control, and the objective function J_2, which evaluates the complexity of further parametric optimization, are calculated based on expressions (8) and (10), respectively. The optimal (boundary) values of the complex objective function J_C and the criterion J_1 are chosen: $J_{Copt} = 0.1$; $J_{1opt} = 0.05$. The weighting factor for criterion J_2 is taken equal to 0.00166 ($k_{J2} = 0.00166$).

Before carrying out the procedure of searching for optimal membership functions for the fuzzy controller of the robot's ACS with the rule base given in Table 2, and

Table 2 Fragment of the RB of the fuzzy controller for the robot's ACS

Rule number	Linguistic terms of input variables			Weight coefficients of rules consequents		
	$K_P\varepsilon_V$	$K_D\frac{d\varepsilon_V}{dt}$	$K_I\int\varepsilon_V dt$	k_{1r}	k_{2r}	k_{3r}
1	BN	N	N	61.29	46.46	52.72
5	BN	Z	Z	93.84	78.8	68.41
18	SN	P	P	46.48	72.72	45.52
23	Z	Z	Z	59.87	52.6	27.37
30	SP	N	P	11.36	32.24	88.78
36	SP	P	P	30.43	60.62	38.97
41	BP	Z	Z	30.28	18.63	70.88
45	BP	P	P	9.07	23.97	42.28

triangular LTMFs (Fig. 8) the initial values of the objective functions J_C, J_1, and J_2 had the following values: $J_C = 0.112$; $J_1 = 0.057$; $J_2 = 33$.

At *Stage 8*, an iterative global search for the optimal vector of membership functions is carried out based on the following three (adapted to the specifics of this problem) bioinspired global optimization algorithms: genetic algorithm, the algorithm of artificial immune systems, biogeography-based optimization algorithm.

When carrying out the procedure of searching for optimal membership functions at the stage of initialization of bioinspired evolutionary algorithms, the main parameters of GA, AIS, and BBO were selected experimentally for this particular problem. In particular, for the genetic algorithm, an initial population N_0 was created from l_{max} = 100 chromosomes. Proportional selection [65, 66] was chosen as the selection operator, single-point crossover as the crossover operator, and simple mutation as the mutation operator [67, 68]. In this case, the values of the probabilities of crossover P_C and mutation P_M are given as: $P_C = 0.25$; $P_M = 0.1$.

For the algorithm of artificial immune systems, an initial population N_0 of l_{max} = 100 immune cells was created. The uniform cloning operator [69] is used as the cloning operator. In turn, the number of memory cells $N_m = 10$, the number of cells in the population with the worst affinity $N_w = 50$, the parameter of the cloning operator $N_c = 5$, the parameter of the mutation operator $r = 2$.

For the BBO algorithm, an ecosystem of $l_{max} = 100$ habitats (islands) was created at the initialization stage. The dependences of species migration on the number of species on the islands $\lambda(N)$ and $\nu(N)$ are linear, with $\lambda_{max} = \nu_{max} = 1$ [70, 71]. Mutation operator coefficient $r = 0.1$, the maximum possible number of species on the island, corresponding to the optimal value of the habitat suitability index f_{opt}, $N_{max} = 10$.

When implementing the above 3 algorithms, it is advisable to calculate the fitness function, the affinity function, and the habitat suitability index f based on the complex objective function J_C using the expression (12). As a criterion for the completion of optimization for each of the 3 algorithms given, the execution of the maximum number of iterations $Z_{max} = 100$ was chosen.

Procedures of the iterative search for optimal membership functions at the 8th stage of the implementation of the information technology for the presented FC of the speed ACS were carried out, in turn, using each of the 3 algorithms (GA, AIS, BBO). Herewith, when calculating the values of the complex objective function J_C (7) at each Z-th iteration of the IIT, the simulation of transients of the robot's ACS was carried out in all possible operation modes (under the action of various input and disturbing influences) to effectively search for the optimal membership functions of all FC terms.

Since bioinspired evolutionary algorithms are stochastic and when conducting each computational experiment under the same conditions and with the same controlled parameters (the number of individuals in the population l_{max}, the number of memory cells N_m, the coefficient of the mutation operator r, etc.) can give different results, then for an objective comparative analysis of their effectiveness, iterative procedures with the help of each algorithm should be carried out repeatedly followed by averaging the results obtained [72–75]. In this case, iterative procedures of finding the optimal LTMFs were carried out 10 times using each of the 3 selected evolutionary algorithms (GA, AIS, BBO), followed by averaging the obtained values of the complex objective function J_C at each Z-th iteration.

Figure 9 shows the average curves of change in the complex objective function (7) in the process of searching for optimal membership functions **S** at *Stage 8* based on the considered algorithms (GA, AIS, BBO).

At the final (9th) stage of the implementation of the information technology, the results of the search for optimal membership functions (using various bioinspired global optimization algorithms applied at *Stage 8*) were analyzed, and the best variant of the vector **S** was selected. Also, the effectiveness study of the fuzzy ACS for MR was conducted based on the results of the structural optimization of the FC membership functions. The average results of the experiments obtained in the process

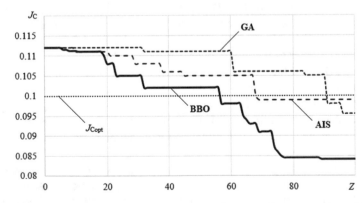

Fig. 9 Curves of changes in the complex objective function (7) in the process of searching for the optimal membership functions of the FC

Table 3 Averaged results of experiments obtained during the implementation of the IIT

Search algorithm	Z_{JCopt}	J_{Cmin}	J_{1min}	J_{2min}
GA	91	0.0955	0.049	28
AIS	68	0.099	0.0525	28
BBO	57	0.0841	0.041	26

of searching for optimal membership functions using each of the three different algorithms are presented in Table 3, where the following designations are accepted: Z_{JCopt} is the number of iterations in the implementation of the algorithm required to achieve the optimal value of the complex objective function J_C; $J_{Cmin}, J_{1min}, J_{2min}$ are minimum values of the complex objective function J_C and its components J_1, J_2, achieved in the process of optimizing the LTMFs.

As can be seen from Fig. 9 and Table 3, the proposed optimization IIT based on bioinspired evolutionary algorithms (GA, AIS, and BBO) makes it possible to successfully search for the optimal membership functions of linguistic terms for the FC of the mobile robot's ACS to improve the efficiency of the speed controlling, as well as to reduce the number of parameters during further parametric optimization of LTs.

The use of the IIT proposed by the authors based on the combination of GA, AIS, and BBO made it possible to reduce the value of the objective function J_1 by 14%, 7.9%, and 28%, respectively, compared with the initial FC with triangular membership functions of linguistic terms. In addition, for the optimal membership functions found using the developed search technology based on GA, AIS, and BBO, the number of parameters of the linguistic terms of the FC input variables decreased by 5, 5, and 7, respectively.

The vectors S_{GA}, S_{AIS}, S_{BBO} formed using the GA, AIS, and BBO algorithms, which ensure the achievement of the set boundary values of the complex objective function J_{Copt} of the fuzzy ACS, have a different structure:

$$S_{GA} = \{ZFN, TrpFN, Gs2FN, Gs1FN, TrFN, Gs1FN,$$
$$GbFN, Gs1FN, Gs1FN, Gs1FN, SgFN\};$$
$$S_{AIS} = \{TrFN, GbFN, GbFN, TrFN, SFN, ZFN,$$
$$TrFN, Gs1FN, ZFN, TrF, SFN\};$$
$$S_{BBO} = \{Gs1FN, TrFN, TrpFN, Gs1FN, Gs1FN, Gs1FN,$$
$$TrFN, Gs1FN, Gs1FN, Gs1FN, Gs1FN\}.$$

In turn, the appearance of the linguistic terms of FC with optimized sets of membership functions corresponding to the S_{GA}, $S_{AIS,}$ and S_{BBO} vectors is shown in Figs. 10, 11 and 12.

To solve this specific problem, the most effective is the use of the BBO algorithm at *Stage 8*, during the implementation of which it was possible to achieve the optimal value of the complex objective function of the fuzzy ACS ($J_C \leq 0.1$) in the least

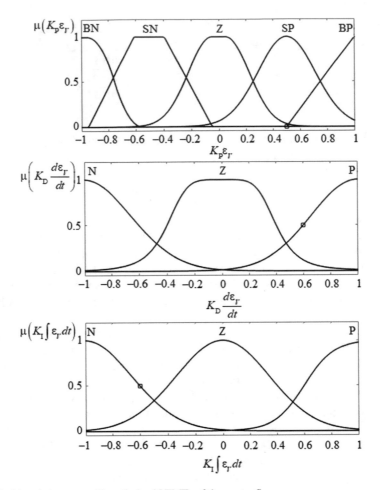

Fig. 10 Linguistic terms with optimized LTMFs of the vector S_{GA}

number of iterations ($Z_{JCopt} = 57$). Also, when implementing this algorithm at the 89th iteration (Fig. 9), the smallest value of the complex objective function was achieved ($J_{Cmin} = 0.0841$).

Moreover, the minimum values of the criteria J_1 and J_2 achieved during the optimization process using the BBO algorithm are also the smallest ($J_{1min} = 0.041$, $J_{2min} = 26$) compared to the values found using other algorithms (GA and AIS).

Thus, the optimal LTMF vector S_{opt} for this fuzzy ACS is the S_{BBO} vector found using the BBO algorithm at the 8th stage of the implementation of the proposed intelligent information technology ($S_{opt} = S_{BBO}$).

Figure 13 shows the transient characteristics of the acceleration of the mobile robot during movement along an inclined ferromagnetic surface under the following simulation conditions: set value of speed $V_S = 0.2$ m/s; angle of inclination of the

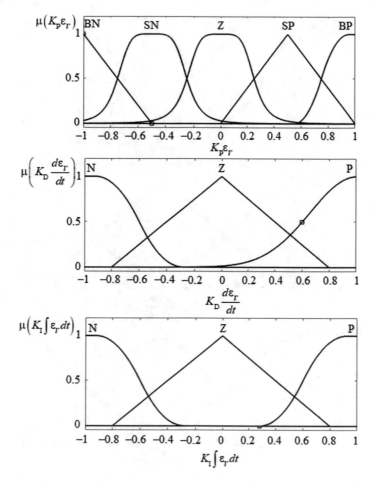

Fig. 11 Linguistic terms with optimized LTMFs of the vector S_{AIS}

working surface $\gamma = 60°$; the presence of a permanent disturbing effect in the form of a load force of a technological operation $F_D = 900$ N.

The given graphs of transients of speed change (Fig. 13) were obtained by modeling the fuzzy ACS with FC: 1—with initial triangular LTMFs; 2—with optimal LTMFs formed by implementing the proposed information technology based on the BBO algorithm.

Table 4 presents a comparative analysis of the main quality indicators [76–78] of the speed automatic control system for transients of the mobile robot movement along an inclined ferromagnetic surface (Fig. 13).

As can be seen from Fig. 13 and Table 4 the ACS of the MR speed with optimal membership functions of linguistic terms of FC input variables, formed using the structural optimization IIT proposed by the authors based on a biogeographical

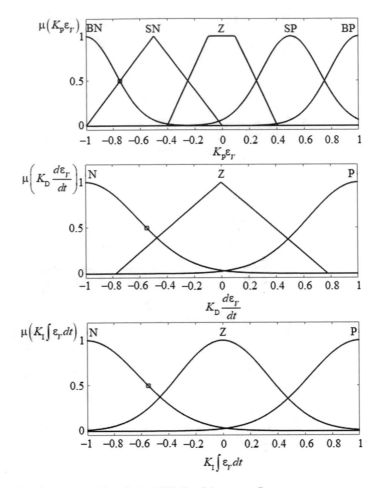

Fig. 12 Linguistic terms with optimized LTMFs of the vector S_{BBO}

algorithm, has higher control quality indicators than ACS with pre-selected triangular type LTMFs. It should also be noted that to find the optimal vector S_{opt} of the fuzzy controller using this information technology, it did not require significant computational and time costs ($Z_{JCopt} = 57$), which confirms its rather high efficiency.

If it is necessary to carry out further parametric optimization [79–83] of the linguistic terms of the considered FC (for further improvement of the quality indicators of the MR control system), the vector of optimized parameters for the found optimal LTMFs will have 7 variables less than the same vector for triangular LTMFs, that significantly simplifies the parametric optimization procedure. Also, the further procedure of the software and hardware implementation of the optimized fuzzy controller for the mobile robot control system will be simplified.

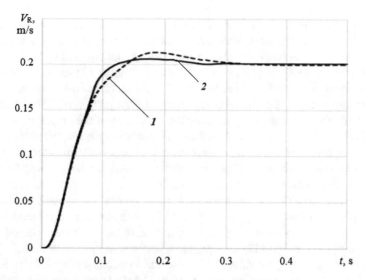

Fig. 13 Transient characteristics of MR acceleration with speed ACS

Table 4 Comparative analysis of quality indicators of ACS with different LTMFs

Quality indicators	Quality indicators for the ACS of MR speed	
	LTMFs of triangular type	Optimal LTMFs
Rise time t_r, s	0.139	0.128
Regulating time t_{rmax}, s	0.31	0.236
Overshoot σ_{max}, %	6.5	3.2

6 Conclusions

This chapter presents the development and research of intelligent information technology for structural optimization of fuzzy systems based on the evolutionary search of the optimal membership functions. The obtained IIT makes it possible to form optimal LTMFs when solving compromise problems of multi-criteria structural optimization of various FSs (control, decision making, etc.) to increase their efficiency and reduce the degree of complexity of further parametric optimization. A specific feature of the proposed information technology is the implementation of iterative search procedures based on a combination of different (two or more) bioinspired evolutionary algorithms, followed by the analysis of the results obtained and selection of the best variant of the LTMF vector corresponding to the global optimum of the problem being solved. In particular, this chapter considers procedures of an iterative global search for the optimal vector of membership functions using such (adapted to the specifics of the problem being solved) bioinspired evolutionary algorithms as genetic, artificial immune systems, and biogeographic.

To study the effectiveness of the proposed IIT, in this chapter, we searched for optimal membership functions of the fuzzy controller for the speed ACS of the multi-purpose caterpillar mobile robot able to move along inclined and vertical ferromagnetic surfaces, with the implementation of this technology based on 3 bioinspired evolutionary algorithms: GA, AIS, and BBO. The analysis of the obtained results of computer simulation shows that the application of the proposed IIT for finding the optimal LTMFs can significantly increase the efficiency of the processes of controlling the MR's speed, as well as reduce the total number of parameters at further parametric optimization of linguistic terms. Herewith, for solving this specific problem, the BBO algorithm is the most effective, since it allowed to achieve the optimal value of the complex objective function of the fuzzy ACS ($J_C \leq 0.1$) in the least number of iterations ($Z_{JCopt} = 57$). Also, using this IIT based on BBO, the smallest value of the complex objective function $J_{Cmin} = 0.0841$ (at the 89th iteration) was found, which is 1.33 times less than the value of J_C for the initial fuzzy ACS with pre-selected LTMFs of triangular type ($J_C = 0.112$). In turn, when implementing this modification of the information technology, the value of the criterion J_1 decreased by 1.39 times ($J_{1min} = 0.041$), and the number of parameters of linguistic terms decreased by 7 ($J_{2min} = 26$). As a result, the ACS of the MR speed with optimal membership functions of linguistic terms, found using the proposed by the authors IIT based on a combination of bioinspired evolutionary algorithms (with the best results according to the BBO algorithm), has higher control quality indicators compared to fuzzy ACS with preselected LTMFs of triangular type, namely: the system performance increased by 1.3 times, the overshoot decreased by 2 times.

Thus, the research results presented in this chapter confirm the high efficiency of the information technology developed by the authors for searching the optimal LTMFs, as well as the feasibility of its application for the structural optimization of fuzzy systems and devices of various types, configurations, and purposes.

References

1. Zadeh, L. A., Abbasov, A. M., Yager, R. R., Shahbazova, S. N., & Reformat, M. Z. (Eds.). (2014). *Recent developments and new directions in soft computing, STUDFUZ 317*. Springer.
2. Jamshidi, M., Kreinovich, V., & Kacprzyk, J. (Eds.). (2013). *Advance trends in soft computing*. Springer.
3. Kondratenko,Y. P., Korobko, O. V., & Kozlov, O.V. (2016). Synthesis and optimization of fuzzy controller for thermoacoustic plant. In Zadeh L. A. et al. (Eds.), *Recent developments and new direction in soft-computing foundations and applications, studies in fuzziness and soft computing* (Vol. 342, pp. 453–467). https://doi.org/10.1007/978-3-319-32229-2_31.
4. Zadeh, L. A. (1994). The role of fuzzy logic in modeling, identification and control. *Modeling Identification and Control, 15*(3), 191–203.
5. Mamdani, E. H. (1974). Application of fuzzy algorithms for control of simple dynamic plant. *Proceedings of IEEE, 121*, 1585–1588.
6. Kosko, B. (1994). Fuzzy systems as universal approximators. *IEEE Transactions on Computers, 43*(11), 1329–1333.

7. Kondratenko, Y. P., & Kozlov, O. V. (2012). Mathematic modeling of reactor's temperature mode of multiloop pyrolysis plant. *Modeling and Simulation in Engineering, Economics and Management, Lecture Notes in Business Information Processing, 115,* 178–187.

8. Kondratenko, Y. P., Korobko, O. V., Kozlov, O. V. (2012). Frequency tuning algorithm for loudspeaker driven thermoacoustic refrigerator optimization. In K. J. Engemann, A. M. Gil-Lafuente, & J. M. Merigo (Eds.), *Lecture notes in business information processing: Modeling and simulation in engineering, economics and management* (Vol. 115, 270–279). Springer. https://doi.org/10.1007/978-3-642-30433-0_27.

9. Kondratenko, Y. P., & Kozlov, O. V. (2016). Mathematical model of ecopyrogenesis reactor with fuzzy parametrical identification. In L. A. Zadeh et al. (Eds.), *Recent developments and new direction in soft-computing foundations and applications, studies in fuzziness and soft computing* (Vol. 342, pp. 439–451). Springer. https://doi.org/10.1007/978-3-319-32229-2_30.

10. Kondratenko, Y. P., Kozlov, O. V. & Korobko, O. V. (2018). Two modifications of the automatic rule base synthesis for fuzzy control and decision making systems. In J. Medina et al. (Eds), *Information Processing and Management of Uncertainty in Knowledge-Based Systems: Theory and Foundations, 17th International Conference, IPMU 2018, Cadiz, Spain, Proceedings, Part II, CCIS* (Vol. 854, pp. 570–582). Springer International Publishing AG. https://doi.org/10.1007/978-3-319-91476-3_47.

11. Simon, D. (2013). *Evolutionary optimization algorithms: Biologically inspired and population-based approaches to computer intelligence.* Wiley.

12. Kondratenko, Y., Khalaf, P., Richter, H., & Simon, D. (2019). *Fuzzy real-time multi-objective optimization of a prosthesis test robot control system.* In Y. P. Kondratenko, A. A. Chikrii, V. F. Gubarev, & J. Kacprzyk (Eds.), *Advanced control techniques in complex engineering systems: Theory and applications. dedicated to Professor Vsevolod M. Kuntsevich.* Studies in systems, decision and control (Vol. 203, pp. 165–185). Springer Nature Switzerland AG. https://doi.org/10.1007/978-3-030-21927-7_8.

13. Abbasian, R., & Mouhoub, M. (2011). An efficient hierarchical parallel genetic algorithm for graph coloring problem. In *Proceedings of The 13th Annual Conference on Genetic and Evolutionary Computation, ACM, Dublin, Ireland* (pp. 521–528).

14. Zhu, J., Lauri, F., Koukam, A., & Hilaire, V. (2014). *Fuzzy logic control optimized by artificial immune system for building thermal condition.* In P. Siarry, L. Idoumghar, & J. Lepagnot (Eds.), *Swarm intelligence based optimization. ICSIBO 2014.* Lecture notes in computer science (vol. 8472, pp. 42–49). Springer.

15. Melendez, A., & Castillo, O. (2013). Evolutionary optimization of the fuzzy integrator in a navigation system for a mobile robot. In *Recent advances on hybrid intelligent systems* (pp. 21–31).

16. Smiley, A., & Simon, D. (2016). Evolutionary optimization of atrial fibrillation diagnostic algorithms. *International Journal of Swarm Intelligence, 2*(2/3/4), 117–133.

17. Kondratenko, Y. P., & Kozlov, A. V. (2019). Generation of rule bases of fuzzy systems based on modified ant colony algorithms. *Journal of Automation and Information Sciences, 51*(3), 4–25; Begel House Inc: https://doi.org/10.1615/JAutomatInfScien.v51.i3.20.

18. Hampel, R., Wagenknecht, M., & Chaker N. (2000). Fuzzy control: Theory and practice (p. 410). Physika-Verlag .

19. Sadegheih, A., & Drake, P. R. (2001). Network optimization using linear programming and genetic algorithm, neural network world. *Neural Network World, 11*(3), 223–233.

20. Kacprzyk, J. (1997). *Multistage fuzzy control: A prescriptive approach.* Wiley.

21. Kozlov, O., Kondratenko, G., Gomolka, Z., & Kondratenko, Y. (2019). Synthesis and optimization of green fuzzy controllers for the reactors of the specialized pyrolysis plants. In V. Kharchenko, Y. Kondratenko, J. Kacprzyk (Eds.), *Green IT engineering: social, business and industrial applications, studies in systems, decision and control* (Vol 171, 373–396) Springer. https://doi.org/10.1007/978-3-030-00253-4_16.

22. Kondratenko, Y. P., & Simon, D. (2018). Structural and parametric optimization of fuzzy control and decision making systems. In L. Zadeh, R. Yager, S. Shahbazova, M. Reformat, & V. Kreinovich (Eds.), *Recent developments and the new direction in soft-computing foundations*

and applications, studies in fuzziness and soft computing (Vol. 361, pp. 273–289). Springer. https://doi.org/10.1007/978-3-319-75408-6_22.

23. Ovreiu, M., & Simon, D. (2010). Biogeography-based optimization of neuro-fuzzy system parameters for diagnosis of cardiac disease. In *Proceedings of Genetic and Evolutionary Computation Conference* (pp. 1235–1242).

24. Nabi, A, & Singh, N. A. (2016). Particle swarm optimization of fuzzy logic controller for voltage sag improvement. In: Proceedings of 2016 3rd International Conference on Advanced Computing and Communication Systems (ICACCS) (Vol. 01, pp. 1–5).

25. Alves, R. T., Delgado, M. R., Lopes, H. S., & Freitas, A. A. (2004). An artificial immune system for fuzzy-rule induction in data mining. In X. Yao et al. (Eds.), *Parallel Problem Solving from Nature - PPSN VIII. PPSN 2004.* Lecture Notes in Computer Science (Vol. 3242, pp. 1011–1020). Springer.

26. Castillo, O., Ochoa, P., & Soria, J. (2016). Differential evolution with fuzzy logic for dynamic adaptation of parameters in mathematical function optimization. In P. Angelov, S. Sotirov (Eds.), *Imprecision and uncertainty in information representation and processing. studies in fuzziness and soft computing* (Vol. 332, 361–374). Springer.

27. Khan, S., et al. (2008). Design and implementation of an optimal fuzzy logic controller using genetic algorithm. *Journal of Computer Science, 4*(10), 799–806.

28. Cordon, O., Gomide, F., Herrera, F., Hoffmann, F., & Magdalena, L. (2004). Ten years of genetic fuzzy systems: Current framework and new trends. *Fuzzy Sets and Systems, 141*(1), 5–31.

29. Liu, X.-H., Kuai, R., Guan, P., Ye, X.-M., & Wu, Z.-L. (2009). Fuzzy-PID control for arc furnace electrode regulator system based on genetic algorithm. In *Proceedings of the Eighth International Conference on Machine Learning and Cybernetics, Baoding* (pp. 683–689).

30. Zhao, J., Han, L., Wang, L., & Yu, Z. (2016) The fuzzy PID control optimized by genetic algorithm for trajectory tracking of robot arm. In *2016 12th World Congress on Intelligent Control and Automation (WCICA), Guilin, China* (pp. 556–559).

31. Chen, Ch., Li, M., Sui, J., Wei, K., Pei, Q.: A genetic algorithm-optimized fuzzy logic controller to avoid rear-end collisions, Journal of advanced transportation, Vol. 50, 1735-1753 (2016).

32. Li, H. F., Feng, Z. G, & Wang, J. (2008). GA based design of fuzzy control law for hypersonic vehicle. In *2008 2nd International Symposium on Systems and Control in Aerospace and Astronautics* (pp. 1–4).

33. Li, L., Zhu, Q., & Gao, Sh. (2006). Design and realization of waterjet propelled craft autopilot based on fuzzy control and genetic algorithms. In *Proceedings of the 2006 IEEE International Conference on Mechatronics and Automation, Luoyang, China* (pp. 1362–1366).

34. Prakash, A., & Deshmukh, S. G. (2011). A multi-criteria customer allocation problem in supply chain environment: An artificial immune system with fuzzy logic controller based approach. *Expert Systems with Applications, 38*(4), 3199–3208.

35. Visconti, A., & Tahayori, H. (2011). Artificial immune system based on interval type-2 fuzzy set paradigm. *Applied Soft Computing, 11*(6), 4055–4063.

36. Hachicha, N., Jarboui, B., & Siarry, P. (2011). A fuzzy logic control using a differential evolution algorithm aimed at modelling the financial market dynamics. *Information Sciences, 181*, 79–91.

37. Minku, F.L., & Ludermir, T. (2005). Evolutionary strategies and genetic algorithms for dynamic parameter optimization of evolving fuzzy neural networks. In *Evolutionary Computation, The 2005 IEEE Congress* (Vol. 3, pp. 1951–1958).

38. Thomas, G., Lozovyy, P., & Simon, D. (2011) Fuzzy robot controller tuning with biogeography-based optimization. In: *Modern Approaches in Applied Intelligence: 24th International Conference on Industrial Engineering and Other Applications of Applied Intelligent Systems, IEA/AIE 2011, Syracuse, NY, USA, June 28 – July 1, Proceedings, Part II* (pp. 319–327).

39. Zhang, M., Jiang, W., Zhou, X., et al. (2019). A hybrid biogeography-based optimization and fuzzy C-means algorithm for image segmentation. *Soft Computing, 23*, 2033–2046.

40. Piegat, A. (2013). Fuzzy modeling and control (Vol. 69). Physica.

41. Kondratenko, Y. P., & Kondratenko, N. Y. (2015). Soft computing analytic models for increasing efficiency of fuzzy information processing in decision support systems. In: R. Hudson (Ed.),

Decision making: processes, behavioral influences and role in business management (pp. 41–78). Nova Science Publishers.

42. Kondratenko, Y. P., Klymenko, L. P., Al Zu'bi & E. Y. M. (2013). *Structural optimization of fuzzy systems' rules base and aggregation models, kybernetes* (Vol. 42, Iss. 5, pp. 831–843). https://doi.org/10.1108/K-03-2013-0053.

43. Ishibuchi, H., & Yamamoto, T. (2004). Fuzzy rule selection by multi-objective genetic local search algorithms and rule evaluation measures in data mining. *Fuzzy Sets and Systems, 141*(1), 59–88.

44. Kondratenko, Y., Kondratenko, G., & Sidenko, I. (2018). Multi-criteria decision making for selecting a rational IoT platform. In *Proceedings of 2018 IEEE 9th International Conference on Dependable Systems, Services and Technologies, DESSERT 2018, Kyiv, Ukraine* (pp. 147–152). https://doi.org/10.1109/DESSERT.2018.8409117.

45. Nazarenko, A. M., & Karpusha, M. V. (2014). Modeling and identification in the problems of multicriteria optimization with linear and quadratic performance criteria under statistical uncertainty. *Journal of Automation and Information Sciences, 46*(3), 17–29.

46. Kondratenko, Y. P., & Al Zubi, E. Y. M. (2009). *The Optimisation Approach for Increasing Efficiency of Digital Fuzzy Controllers, Annals of DAAAM for 2009 & Proceeding of the 20th International DAAAM Symposium. "Intelligent Manufacturing and Automation", Published by DAAAM International, Vienna, Austria* (pp. 1589–1591).

47. Kondratenko, Y., Kondratenko, G., & Sidenko, I. (2019). Multi-criteria decision making and soft computing for the selection of specialized IoT platform. In O. Chertov, T. Mylovanov, Y. Kondratenko, J. Kacprzyk, V. Kreinovich, V. Stefanuk (Eds.), *Recent Developments in Data Science and Intelligent Analysis of Information. Proceedings of the XVIII International Conference on Data Science and Intelligent Analysis of Information, June 4–7, 2018, Kyiv, Ukraine. ICDSIAI 2018, Advances in Intelligent Systems and Computing* (Vol. 836, 71–80). Springer International Publishing. https://doi.org/10.1007/978-3-319-97885-7_8.

48. Puris, A., Bello, R., Molina, D., & Herrera, F. (2011). Variable mesh optimization for continuous optimization problems. *Soft Computing, 16*(3), 511–525.

49. Mehrabian, A. R., & Lucas, C. (2006). A novel numerical optimization algorithm inspired from weed colonization. *Ecological Informatics, 1*, 355–366.

50. Li, J., Li, Y., Tian, S., et al. (2020). An improved cuckoo search algorithm with self-adaptive knowledge learning. *Neural Computing and Applications, 32*, 11967–11997.

51. Kondratenko, Y., et al. (2021). Inspection mobile robot's control system with remote IoT-based data transmission. *Journal of Mobile Multimedia. Special issue "Mobile Communication and Computing for Internet of Things and Industrial Automation"* 17(4), 499–522.

52. Kozlov, O. (2021). Optimal selection of membership functions types for fuzzy control and decision making systems. In *Proceedings of the 2nd International Workshop on Intelligent Information Technologies & Systems of Information Security with CEUR-WS, Khmelnytskyi, Ukraine, IntelITSIS 2021, CEUR-WS* (Vol. 2853, pp. 238–247).

53. Lewis, F. L., Campos, J., & Selmic, R. (2002). *Neuro-fuzzy control of industrial systems with actuator nonlinearities*. SIAM.

54. Duro, R. J., & Kondratenko, Y. P. (Eds.). (2015). *Advances in intelligent robotics and collaboration automation*. Control and Robotics, River Publishers, Denmark.

55. Spong, M. W., Hutchinson, S., & Vidyasagar, M. (2006). *Robot modeling and control*. Wiley.

56. Souto, D., Faiña, A., Deibe, A., Lopez-Peña, F., & Duro, R. J. (2012). A robot for the unsupervised grit-blasting of ship hulls. *International Journal of Advanced Robotic Systems, 9*, 1–16.

57. Longo, D., & Muscato, G. (2004). A small low-cost low-weight inspection robot with passive-type locomotion. *Integrated Computer-Aided Engineering, 11*, 339–348.

58. Xiao, J., Xiao, J., Xi, N., Tummala, R.L., & Mukherjee, R.: Fuzzy controller for wall-climbing microrobots. *IEEE Transactions on Fuzzy Systems, 12*(4), 466–480.

59. Elayaraja, D., & Ramabalan, S. (2014). Fuzzy logic control of low cost obstacle climbing robot. *Applied Mechanics and Materials, 592–594*, 2150–2154.

60. Churavy, C., Baker, M., Mehta, S., Pradhan, I., Scheidegger, N., Shanfelt, S., Rarick, R., & Simon, D. (2008). Effective implementation of a mapping swarm of robots. *IEEE Potentials, 27*(4), 28–33.

61. Kondratenko, Y. P., & Kozlov, A. V. (2019). Parametric optimization of fuzzy control systems based on hybrid particle swarm algorithms with elite strategy. *Journal of Automation and Information Sciences, 51*(12), 25–45; New York: Begel House Inc. https://doi.org/10.1615/JAutomatInfScien.v51.i12.40.

62. Li, J., Wang, J. Q., & Hu, J. H. (2019). Multi-criteria decision-making method based on dominance degree and BWM with probabilistic hesitant fuzzy information. *International Journal of Machine Learning and Cybernetics, 10*, 1671–1685.

63. Kondratenko, Y. P., Kozlov, O. V., Klymenko, L. P., & Kondratenko, G. V. (2014). Synthesis and research of neuro-fuzzy model of ecopyrogenesis multi-circuit circulatory system. In M. Jamshidi, V. Kreinovich, J. Kazprzyk (Eds.), *Advance trends in soft computing*. Series: Studies in fuzziness and soft computing (Vol. 312, pp. 1–14). https://doi.org/10.1007/978-3-319-036 74-8_1.

64. Mazandarani, M., & Xiu, L. (2020). Fractional fuzzy inference system: The new generation of fuzzy inference systems. *IEEE Access, 8*, 126066–126082.

65. Piwonska, A. (2010). Genetic algorithm finds routes in travelling salesman problem with profits. Zeszyty Naukowe Politechniki Bia lostockiej, Informatyka 51–65.

66. Sun, X., Wang, J.: Routing design and fleet allocation optimization of freeway service patrol: Improved results using genetic algorithm. *Physica A: Statistical Mechanics and its Applications, 501*, 205–216.

67. Arakaki, R. K., & Usberti, F. L. (2018). Hybrid genetic algorithm for the open capacitated arc routing problem. *Computers & Operations Research, 90*, 221–231.

68. Ashby, H., & Yampolskiy, R. V. (2011). Genetic algorithm and wisdom of artificial crowds algorithm applied to light up. In *Proceedings of 16th International Conference on Computer Games, AI, Animation, Mobile, Interactive Multimedia, Educational & Serious Games, Louisville, KY, USA* (pp. 27–30).

69. Dasgupta, D. (2006). Advances in artificial immune systems. *IEEE Computational Intelligence Magazine, 1*(4), 40–49.

70. Simon, D. (2008). Biogeography-based optimization. *IEEE Transactions on Evolutionary Computation, 12*(6), 702–713.

71. Sayed, M. M., Saad, M. S., Emara, H. M., & Abou El-Zahab, E. E. (2013). A novel method for type-2 fuzzy logic controller design using a modified biogeography-based optimization. In *2013 IEEE International Conference on Industrial Technology (ICIT), Cape Town* (pp. 28–33).

72. Aseri, N. A. M., Ismail, M. A., Fakharudin, A. S., & Ibrahim, A. O. (2020). Review of the metaheuristic algorithms for fuzzy modeling in the classification problem. *International Journal of Advanced Trends in Computer Science and Engineering, 9*(1.4), 387–400.

73. Oh, S. K., & Pedrycz, W. (2002). The design of hybrid fuzzy controllers based on genetic algorithms and estimation techniques. *Journal of Kybernetes, 31*(6), 909–917.

74. Pedrycz, W., Li, K., & Reformat, M. (2015). Evolutionary reduction of fuzzy rule-based models. In *Fifty Years of Fuzzy Logic and its Applications, STUDFUZ 326* (pp. 459–481). Springer.

75. Nagata, Y., & Soler, D. (2012). A new genetic algorithm for the asymmetric traveling salesman problem. *Expert Systems with Applications, 39*(10), 8947–8953.

76. Kuntsevich, V. M. et al. (Eds). *Control systems: Theory and applications*. Series in automation, control and robotics. River Publishers, Gistrup, Delft.

77. Kondratenko, Y. P., Kuntsevich, V. M., Chikrii, A. A., & Gubarev, V. F. (Eds.) *Advanced control systems: theory and applications*. Series in automation, control and robotics. River Publishers: Gistrup. ISBN: 9788770223416.

78. Topalov, A. et al. (2016). Control processes of floating docks based on SCADA systems with wireless data transmission. In *Perspective Technologies and Methods in MEMS Design, MEMSTECH 2016 - Proceedings of 12th International Conference* (pp. 57–61). https://doi.org/10.1109/MEMSTECH.2016.7507520.

79. Kosanam, S., & Simon, D. (2006). Fuzzy membership function optimization for system identification using an extended Kalman filter. *Fuzzy Information Processing Society* 459–462.
80. Shill, P. C., Maeda, Y., & Murase, K. (2013). Optimization of fuzzy logic controllers with rule base size reduction using genetic algorithms. In *2013 IEEE Symposium on Computational Intelligence in Control and Automation (CICA)* (pp. 57–64). Singapore.
81. Gupta, M., Behera, L., & Venkatesh, K. S. (2010). PSO based modeling of Takagi-Sugeno fuzzy motion controller for dynamic object tracking with mobile platform. In *International Multiconference Computer Science and Information Technology (IMCSIT)* (pp. 37–43).
82. Safaee, B., & Mashhadi, S. K. M. (2016). Fuzzy membership functions optimization of fuzzy controllers for a quadrotor using particle swarm optimization and genetic algorithm. In *Proceedings of 2016 4th International Conference on Control, Instrumentation, and Automation (ICCIA)* (pp. 256–261).
83. Kondratenko, Y., Atamanyuk, I., Sidenko, I., Kondratenko, G., & Sichevskyi, S. (2022). Machine learning techniques for increasing efficiency of the robot's sensor and control information processing. *Sensors, 22*, 1062. https://doi.org/10.3390/s22031062.

Neural and Granular Fuzzy Adaptive Modeling

Alisson Porto and Fernando Gomide

Abstract The chapter addresses adaptive nonlinear systems modeling focusing on the uninorm-based evolving neuro-fuzzy network and the granular evolving min-max fuzzy approaches. Neuro-fuzzy networks have shown to be highly competitive with multilayer neural networks, but their performance have not been yet contrasted with contemporary adaptive granular modeling methods such as evolving min-max modeling. Neuro-fuzzy network and granular evolving min-max modeling are benchmarked against state of the art evolving modeling methods using the Box and Jenkins gas furnace, and Mackey-Glass chaotic time series data. It is shown that evolving granular min-max modeling is a powerful modeling approach because, in addition to accuracy, it benefits from the transparency conveyed by the developed models.

1   Introduction

Stream data processing and learning from stream data require fast and responsive methods to analyze data and to develop system models online. Data-driven modeling of nonlinear, nonstationary systems require adaptive machine learning methods to cope with the continuous, non-ending time-variant nature of the real world system dynamics. Evolving modeling is a modeling approach capable of gradual, continuous, and simultaneous adaptation of the structural components of a model and its parameters [1]. A comprehensive and state-of-the art review of evolving modeling approaches is found in [16]. Evolving granular rule-based and fuzzy tree-based modeling is reviewed in [8, 9].

This chapter uses the uninorm-based evolving neural network introduced in [15] and the granular evolving min-max fuzzy modeling suggested in [14] to model non-

A. Porto
Nitryx Progress Rail, Av. Avelino Silveira Franco, 149, 13105-822 Campinas, Sao Paulo, Brazil

F. Gomide (✉)
School of Electrical and Computer Engineering, University of Campinas, Av. Albert Einstein, 400, 13083-512 Campinas, Sao Paulo, Brazil
e-mail: gomide@unicamp.br

© The Author(s), under exclusive license to Springer Nature Switzerland AG 2023
Y. P. Kondratenko et al. (eds.), *Artificial Intelligence in Control and Decision-making Systems*, Studies in Computational Intelligence 1087,
https://doi.org/10.1007/978-3-031-25759-9_8

linear systems. The goal is to evaluate their performance when modeling two classic benchmarks, the Box and Jenkins gas furnace, and the Mackey-Glass chaotic time series, respectively. Performances of the evolving neuro-fuzzy and granular min-max models are compared with the ones developed by state of the art adaptive modeling approaches. The uninorm-based neural network competes closely and its performance surpasses its classic neural modeling counterparts, but it has not been systematically evaluated and compared with contemporary evolving methods. Similarly, granular min-max fuzzy modeling have shown to perform better than many adaptive machine learning methods [3, 14], but has not been contrasted with neural modeling.

After this introduction, the chapter proceeds as follows. The next section reviews the evolving neuro-fuzzy network and granular min-max fuzzy modeling methods. Section 3 addresses the Box and Jenkins gas furnace identification, and the Mackey Glass chaotic time series forecasting problems. Gas furnace modeling and chaotic time series are benchmarks widely adopted in the literature to evaluate the performance of modeling methods and forecasting algorithms. The evolving neuro-fuzzy and granular min-max fuzzy modeling are evaluated and compared with state of the art evolving modeling methods using the root mean square error index. Section 4 concludes the chapter summarizing its findings and suggesting issues for future work.

2 Adaptive Neuro-Fuzzy and Granular Min-Max Modeling

2.1 Uninorm-Based Evolving Neural Network Modeling

Neuro-fuzzy networks are a framework to build models based on fuzzy set, neural network, and machine learning methods. First work in the area dates to early nineties [13], occasion in which the idea of logic neurons derived from the notion of t-norms and t-conorms was raised. For instance, the framework has shown to be very powerful to model multistage decision making and control problems in fuzzy environments, particularly when one needs to manage objective and subjective goals, and fuzzy constraints on decisions and states [6].

In fuzzy set theory, uninorms (denoted u and U) are a generalization of t-norms and t-conorms. While a t-norm has the identity element $e = 1$, and a s-norm the identity element $e = 0$, the identity element of a uninorm is anywhere in the unit interval, that is, $e \in [0, 1]$ [12]. There are many ways to construct uninorms. In this chapter constructors of uninorms are as follows [18]:

$$
a \text{ u } w = \begin{cases} e + (1 - e) \left(\frac{(a-e)}{(1-e)} \text{s} \frac{(w-e)}{(1-e)} \right) & \text{if } a, w \in [e, 1] \\ e \left(\frac{a}{e} \text{t} \frac{w}{e} \right) & \text{if } a, w \in [0, e) \\ \max(a, w) (\text{or } \min(a, w)) & \text{otherwise} \end{cases} \tag{1}
$$

where t and s denote a t-norm and a t-conorm, respectively. Here in this chapter t is the algebraic product (a t $b = ab$), and s is the algebraic sum (a s $b = a + b - ab$), $a, b \in [0, 1]$.

The output of an unineuron processing unit with n inputs is of the form

$$z = \mathbf{U}_{l=1}^{n} \lambda_{lj} \text{ u } w_{lj}, \tag{2}$$

where λ_{lj} is the membership degree of the lth input variable in the jth fuzzy set of its domain partition, and w_{lj} the corresponding weight. This form produces neuro-fuzzy networks which are similar to fuzzy rule-based models [3]. Alternatively, the one adopted in this chapter, λ can be viewed as the membership degree of an n dimensional input in a n-dimensional fuzzy set called cloud [17]. In this case output z of the unineuron becomes [15]

$$z = \lambda \text{ u } w \tag{3}$$

Therefore, the ith unineuron of a neuro-fuzzy network with L^t units in the hidden layer produces the output

$$z_i = \lambda_i \text{ u } w_i \tag{4}$$

whose membership degree λ_i is computed using

$$\lambda_i^t = \frac{\gamma_i^t}{\sum_{j=1}^{L^t} \gamma_j^t}, \quad i = 1, ..., L^t \tag{5}$$

Here γ_i^t is the local density of the n-dimensional input data $\mathbf{x}^t$ in the ith cloud. At each step t, γ_i^t is given by:

$$\gamma_i^t = \frac{1}{1 + ||\mathbf{x}^t - \boldsymbol{\mu}_i^t||^2 + \Sigma_i^t - ||\mathbf{x}^t||^2}, \tag{6}$$

where $\boldsymbol{\mu}_i^t = ((M_i^t - 1)/(M_i^t))\boldsymbol{\mu}_i^{t-1} + \mathbf{x}^t/(M_i^t)$, $\boldsymbol{\mu}_i^1 = \mathbf{x}^1$, M_i^t is the number of input data of the i-th cloud at step t. The dispersion Σ_i^t is computed using:

$$\Sigma_i^t = \frac{M_i^t - 1}{M_i^t} \Sigma_i^{t-1} + \frac{1}{M_i^t} ||\mathbf{x}^t||^2, \quad \Sigma_i^1 = ||\mathbf{x}^1||^2 \tag{7}$$

The neuro-fuzzy network adaptation process requires global density information, which proceeds similarly as for local densities, but considering all input data instead. It is computed recursively using

$$\Gamma^t = \frac{1}{1 + ||\mathbf{x}^t - \boldsymbol{\mu}_G^t||^2 + \Sigma_G^t - ||\mathbf{x}^t||^2}, \tag{8}$$

where $\boldsymbol{\mu}_G^t = ((t - 1)/t)\boldsymbol{\mu}_G^{t-1} + \mathbf{x}^t/t$, $\boldsymbol{\mu}_G^1 = \mathbf{x}^1$ and

$$\Sigma_G^t = \frac{t-1}{t}\Sigma_G^t - 1 + \frac{1}{t}||\mathbf{x}^t||^2, \quad \Sigma_G^1 = ||\mathbf{x}^1||^2 \tag{9}$$

Define

$$Cloud_i = \arg\max_i(\gamma_i^t). \tag{10}$$

It is worth to recall that the network has a feedforward structure governed by the number of clouds formed by a granulation of the input data space. Because to each cloud corresponds an unineuron, the number of units in the hidden layer at step t, L^t, is the same as the number of clouds. Therefore, the structure of the network is adapted whenever the number of clouds are. Figure 1 shows the evolving neuro-fuzzy network, denoted by eHFN, structure.

The output layer is composed by m neural processing units with sigmoidal activation function whose jth output is

$$\hat{y}_j = f\left(\sum_{i=1}^{L^t} r_{ji} z_i\right), \tag{11}$$

where r_{ji} is the connection weight between the j-th output neuron and the ith unineuron of the hidden layer.

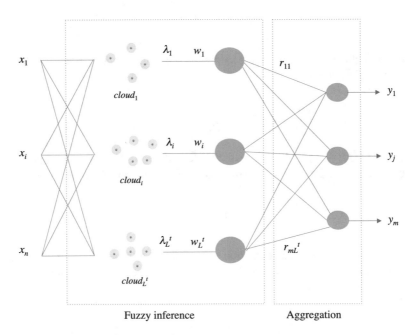

Fig. 1 Neuro-fuzzy network structure

Hidden layer weights w_i and unineurons identity values e_i are chosen randomly during initialization and kept fixed afterwards. This is the same approach as the one adopted in extreme learning machines [3]. The output layer weights $\mathbf{R}^t = [r_{jl}^t]$ are updated using the recursive least squares (RLS)

$$\mathbf{p}^t = \mathbf{Q}^{t-1}\mathbf{z}^t\{\psi + (\mathbf{z}^t)^T\mathbf{Q}^{t-1}(\mathbf{z}^t)\}^{-1}, \tag{12}$$

$$\mathbf{Q}^t = (\mathbf{I}_{L^t} - \mathbf{p}^t(\mathbf{z}^t)^T)\psi^{-1}\mathbf{Q}^{t-1}, \tag{13}$$

$$\mathbf{R}^t = \mathbf{R}^{t-1} + (\mathbf{p}^t)^T(\mathbf{f}^{-1}(\mathbf{y}^t) - \mathbf{R}^{t-1}\mathbf{z}^t), \tag{14}$$

where $\mathbf{f}^{-1}(\mathbf{y}^t) = \log(\mathbf{y}^t) - \log(1 - \mathbf{y}^t)$ and $\mathbf{y}^t = [y_1, \ldots, y_m]$. Matrix $\mathbf{Q}$ starts with $\omega\mathbf{I}_{L^t}$, $\omega = 1000$ and $\mathbf{I}_{L^t}$ is an $L^t \times L^t$ identity matrix. Figure 2 summarizes the uninorm-based evolving modeling method. Uninorm-based fuzzy neural networks are universal approximators [10]. The detailed neuro-fuzzy network adaptation procedure is as follows:

read first input data
initialize global density $\mu_G^1 \leftarrow \mathbf{x}^1$; $\Sigma_G^1 \leftarrow ||\mathbf{x}^1||^2$; $\Gamma^1 \leftarrow 1$
initialize first cloud $\mu_1^1 \leftarrow \mathbf{x}^1$; $M_1^1 \leftarrow 1$; $\Sigma_1^1 \leftarrow ||\mathbf{x}^1||^2$; $\gamma_1^1 \leftarrow 1$; $\mathbf{X}_1^f \leftarrow \mathbf{x}^1$; $\gamma_1^f \leftarrow \gamma_1^1$; $\Gamma_1^f \leftarrow \Gamma^1$
set initial network structure: $L^1 \leftarrow 1$;
while there are input data **do**
  read $\mathbf{x}^t$
  output $\hat{\mathbf{y}}^t$
  read $\mathbf{y}^t$
  update global density (8)
  compute local density for each cloud (6)
  if $\Gamma^t > \Gamma_i^f \; \forall i | i = [1, L^t]$ **then**
    criate new cloud i; $\mu_i^t \leftarrow \mathbf{x}^t$; $M_i^t \leftarrow 1$; $\Sigma_i^t \leftarrow ||\mathbf{x}^t||^2$; $\gamma_i^t \leftarrow 1$; $\mathbf{X}_i^f \leftarrow \mathbf{x}^t$; $\gamma_i^f \leftarrow \gamma_i^t$; $\Gamma_i^f \leftarrow \Gamma^t$; $L^t \leftarrow L^t + 1$
  else
    find $Cloud_i$ using (10)
    update $Cloud_i$ $M_i^t \leftarrow M_i^t + 1$;
    if $\gamma_i^t > \gamma_i^f$ **and** $\Gamma^t > \Gamma_i^f$ **then**
      update focal $\mathbf{X}_i^f \leftarrow \mathbf{x}^t$; $\gamma_i^f \leftarrow \gamma_i^t$; $\Gamma_i^f \leftarrow \Gamma^t$
    end if
  end if
  compute membership degree using (5)
  update weights $\mathbf{R}^t$ (12)–(14)
end while

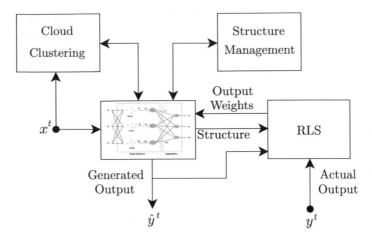

Fig. 2 Uninorm-based evolving neural network modeling

2.2 Granular Evolving Min-Max Modeling

Granular evolving min-max modeling, denoted by eFMM for short, uses hyperboxes
to granulate the input data space [14]. A hyperbox is a n-dimensional rectangle
defined by a minimum and a maximum points. It is formally defined as

$$B_i = \{X, W_i, V_i, b_i(x, W_i, V_i, c_i)\} \tag{15}$$

where B_i is the ith hyperbox, X is the n-dimensional input data space, b_i is the
membership function associated with B_i, $x \in X$ is an input data point, and c_i, W_i, V_i
are the center, maximum and minimum points, respectively.

The eFMM is a fuzzy rule-based modeling approach that assigns a functional
fuzzy rule to each hyperbox. In this chapter, eFMM uses affine functions in the
consequents of the fuzzy rules. Thus, the collection of hyperboxes define a fuzzy
model consisting of R functional fuzzy rules of the form:

$$R_i : \text{If } x \text{ is } B_i \text{ then } \hat{y}_i = \theta_{0i} + \sum_{j=1}^{n} \theta_{ji} x_j \tag{16}$$

where R_i is the ith local fuzzy rule, $\hat{y}_i$ is the rule output, and θ_{ji}, $i = 1, \ldots, R$, $j =
1, \ldots, n$ are the consequent parameters, the coefficients of the consequent function.
The set of rules produce the model output as a weighted combination of the R local
models. The contribution of each local model to the output is proportional to the
activation level of each rule. The eFMM model output is:

$$\hat{y}^t = \sum_{i=1}^{R} \psi_i \theta_i \bar{x}^t, \quad \psi_i = \frac{b_i}{\sum_{l=1}^{R} b_l} \tag{17}$$

where $\theta_i = [\theta_{0i}, \ldots, \theta_{ni}]$, $\bar{x}^t = [1, x_1, \ldots, x_n]^T$ is the augmented input data vector at t, and ψ_i is the normalized firing degree of ith rule.

Rules antecedents are the product of component-wise Gaussian membership functions:

$$b_i = \prod_{j=1}^{n} \bar{b}_{ji} \tag{18}$$

$$\bar{b}_{ji} = \exp\left(\frac{-(x_j - c_{ji})^2}{2\sigma_{ji}^2}\right) \tag{19}$$

$$\sigma_{ji} = \min(w_{ji} - c_{ji}, c_{ji} - v_{ji}) \tag{20}$$

where $\bar{b}_{ji}$ is the jth component of the i-th rule membership function, σ_{ji} is the width, x_j is the j-th component of the n-dimensional input data x, and w_{ji}, v_{ji}, c_{ji} are the components of the ith rule maximum, minimum, and center points, respectively. Figure 3 shows the local dispersion expressed by (20).

After constructing the antecedents of the rules, the eFMM estimates the local affine functions parameters θ_i using the recursive least squares (RLS) with forgetting factor. Similarly as in the eHFN, the forgetting factor emphasizes newer data points when estimating the local models to enhance adaptability in nonstationary environments.

When a data sample is input, eFMM decides if a new rule must be created, or if an existing one should be modified. When a rule is modified, the first step attempts to expand a hyperbox to encompass the new data point. This is done by displacing the maximum (W) and minimum (V) points to accommodate the data within the hyperbox borders using:

$$W_i^t = \max(W_i^{t-1}, x^t) \quad V_i^t = \min(V_i^{t-1}, x^t) \tag{21}$$

Fig. 3 Hyperbox B_i and its local dispersion

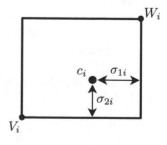

Expansion allows hyperboxes to incorporate knowledge about the underlying process that generates the data stream. There is, however, a need to control expansion to prevent a hyperbox from incorporating data samples with dissimilar properties. Expansion must be such that:

$$\max(w_{ji}^t, x_j^t) - \min(v_{ji}^t, x_j^t) \leq \delta_{ji} \qquad (22)$$

where w_{ji}^t and v_{ji}^t are the maximum and minimum points components at step t, and δ_{ji} is a parameter. When a rule B_i is created, the parameters δ are set as $\delta_{ji} = \delta_0$ for $j = 1, 2, \ldots, n$, where δ_0 is a user supplied value. Parameters δ are automatically adjusted during the adaptation process [14].

The eFMM has a rule base management mechanism to monitor redundant and outdated rules, to enhance robustness, and to prevent the rule base from growing boundlessly. A threshold $\epsilon \in [0.01, 0.5]$ decides if a rule should be deleted or not. Redundant rules are merged automatically. See [14] for details on merging of redundant rules. Figure 4 summarizes the granular evolving min-max modeling method. The detailed adaptation procedure is as follows:

> choose δ_0 and ϵ
> read first input data x^1
> set $W_i = V_i = c_i = x^1$
> **while** there are input data **do**
>   read x^t
>   output $\hat{y}^t$ using current rule base
>   **for** $i = 1, 2, \ldots, R$ **do**
>   compute ith rule firing degree using (18)
>   find the rule i with the highest firing degree and not tested for condition (22) yet;
>   test rule B_i for condition (22);
>   **if** condition (22) holds **then**
>   expand B_i to include x^t using (21).
>   update B_i consequent parameters θ_i;
>   update δ_i
>   **end if**
>   **end for**
>   **if** no rule meets condition (22) **then**
>   create a new rule
>   **end if**
>   check rule base for redundant, similar and outdated rules
> **end while**

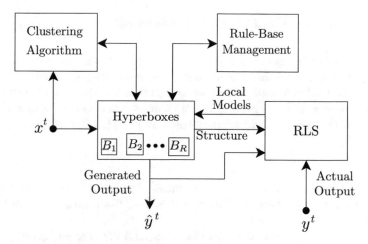

Fig. 4 Granular evolving min-max modeling

3 Results

This section summarizes the computational experiments done to evaluate and to compare the performance of the uninorm-based evolving neuro-fuzzy network and granular evolving min-max fuzzy modeling with state-of-the art evolving modeling approaches. The data concern two classic challenging benchmarks, the Box and Jenkins gas furnace identification problem [4], and the Mackey-Glass chaotic time series forecasting [11], respectively. These benchmarks have been extensively studied in the area of system identification, machine learning, and computational intelligence. Here the models assume the same inputs as those adopted in the literature. Performance is measured using the root mean square error (RMSE) (23)

$$\text{RMSE} = \sqrt{\frac{1}{T} \sum_{t=1}^{T} (\hat{y}^t - y^t)^2} \tag{23}$$

where $\hat{y}^t$ denotes the model output, and y^t denotes the actual output at step t, respectively. Simulations assume that data samples are input and processed as a stream. Performance evaluation computes the RMSE considering all data produced by the models up to the simulation horizon T. All data are normalized within $[0.1, 0.9]$ range to properly compare the prediction performances against those reported in [15]. The number of neurons in the hidden layer, or number of rules in the rule base, are those reached at the end of the simulation, that is, at step $t = T$.

3.1   Box and Jenkins Gas Furnace Identification

The Box and Jenkins gas furnace identification concerns a model whose input is the oxygen-methane gas flow rate u^t, and the output is the concentration of carbon dioxide y^t emitted as a result of the combustion process. The aim is to estimate the value of output y^t at step t using the immediate past value of the output y^{t-1}, and the value of the input at the fourth previous step u^{t-4} [17]. Therefore, the model is of the form:

$$\hat{y}^t = f_{BJ}(y^{t-1}, u^{t-4}) \tag{24}$$

The data set has 290 input-output samples. The RMSE was computed using normalized data. The forgetting factor is $\psi = 0.9$. The parameters of eFMM are $\epsilon = 0.01$, $\delta_0 = 0.9$, and $\gamma = 0.85$.

Table 1 shows the average number of rules, and the average RMSE produced by the granular evolving min-max modeling eFMM. The corresponding values for the uninorm-based evolving neuro-fuzzy modeling eHFN and remaining modeling methods are those reported in [15]. Originally, DENFIS was developed in [7], eTS in [2], ANYA in [17], and ELM in [5].

The eHFN, eFMM, and ANYA produce similar average RMSE values, but DENFIS performs slightly better. The best RMSE value that eHFN achieved was 0.0232 whereas DENFIS reached 0.0190 [15]. On the other hand, eFMM produces the most parsimonious model: while DENFIS develops 12 rules and eHFM averages 7 unineurons, eFMM develops at most two rules only. Fig. 5 illustrates the performance of the eFMM model estimates.

3.2   Mackey-Glass Time Series

Mackey-Glass is a synthetic time series data generated by the equation

$$\frac{dx}{dt} = \frac{0.2x^{t-\tau}}{1 + (x^{t-\tau})^{10}} - 0.1x^t \tag{25}$$

Table 1   Box and Jenkins modeling performance

Model	Number rules/neurons	RMSE
DENFIS	12	0.0190
eTS	12	0.0796
ANYA	7	0.0393
ELM	20	0.0931
eHFN	7	0.0398
eFMM	2	0.0358

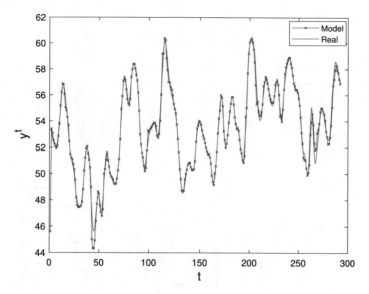

Fig. 5 Box and Jenkins eFMM model output

The time series value at each step t is obtained via numerical integration of (25) with step size 0.1, initial condition $x(0) = 1.2$, and $\tau = 17$. It is well known that the series is chaotic for this value of τ. The goal is to forecast the value of x^t at $t = 85$ steps ahead, that is, $\hat{y}^t = x^{t+85}$ using past values of the series. As indicated in the literature of time series forecasting and machine learning [2], to compare the performance with previous work reported often the forecast model should use the current, the sixth, the twelfth and the eighteenth previous values. Thus, similarly as in the literature, the model here takes the form

$$x^{t+85} = f_{MG}\left(x^t, x^{t-6}, x^{t-12}, x^{t-18}\right) \tag{26}$$

The values of the parameters of eFMM are $\epsilon = 0.5$, $\delta_0 = 0.9$ and $\gamma = 0.8$. Table 2 summarizes the results. Interestingly, in this case both, eHFN and eFMM, surpass the remaining approaches. Between eHFN and eFMM, clearly eFMM exhibits the best accuracy and produces the simplest model. Figure 6 shows the forecasting performance of eFMM.

It is interesting to note that eFMM produced considerably fewer rules when compared with the remaining methods in both experiments addressed in this chapter. One of the reasons for this is the choice of the value for δ_0. As condition (22) reveals, the parameter δ controls rule expansion, and it is made $\delta = \delta_0$ whenever a new rule is created. The user may choose any value in the [0, 1] range for δ_0; the higher the value, the fewer rules eFMM tends to create (because condition (22) will be more easily attained, which allows expansion of a rule instead of creation of a new rule). Both Box and Jenkins identification and Mackey-Glass forecast use the value $\delta_0 = 0.9$.

Table 2 Mackey-glass forecasting performance

Model	Number rules/neurons	RMSE
DENFIS	25	0.0730
eTS	24	0.0779
ANYA	13	0.1081
ELM	20	0.2096
eHFN	13	0.0435
eFMM	1	0.0167

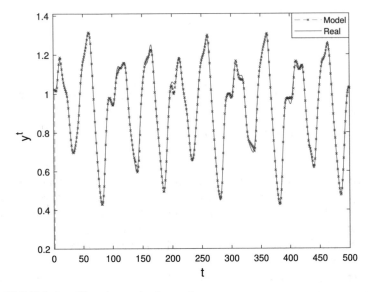

Fig. 6 eFMM Mackey-Glass time series forecasting

Further experiments were done using different values of δ_0, which could led the eFMM to create up to 10 rules (when $\delta_0 = 0.2$) in the Mackey-Glass experiment, but the forecast performance deteriorates considerably. Another reason for the small number of rules the eFMM develops was the choice of $\epsilon = 0.5$. This value allows the eFMM to delete outdated rules more often. Figure 7 shows the number of rules in a Mackey-Glass forecasting experiment. Notice that eFMM creates a rule at $t = 0$ and a new rule at four subsequent time steps during the simulation, but whenever a new rule was created, an existing one was rapidly deleted.

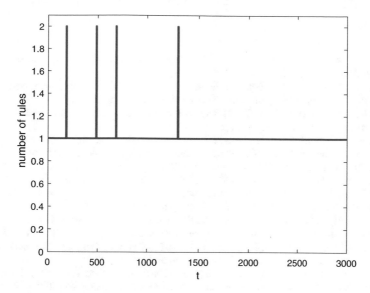

Fig. 7 Number of eFMM rules for Mackey-glass forecasting

4 Conclusion

This chapter addressed the uninorm-based evolving neuro-fuzzy network and the granular evolving min-max fuzzy modeling of nonlinear systems. The purpose was to benchmark their performance using the Box and Jenkins gas furnace identification, and Mackey-Glass chaotic time series forecasting. Both, the uninorm-based evolving neuro-fuzzy network and the evolving granular min-max fuzzy modeling have shown to be highly competitive with their contemporary neural and rule-based evolving counterparts. In particular, the evolving granular min-max fuzzy modeling has shown to be very powerful to model nonlinear systems with stream data because it produces models which satisfactorily trade-off accuracy and model complexity when compared with state of the art evolving models. This is a highly desirable feature especially in online and real time applications. Future work shall proceed to underpin the nature of the approximation and model complexity features of the evolving granular min-max fuzzy modeling and its variations.

Acknowledgements The authors are grateful to the Brazilian National Council for Scientific and Technological Development (CNPq) for its support via a fellowship, and grant 302467/2019-0, respectively. The helpful and constructive comments of the anonymous reviewers are also kindly acknowledged.

References

1. Angelov, P., Filev, D., & Kasabov, N. (2010). *Evolving intelligent systems: Methodology and applications*. Wiley IEEE Press.
2. Angelov, P., & Zhou, X. (2006). *Evolving fuzzy systems from data streams in real-time* (pp. 29–35). Intenational Symposium on Evolving Fuzzy Systems.
3. Bordignon, F., & Gomide, F. (2014). Uninorm based evolving neural networks and approximation capabilities. *Neurocomputing, 127*, 13–20.
4. Box, G., & Jenkins, G. (1990). *Time series analysis, forecasting, and control*. Holden Day.
5. Huang, G., Zhu, Q., & Siew, C. (2004). Extreme learning machine: a new learning scheme of feedforward neural net- works. *Proceedings of IJCNN, Budapest, Hungary, 2*, 985–990.
6. Kacprzyk, J., Romero, R., & Gomide, F. (1999). Involving objective and subjective aspects in multistage decision making and control under fuzziness: Dynamic programming and neural networks. *International Journal of Intelligent Systems, 14*(1), 79–104.
7. Kasabov, N. (2002). Denfis: Dynamic evolving neural-fuzzy inference system and its application for time-series prediction. *IEEE Transactions on Fuzzy Systems, 10*(2), 144–154.
8. Leite, D., Ballini, R., Costa, P., & Gomide, F. (2012). Evolving fuzzy granular modeling from nonstationary fuzzy data streams. *Evolving Systems, 3*(2), 65–79.
9. Lemos, A., Caminhas, W., & Gomide, F. (2013). Evolving intelligent systems: Methods, algorithms and applications. *Emerging Paradigms in Machine Learning, 490*, 117–159.
10. Lemos, A., Kreinovich, V., Caminhas, W., & Gomide, F. (2011). Universal approximation with uninorm-based fuzzy neural networks. In *2011 annual meeting of the North American fuzzy information processing society* (pp. 1–6)
11. Mackey, M., & Glass, L. (1977). Oscillation and chaos in physiological control systems. *Science, 197*, 287–289.
12. Pedrycz, W., & Gomide, F. (2007). *Fuzzy systems engineering: Towards human-centric computing*. Wiley IEEE Press.
13. Pedrycz, W., & Rocha, A. (1993). Fuzzy set-based models of neurons and knowledge-based networks. *IEEE Transactions on Fuzzy Systems, 1*(4), 254–266.
14. Porto, A., Gomide, F.: Granular evolving min-max fuzzy modeling. In *Proceedings of EUSFLAT 2019*. Czeck Republic
15. Rosa, R., Gomide, F., & Ballini, R. (2013) Evolving hybrid neural fuzzy network for system modeling and time series forecasting. In *12th international conference on machine learning and applications ICMLA* (pp. 378–383)
16. Skrjanc, I., Iglesias, J., Sanchis, A., Leite, D., Lughofer, E., & Gomide, F. (2019). Evolving fuzzy and neuro-fuzzy approaches in clustering, regression, identification, and classification: A survey. *Information Sciences, 490*, 344–368.
17. Yager, R., & Angelov, P. (2011). A new type of simplified fuzzy rule-based system. *International Journal of General Systems, 41*(2), 163–185.
18. Yager, R., & Rybalov, A. (1996). Uninorm aggregation operators. *Fuzzy Sets and Systems, 80*(1), 111–120.

State and Action Abstraction for Search and Reinforcement Learning Algorithms

Alexander Dockhorn📵 and Rudolf Kruse📵

Abstract Decision-making in large and dynamic environments has always been a challenge for AI agents. Given the multitude of available sensors in robotics and the rising complexity of simulated environments, agents have access to plenty of data but need to carefully focus their attention if they want to be successful. While action abstractions reduce the complexity by concentrating on a feasible subset of actions, state abstractions enable the agent to better transfer its knowledge from similar situations. In this article, we want to identify the different techniques for learning and using state and action abstractions and compare their effect on an agent's training and its resulting behavior.

Keywords Action abstraction · State abstraction · Reinforcement learning · Search-based algorithms · Computational intelligence in games

1 Introduction

Decision-making plays an integral part in artificial intelligence (AI) agents. Being confronted with a decision-making problem, possible decisions must be identified first. After that, an agent needs to construct possible action plans and gather information on them. As a result, the agent can weigh the evidence to choose among the available options. After having executed the selected actions, the agent will be able to review its decision based on the observed consequences.

In a reinforcement learning context, we are confronted with a sequential decision-making problem. Here, the agent stays in interaction with its environment and continuously selects actions to maximize its reward over time. Over the course of multiple

A. Dockhorn (✉)
Gottfried Wilhelm Leibniz University, 30167 Hannover, Germany
e-mail: dockhorn@tnt.uni-hannover.de

R. Kruse
Otto Von Guericke University, 39106 Magdeburg, Germany
e-mail: rudolf.kruse@ovgu.de

Y. P. Kondratenko et al. (eds.), *Artificial Intelligence in Control and Decision-making Systems*, Studies in Computational Intelligence 1087,
https://doi.org/10.1007/978-3-031-25759-9_9

181

interactions, the agents gather more information and may revise their previous decisions in later time steps. Furthermore, the learning agent might have to face a large variety of situations and have many actions to choose from for handling any given situation. Thus, effectively re-using the gathered evidence from previous time-steps is key to optimizing the agent's performance over time.

Both, search and reinforcement learning algorithms, can struggle to perform well at training or run-time in either of the two settings. A popular method for improving the agent's performance is to implement abstractions [80] of the problem at hand. We naturally do this, when designing AI agents with a certain task in mind, by providing it with the information sources and actions we deem to be relevant, thereby restricting the agent to the envisioned setting. Thereby, abstractions help the agent to focus its attention and narrow down the number of relevant alternatives to choose from.

Learning abstractions from experience becomes especially relevant when aiming for more general AI agents [50]. Many AI agents are presented with exactly the information they need to solve the given task. However, detecting the relevant information on your own and abstracting relevant concepts is a non-trivial task but required for becoming proficient in solving more general problems [36, 91].

This survey shall provide an overview of common abstraction methods for state and action spaces of Markov Decision Processes (MDP) [32]. Both, reinforcement learning and search-based algorithms, can make use of such abstractions, to speed up the decision process and thereby often improve the agent's performance. In Sect. 2 we will summarize preliminaries on MDPs, search, and reinforcement learning algorithms. Section 3 covers methods for state abstraction, whereas Sect. 4 will focus on action abstractions. The work will be concluded in Sect. 5 followed by propositions for future research and applications.

2   Preliminaries

The following sections will summarize preliminaries on MDPs (Sect. 2.1), reinforcement learning (Sect. 2.2), and search algorithms (Sect. 2.3). On this basis, we will summarize the research on state and action abstraction in later Sections.

2.1   Markov Decision Processes

To discuss abstractions in more detail, we first review Markov Decision Processes [32] as a general formalization of a reinforcement learning problem. An MDP is described by a tuple:

$$M = (S, A, T, R) \tag{1}$$

where S is the state space, A the action set, T is the state transition function, and R is the reward function. The state space S consists of all the environment states s that the agent can observe. Action set A consists of all actions a the agent can use to interact with its environment. The state transition function $T(s, a, s')$ represents a probability function, mapping the current state and the agent's action to the probability of the next state s'. Finally, the reward function $R(s, a, s)$' provides the agent with a numerical reward when transitioning from state s to s' via action a.

In the context of an MDP, the agent is in constant interaction with its environment. Every time-step t, the agent observes the state of the environment and chooses an action to execute. As a reaction, the environment will change its state and provide the agent with a reward. The agent's goal is to choose actions such that it maximizes the received reward over time.

In terms of the applications, we will discuss in this work, we want to highlight combinatorial state and action spaces. Combinatorial states are made of multiple sensory inputs $s = (s_1, \ldots, s_n)$. Sensors may be part of the same modality (e.g., pixels of a camera-image, or receive data from multiple modalities (e.g., audio, video, tactile sensors, etc.). Making efficient use of multiple sensors and multiple modalities may require extensive pre-processing [7] or fusion algorithms [9]. A similar principle applies to combinatorial action spaces in which each action consists of multiple action components $a = (a_1, \ldots, a_m)$. Those can be commonly observed in robotics and multi-unit games in which multiple independent entities need to be controlled during every tick of the environment.

2.2   Reinforcement Learning

The theory of reinforcement learning algorithms finds its origin in psychology. Conditioning, a form of learning stimulus–response associations, describes the coupling of a neutral stimulus with an unconditional stimulus. Here, the neutral stimulus becomes a conditioned stimulus and triggers a comparable response in the subject (cf. [31]). Reinforcement learning algorithms, adopt this principle to train an agent in arbitrary tasks through continuous feedback. Using a multi-armed bandit, the associative version of the learning problem can be described [35]. Here, the agent learns the average value when using an action without changing the state of the underlying system. If the state is mutable, modeling by Markov decision processes is resorted to [6]. In this, the agent also receives feedback in the form of a numerical reward while at the same time changing the state of its environment.

Here, the learning process is fundamentally different from supervised learning. While the agent is provided with a data set of stimulus–response observations, in reinforcement learning the agent independently explores its environment. Hereby, a balance between the exploration of unknown actions and the exploitation of promising actions must be found continuously. Learning procedures of reinforcement learning can thus be divided into on- and off-policy procedures. While on-policy methods always select the currently best-known action, off-policy methods allow the exploration of actions that have not yet been considered optimal.

Classical methods, also known as tabular methods, such as Temporal Difference Learning [87] or the Monte Carlo Method [79] determine the expected reward of each action or state-action pair based on repeated interaction with the system under observation. If the underlying system is fully known, these values can be determined by Dynamic Programming [37]. In these classical methods, decisions are made based on the learned expected reward. Here, this value must be learned for each action (non-associative), or each state-action pair (associative). This property makes classical learning methods unusable, especially for large state and action spaces.

The use of neural networks in the context of deep reinforcement learning [58], has brought fundamental changes for tackling the learning problem. While before, the observation of the agent and thus the possible state space had to be designed by hand, the use of neural networks allows the internal reshaping of unstructured state observations. Here, training the neural network creates an approximation of the reward signal or an action probability to maximize the reward signal. In contrast to tabular methods, the network-internal transformation of the input signal can also approximate the reward of previously unobserved states. While often done implicitly, later sections will show how such a network behavior can be actively approached to learn state and action abstractions.

2.3   Search/Optimization-Based Decision-Making

In contrast to reinforcement learning, search and optimization-based decision-making tries to determine the best action at run-time. Therefore, the agent usually relies on a forward model to determine the outcome of planned actions and optimize an action sequence. In cases where the forward model cannot be accessed, it can be approximated from observation [15–17].

Exhaustive search methods try to simulate all possible action sequences in a structured manner. Classic tree-search algorithms build a tree starting from the current state as the root node. Each simulated action is represented as an edge, connecting the state in which the action has been initialized with the resulting next state. Methods such as breadth-first search and depth-first search [20] do so without any optimization. More efficient algorithms, such as the minimax algorithm [62] and alpha–beta pruning [80] skip infeasible subtrees and have been successfully applied to games with large state spaces (e.g., chess [21]). Nevertheless, they have shown to be infeasible for games such as Go which features a state of approximately 10^{170} states [89].

Due to the large branching factor and the sheer number of possible states, exhaustive search often takes too much time to compute a result. Instead, heuristic search methods can be used to approximate the optimal action given a sample of the gam tree.

Methods such as flat Monte Carlo [49] and Monte Carlo Tree Search (MCTS) [12] achieve this by simulating the outcome of multiple action sequences. To focus the search, these methods balance the exploration and exploitation of available options [49], whereas during exploration we aim to find good actions that have not been analyzed yet and during exploitation, we further analyze parts of the game tree that has shown good results so far.

In contrast to tree-search methods, optimization-based methods try to directly optimize an action sequence. Algorithms such as the Rolling Horizon Evolutionary Algorithm (RHEA) [33, 34] do so, by using an evolutionary algorithm to improve multiple candidate action sequences throughout several generations. Once the algorithm terminates, the first action of the best sequence will be applied.

In contrast to reinforcement learning, search- and optimization-based methods can be used without any prior training since necessary evaluations are done at run-time. This made them especially relevant in more general domains, such as general game-playing [57, 76] and general strategy game-playing [27, 74], in which the agent needs to act in a previously unknown environment. Nevertheless, they struggle with optimizing long action sequences and high branching factors. For this reason, they are commonly applied in environments with simple state and action spaces. The performance for more complex scenarios can be improved by methods of state and action abstraction, which will be detailed in the following sections.

3 State Abstractions

In a state abstraction, we intend to reduce the complexity of the decision-making task, by mapping from the history of previously visited states H to a state in the compact state space S' [60]. In early papers, the abstraction has solely been based on a mapping of the current state $s \in S$ to another state in the compact state space $s' \in S'$, whereas $S' \ll S$. Those methods have been designed for MDPs of the first degree, which makes the mapping independent of previously visited states. In either way, the decision space become more compact, and the agent's effort is reduced. For simplicity, we will constrain the following discussion on MDPs of the first degree as well. The following categories of state abstractions have been proposed by Konidaris [50] and served as a guideline to structure this review. We extend his work with an updated summary of the state-of-the-art and an overview of state abstraction in highly structured state spaces as they can be commonly observed in path planning. An overview of discussed methods is presented in Fig. 1, whereas each type of method will be discussed in the following sections, respectively.

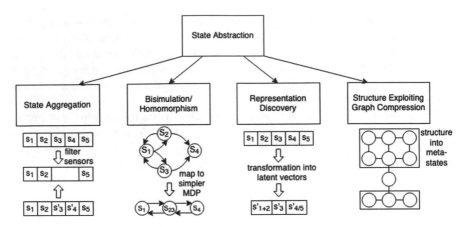

Fig. 1 Overview of the types of state abstraction algorithms

3.1 Bisimulation Approaches

Bisimulation approaches aim to construct a minimized MDP with similar properties as the original one. Such model minimization approaches are often variations
of automaton minimization algorithms. The state space abstraction by Boutilier and
Dearden [11] solves a factored MDP by constructing a minimized version based on
the impact of propositions on the utility. Due to the existence of multiple factor-
izations per MDP, the proposed approach is non-deterministic and can result in
abstractions of varying quality and size. In any case, the minimization is adequate
and computable in polynomial time. The approach by Dean and Givan [22] also
uses a factored representation to find the coarsest homogenous refinement of any
partition of the MDP's state space. The resulting reduced MDP is minimal in a well-
defined sense and can be used to find an optimal policy in the original MDP. Sadly,
this process is NP-hard. Simplifications of this process can run in polynomial time
but cannot produce an optimal minimization. In contrast to exact abstractions, an
approximate abstraction can often be achieved in polynomial time [23, 30]. Thereby,
algorithms balance the coarseness of the abstraction with the quality of the obtained
solution from the abstract MDP.

The high complexity of bisimulation approaches often makes them infeasible
for very large state spaces. Approximate solutions can reduce the computational
complexity but may cause the agent to learn an invalid policy. Often the approximation of the MDP can be iteratively improved at the cost of additional computations.
The problem is typically studied in the context of model-based reinforcement learning
[40], in which algorithms aim for a good balance between building an internal model
and improving their policy. Another solution to the problem is the search for local
abstractions of the MDP. In contrast, to the approaches above, local abstractions only
model a subgraph of the original MDP. Due to the reduced size of the abstracted graph,
such abstractions can be found much faster. Jiang et al. have used this approach in

the context of a UCT (MCTS using upper confidence bounds as a selection criterion) to efficiently use the data generated from previously simulated trajectories. Given a small budget of simulations, the local abstractions have been shown to improve the performance of UCT.

3.2 State Aggregation Approaches

In homomorphic state abstraction, we construct a more compact state space S' and a mapping from S to S'. The simplest way to do so is by discarding variables from our state observation [38]. Thereby, we reduce the dimensionality of the state space and often considerably shrink its size. Identifying suitable variables to discard can either be done in communication with domain experts or as a result of feature analysis [37–39]. For example, the approach by Jong and Stone [43] measures the relevance of a variable in terms of its impact on the optimal policy. In case a variable does not significantly change the decisions suggested by the policy, we can safely remove it from our representation. Interesting about this method is that it preserves convergence to the optimal policy, which cannot be ensured in expert-driven abstractions. In contrast to filtering approaches, feature selection can also be done in a bottom-up manner, i.e., starting with no features and adding the ones that improve the model quality the most [63].

Instead of ignoring/selecting state variables to project a state into a subspace, we can also group states into clusters [41–43]. Given the agent's previous trajectories, we can try to identify states with a similar policy, transition, or reward function to create a more compact clustered representation. The aggregation of states into a soft/fuzzy partition can result in more flexibility and model the agent's confidence in the current clustering [84]. This approach will become relevant once again in Sect. 4.3.

Whereas previous methods have mostly focused on reinforcement learning, state aggregation can also be found in search-based algorithms. Most commonly, stochastic search algorithms aim to group simulations based on their trajectories to evaluate visited states more efficiently [29, 39]. Used approaches are similar to the ones presented above but the feature selection is commonly applied in between iterations of the search, may change over time, and therefore adapt to the current local context of the agent's state space.

3.3 Representation Discovery Approaches

While the previous section discussed methods for reducing the complexity of an existing state space into a representative subspace, representation discovery approaches are used to intrinsically learn a state representation based on the original input. Also homomorphic, representation discovery methods often construct abstracted state spaces that bear not much resemblance but share properties of the original state space. In recent work, deep neural networks, specifically, autoencoders [54], have been used to process raw sensory data. Thereby, the network is learning a compact feature vector in its hidden layers that are subsequently used for decision-making. While being hard to interpret for the human observer in the general domain, analysis of convolutional neural networks for image classification has shown that hidden layers represent increasingly complex features of learned image classes [69]. Resulting latent spaces can often be used as highly compact state abstractions.

A special form of convolution is implemented by graph convolutional networks [48]. Those directly operate on graph structures and can therefore be used to create latent representations of MDPs. A study by Jiang et al. [41] has shown their efficient use in reinforcement learning. Finally, a work by Liu et al. [59] has used meta descriptive statistics (MDS) as supplementary state representations. They compared the MDS approach with abstractions obtained from autoencoders and graph convolutional networks as well as combinations of all three methods. In their evaluation, graph convolutional networks resulted in the best performance among the single methods, whereas a combination of MDS and autoencoders performed best overall.

3.4 Graph Compression Using High-Level Knowledge

An interesting application and challenging research domain of state abstraction algorithms is pathfinding. Here, the agent is tasked to explore an environment and find a (near-)optimal path to a given objective. Thereby, the agent either uses a model of its surroundings or needs to create one while exploring the environment. In either case, the time for planning shall be kept as low as possible while also minimizing the length of the resulting path. Sadly, optimizing for one often worsens the performance of the other since abstractions can reduce the planning time, but in many cases only result in near-optimal solutions [15].

Generally, heuristics can be used to guide the agent's search process [80]. In case the same environment is used for multiple pathfinding queries, it might be useful to preprocess the graph model to solve the queries in a more efficient abstract model of the environment. Many algorithms do so by exploiting the highly regular structure of grid-based path-finding problems. In contrast to arbitrary MDPs, the agent's actions and their meaning remain the same for every state. This allows the usage of simple abstraction heuristics, such as the one used in HPA* [10]. Here, the graph is first decomposed into a map of disjoint square sectors. To plan a path from the current position to the target, we first plan a path from one square sector to another and later refine the path in the original grid map. A multi-layered abstraction and refinement process has been implemented in Partial Refinement A* (PRA*) [85]. Sturtevant later proposed a combination of HPA* and PRA* to decrease the agent's memory consumption [86]. The efficiency of real-time heuristic search has been further improved by interweaving it with automatic state abstraction. The work by Bulitko et al. [14] has shown huge performance gains over non-abstracting alternatives.

4 Action Abstractions

In terms of large action spaces, algorithms for decision-making struggle with identifying the best actions during the exploration of a vast number of alternatives. Like large state spaces, large action spaces lengthen the training time of reinforcement learning algorithms and the time to explore the game tree in search-based algorithms. The idea of action abstraction is to either narrow the search on a subset of promising candidate actions (Sect. 4.1) or identify valuable action sequences to be used (Sect. 4.2). Each of the two approaches (cf. Figure 2) will be discussed in the following sections, respectively.

4.1 Script-Based Action Abstraction

Script-based action abstraction reduces the size of the action space by removing inferior actions or focusing the search on a selection of good candidates. While this approach loses granularity and is dependent on the accuracy of the candidate set, it can drastically speed up the agent's search process. A simple concept to retrieve candidate actions is the use of scripts. A script is a subroutine that returns a candidate action for any current state of the environment.

$$Script : S \to A \tag{2}$$

Thereby, a single script is similar to a deterministic policy. In the context of strategy game AI, scripts have been implemented by a set of rules that focus on a certain strategy, e.g., attacking the closest opponent in the range [28]. Such games often involve multiple entities that can be independently controlled by the agent by selecting an entity's action for each of them. An increasing number of such entities (e.g., units, buildings, etc.) results in an exponentially growing combinatorial action space [19, 71]. Due to the independence of multiple entities, scripts are often

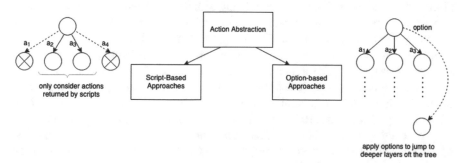

Fig. 2 Overview of the Types of Action Abstraction Algorithms

constrained to a single entity and its respective actions in the context of multi-unit strategy games. Similar implementations could be thought of for robots for which the agent controls multiple joints at the same time. In both cases, the output of a script can be restricted to a single component under the agent's control, and therefore, can also be defined by:

$$EntityScript : S \times Entity \rightarrow A_{Entity} \tag{3}$$

The agent's response is then determined by assigning one script to each entity and building the combinatorial action out of the responses by all the scripts.

4.2 Portfolio-Based Search Algorithms

While the use of a single script with high performance can already yield a successful agent, the combination of multiple scripts has been shown to improve the overall performance of the agent [28, 74]. Let a portfolio be given by a set of scripts, whereas the abstracted action space is the union of all script responses for the current state of the environment. Portfolio-based algorithms vary in the way they encode and optimize the script selection/assignment. For instance, Portfolio Greedy Search (PGS) [18] has been designed for combinatorial action spaces and optimizes the script assignment to each independent entity. For this purpose, a hill-climbing procedure is used to search for the best combination. In presence of an opponent, the opponent's and the player's script choices are updated iteratively. Moraes et al. [68] have shown that the devised hill-climbing procedure can suffer from non-convergence. To overcome this issue, they proposed a nested greedy search (NGS) in which the best opponent's response is computed in each iteration. Nevertheless, this comes at the drawback of being unable to evaluate many actions due to the increased time complexity.

Other search and optimization algorithms have been used for improving the agent's performance. Portfolio Online Evolution (POE) [90] replaces the hill-climbing procedure with an evolutionary algorithm [53]. Thereby, a candidate solution encodes the script assignment for each controlled entity. Its fitness is determined by simulating the outcome of continuously retrieving actions from a given script assignment. Further on, mutation of a single candidate solution (e.g., randomly replacing a script assignment) and crossover on multiple of these candidate solutions (e.g., uniform crossover of script assignment) is used to search for candidate solutions of higher fitness. In a similar manner to POE, the Portfolio Rolling Horizon Evolutionary Algorithm (PRHEA) [28] uses evolution to optimize script assignments of fixed length (so-called horizon). Each turn, the agent applies the first script of the best candidate solution before reusing the population to initialize the search for the next iteration. While this limits the search depth of the agent, it has shown efficiency in optimizing script assignments in multiple strategy games.

Another layer of abstraction is introduced by Stratified Strategy Selection [56] and Cluster-based UCT [44], which first groups the entities into types or clusters and then assigns one script to each type. Given contextual knowledge of the environment, such layered abstractions can improve the efficiency of the agent's search. Some environments may not allow the enforcement of a type system on all entities or define a suitable abstraction for some of the entities under the agent's control. In these cases, an asymmetric action abstraction allows controlling entities at different layers of abstraction. While some may be safely abstracted or even grouped into types, others will be restricted to using their original action space. Moraes and Lelis [67] have proposed the methods Greedy Alpha–Beta Search and Stratified Alpha–Beta Search to search in such asymmetric action abstractions. Additionally, they have shown that the optimal strategy derived using an asymmetric action abstraction is at least as good as using a uniform abstraction.

Another view on the search for combinatorial actions is offered by Naïve MCTS [71], which solves it by a combinatorial multi-armed bandit. Hereby, a combination of MCTS and naïve sampling, which considers each entity's action as an independent contribution to the combinatorial action's reward, is used to optimize the actions for each entity. While the naïve assumption might converge to a local optimum, the guided naïve MCTS proposed by Yang and Ontanon [92] only uses the scripts as recommenders during exploration and keeps the original action space in the tree.

Since previously discussed algorithms mostly differ in the way they structure and sample the script space, a unification of PGS, NGS, POE, SSS, and Naïve MCTS has been proposed by Lelis [55]. General Combinatorial Search for Exponential Action Spaces (GEX) generalizes these algorithms by splitting the search process into subprocesses consisting of select and expand, macro-arm sampling, evaluation, and value propagation. Thereby, more instances of portfolio-based search algorithms can be easily generated by choosing implementations of each subprocess.

The usage of deterministic scripts or their portfolios can result in repetitive behavior. To avoid this, the algorithm puppet search [4, 5] introduced scripts with choice points, which query the state's properties to select an action. A search or optimization process can further be used to modify the thresholds used in these choice points to modify the agent's behavior. This can result in a more reactive game-play experience while adjusting the number of choice points can be an efficient way in tuning the strength of the agent. Other strategies for diverse game-play include the optimization of the portfolio composition [28] as well as the adjustment of search parameters [75]. The simplicity of said approaches makes them approachable to game designers who want to achieve more variety in the agent's gameplay.

4.3   Constructing Higher Level Actions

Another type of action abstraction can be found in the options framework [88]. Options are a temporal action abstraction of an MDP and represent closed-loop policies for taking actions over a period of time. They can be considered high-level actions that involve a whole sequence of actions. Each option o is given by a tuple

$$o = (I_o, \pi_o, \beta_o) \tag{4}$$

where $I_o \subseteq S$ is the initiation set that describes potential starting states in which the option can be applied, $\pi_o : S \rightarrow A$ is the option policy used for retrieving actions while the option is active, and $\beta_o : S \rightarrow [0, 1]$ is the terminal condition which is returning 1 in case the option has to be stopped after reaching a state s.

During search or execution, agents can choose to apply available options as they would apply an action. In contrast to script-based approaches, options increase the branching factor of the search tree by adding additional choices. At the same time, they represent a shortcut to deeper layers of the tree, since if an option is once chosen will remain to be active. Nevertheless, it is important to only add options if they have the potential to reduce the agent's training or search time. Otherwise, poorly chosen options have been shown to slow down the learning or search process [42].

A simple option design that has successfully been used in search-based algorithms are macro actions [77], e.g., repeating a chosen action for multiple time-steps [73, 78]. Similarly, more intricate sub-routines or repeating sub-policies can be designed or extracted from successful runs/play-traces [26]. Alternatively, options can be learned e.g., by subgoal identification, whereas a subgoal decomposes the learning problem into two smaller problems [42]. Ways for finding subgoals include the identification of high reward, or high novelty states [64, 82], bottlenecks in the state space, or by taking other graph connectivity measures into account [79–81].

Continuous domains, in which the agent is unlikely to be in the same state multiple times, pose special challenges for learning options. Due to the missing repetition of the task and a single never-ending episode, the aim is to learn skills that can be used later. Given a formal problem description, this can be done using problem decomposition [46, 81], in which a task is simplified into multiple sub-tasks.

In case this is not possible, option learning needs to generalize the initial and target states of an option to a set of states which are similar to each other [51]. The similarity of states is exploited in their skill chaining algorithm. Once the goal is reached, the agent starts with searching for an option that starts in the neighborhood of the target state and therefore has a high likelihood of reaching said state. Next, we search for an option that has a high likelihood to reach a state of the previous option's initiation set. The process is repeated and thereby multiple options are chained into a high-level skill. Such skills are especially interesting for lifelong reinforcement learning tasks, in which the agent faces a set of related tasks in environments that are similar to each other [1]. Here, it is even more important to identify options that transfer well between the different tasks and do not bloat up the search space. A work by Brunskill and Li [13] shows that transferrable options can be learned and these are ϵ-optimal for all the tasks the agent may encounter.

5 Conclusion and Future Directions

The papers presented in this survey have shown that abstractions can be used to reduce the complexity of many tasks. Both state and action abstractions not just help to reduce the agent's training time, but also improve the agent's performance and can make it more robust to changes in its environment. At the same time, it is important to study the theoretic bounds of these abstractions to ensure near-optimal behavior. While many papers have focused on a formal analysis of the environment or the a priori optimization of the abstraction, more work needs to be spent on identifying and exploiting suitable abstractions at run-time. Especially, highly dynamic environments may require the agent to add or drop abstractions depending on its current situation.

As a result, the agent needs to act under uncertainty about the accuracy and suitability of its learned abstractions. To our surprise, this uncertainty and its impact on the decision-making problem has not been the focus of attention yet. We believe that fuzzy decision-making techniques [8] would be key to finding a suitable abstraction or forming a decision by consensus over multiple imperfect abstractions [45]. Additionally, fuzzy models may play an integral part in transferring learned abstractions to new environments, effectively allowing to reuse the model for similar MDPs while keeping track of its reliability [70]. Finally, the rising complexity of tackled decision-making problems results in an increased demand for explainable models [2] that either support the user to understand the learned abstraction [72] or the decisions made by the agent [16].

References

1. Abel, D., Jinnai, Y., Guo, Y., Konidaris, G., & Littman, M. L. (2018). Policy and value transfer in lifelong reinforcement learning. In *Proceedings of the 35th International Conference on Machine Learning* (pp. 1–10). PMLR, Stockholm.
2. Alonso Moral, J. M., Castiello, C., Magdalena, L., & Mencar, C. (2021). *Explainable fuzzy systems*. Springer International Publishing, Cham. https://doi.org/10.1007/978-3-030-71098-9.
3. Apeldoorn, D., & Dockhorn, A. (2020). Exception-tolerant hierarchical knowledge bases for forward model learning. *IEEE Transactions on Games* 1–14. https://doi.org/10.1109/TG.2020.3008002.
4. Barriga, N. A., Stanescu, M., & Buro, M. (2015). Puppet search: Enhancing scripted behavior by look-ahead search with applications to real-time strategy games. In *Proceedings of the AAAI Conference on Artificial Intelligence and Interactive Digital Entertainment* (pp. 9–15).
5. Barriga, N. A., Stanescu, M., & Buro, M. (2018). Game tree search based on nondeterministic action scripts in real-time strategy games. *IEEE Transactions on Games.*, *10*, 69–77. https://doi.org/10.1109/TCIAIG.2017.2717902
6. Bellman, R. (1957). A Markovian decision process. *Indiana University Mathematics Journal.*, *6*, 679–684. https://doi.org/10.1512/iumj.1957.6.56038

7. Berthold, M. R., Borgelt, C., Höppner, F., & Klawonn, F. (2010). *Guide to intelligent data analysis*. Springer. https://doi.org/10.1007/978-1-84882-260-3.
8. Blanco-Mesa, F., Merigó, J. M., & GilLafuente, A. M. (2017). Fuzzy decision making: A bibliometric-based review. *Journal of Intelligent and Fuzzy Systems, 32*, 2033–2050. https://doi.org/10.3233/JIFS-161640
9. Bloch, I., Hunter, A., Appriou, A., Ayoun, A., Benferhat, S., Besnard, P., Cholvy, L., Cooke, R., Cuppens, F., Dubois, D., Fargier, H., Grabisch, M., Kruse, R., Lang, J., Moral, S., Prade, H., Saffiotti, A., Smets, P., & Sossai, C. (2001). Fusion: General concepts and characteristics. *International Journal of Intelligent Systems., 16*, 1107–1134. https://doi.org/10.1002/int.1052
10. Botea, A., Martin, M., Schaeffer, J., & Tg, C. (2004). Near optimal hierarchical path-finding. *Journal of Game Development., 1*, 1–30.
11. Boutilier, C., & Dearden, R. (1994). Using abstractions for decision-theoretic planning with time constraints. *Proceedings of the National Conference on Artificial Intelligence., 2*, 1016–1022.
12. Browne, C. B., Powley, E., Whitehouse, D., Lucas, S. M., Cowling, P. I., Rohlfshagen, P., Tavener, S., Perez, D., Samothrakis, S., & Colton, S. (2012). A survey of Monte Carlo tree search methods. *IEEE Transactions on Computational Intelligence and AI in Games., 4*, 1–43. https://doi.org/10.1109/TCIAIG.2012.2186810
13. Brunskill, E., & Li, L. (2014). PAC-inspired option discovery in lifelong reinforcement learning. In: *31st International Conference on Machine Learning, ICML 2014* (Vol. 2, pp. 1599–1610).
14. Bulitko, V., Sturtevant, N., & Kazakevich, M. (2005). Speeding up learning in reel-time search via automatic state abstraction. *Proceedings of the National Conference on Artificial Intelligence, 3*, 1349–1354.
15. Bulitko, V., Sturtevant, N., Lu, J., & Yau, T. (2007). Graph abstraction in real-time heuristic search. *Journal of Artificial Intelligence Research., 30*, 51–100. https://doi.org/10.1613/jair.2293
16. Chakraborti, T., Sreedharan, S., & Kambhampati, S. (2020). The emerging landscape of explainable automated planning & decision making. In *IJCAI International Joint Conference on Artificial Intelligence. 2021-January* (pp. 4803–4811). https://doi.org/10.24963/ijcai.2020/669.
17. Cheng, Z., & Ray, L. E. (2014). State abstraction in reinforcement learning by eliminating useless dimensions. In *Proceedings - 2014 13th International Conference on Machine Learning and Applications, ICMLA 2014* (pp. 105–110). https://doi.org/10.1109/ICMLA.2014.22.
18. Churchill, D., & Buro, M. (2013). Portfolio greedy search and simulation for large-scale combat in starcraft. In: *IEEE Conference on Computatonal Intelligence and Games, CIG*. https://doi.org/10.1109/CIG.2013.6633643.
19. Churchill, D., & Buro, M. (2015). Hierarchical portfolio search: Prismata's robust AI architecture for games with large search Spaces. In: *Proceedings of the 11th AAAI Conference on Artificial Intelligence and Interactive Digital Entertainment, AIIDE 2015*. Retrieved from 16–22 November 2015.
20. Cormen, T. H., Leiserson, C. E., Rivest, R. L., & Stein, C. (2009). *Introduction to algorithms*. The MIT Press.
21. Costalba, M., Kiiski, J., Linscott, G., & Romstad, T. Stockfish chess engine. http://www.stockfishchess.org/.
22. Dean, T., & Givan, R. (1997) Model minimization in Markov decision processes. In *Proceedings of the Fourteenth National Conference on Artificial Intelligence* (pp. 106–111).
23. Dean, T. L., Givan, R., & Leach, S. (2013). Model reduction techniques for computing approximately optimal solutions for Markov decision processes 124–131.
24. Dockhorn, A. (2020). Prediction-based search for autonomous game-playing.
25. Dockhorn, A., & Apeldoorn, D. (2018) Forward model approximation for general video game learning. In *Proceedings of the 2018 IEEE Conference on Computational Intelligence and Games (CIG'18)* (pp. 425–432). IEEE. https://doi.org/10.1109/CIG.2018.8490411.
26. Dockhorn, A., Doell, C., Hewelt, M., & Kruse, R. (2017). A decision heuristic for Monte Carlo tree search doppelkopf agents. In *2017 IEEE Symposium Series on Computational Intelligence (SSCI)* (pp. 1–8). IEEE. https://doi.org/10.1109/SSCI.2017.8285181.

27. Dockhorn, A., Grueso, J. H., Jeurissen, D., & Perez-Liebana, D. (2020) "Stratega": A general strategy games framework. In J. C. Osborn (Ed.) *Joint Proceedings of the AIIDE 2020 Workshops co-located with 16th AAAI Conference on Artificial Intelligence and Interactive Digital Entertainment (AIIDE 2020)* (pp. 1–7). CEUR Workshop Proceedings, Worcester (2020)
28. Dockhorn, A., Hurtado-Grueso, J., Jeurissen, D., Xu, L., & Perez-Liebana, D. (2021). Portfolio search and optimization for general strategy game-playing. In *2021 IEEE Congress on Evolutionary Computation (CEC)* (pp. 2085–2092). IEEE. https://doi.org/10.1109/CEC45853.2021.9504824.
29. Dockhorn, A., Hurtado-Grueso, J., Jeurissen, D., Xu, L., & Perez-Liebana, D. (2021). Game state and action abstracting monte carlo tree search for general strategy game-playing. In *2021 IEEE Conference on Games (CoG)* (pp. 1–8). IEEE. https://doi.org/10.1109/CoG52621.2021.9619029.
30. Even-Dar, E., & Mansour, Y. (2003). Approximate equivalence of markov decision processes. *Lecture Notes in Artificial Intelligence (Subseries of Lecture Notes in Computer Science)., 2777*, 581–594. https://doi.org/10.1007/978-3-540-45167-9_42
31. Fearing, F., Pavlov, I. P., & Anrep, G. V. (1929). Conditioned reflexes. *An Investigation of the Physiological Activity of the Cerebral Cortex.* https://doi.org/10.2307/1134737
32. Feinberg, E. A., & Shwartz, A. (Eds.) *Handbook of Markov decision processes.* Springer US (2002). https://doi.org/10.1007/978-1-4615-0805-2.
33. Gaina, R. D., Liu, J., Lucas, S. M., & Pérez-Liébana, D. (2017). Analysis of vanilla rolling horizon evolution parameters in general video game playing. In *Lecture notes in computer science* (pp. 418–434). https://doi.org/10.1007/978-3-319-55849-3_28.
34. Gaina, R. D., Lucas, S. M., & Perez-Liebana, D. (2017) Rolling horizon evolution enhancements in general video game playing. In *2017 IEEE Conference on Computational Intelligence and Games (CIG)* (pp. 88–95). IEEE. https://doi.org/10.1109/CIG.2017.8080420.
35. Ghavamzadeh, M., Mannor, S., Pineau, J., & Tamar, A. (2015). Bayesian reinforcement learning: Λ survey. https://doi.org/10.1561/2200000049.
36. Goertzel, B., & Pennachin, C. (Eds.) (2007). *Artificial general intelligence.* Springer. https://doi.org/10.1007/978-3-540-68677-4
37. Goerzen, C., Kong, Z., & Mettler, B. (2010). A survey of motion planning algorithms from the perspective of autonomous UAV guidance. *Journal of Intelligent and Robotic Systems., 57*, 65–100. https://doi.org/10.1007/s10846-009-9383-1
38. Hao, B., Duan, Y., Lattimore, T., Szepesvári, C., & Wang, M. (2020). Sparse feature selection makes batch reinforcement learning more sample efficient.
39. Hostetler, J., Fern, A., & Dietterich, T. (2014). State aggregation in Monte Carlo tree search. *Proceedings of the National Conference on Artificial Intelligence., 4*, 2446–2452.
40. Janner, M., Fu, J., Zhang, M., & Levine, S. (2019). When to trust your model: Model-based policy optimization. *Advances in Neural Information Processing Systems, 32*.
41. Jiang, J., Dun, C., Huang, T., & Lu, Z. (2018). Graph convolutional reinforcement learning
42. Jong, N. K., Hester, T., & Stone, P. (2008). The utility of temporal abstraction in reinforcement learning. In *AAMAS '08: Proceedings of the 7th International Joint Conference on Autonomous Agents and Multiagent Systems* (pp. 299–306). https://doi.org/10.5555/1402383.1402429.
43. Jong, N. K., & Stone, P. (2005). State abstraction discovery from irrelevant state variables. In *IJCAI International Joint Conference on Artificial Intelligence* (pp. 752–757).
44. Justesen, N., Tillman, B., Togelius, J., & Risi, S. (2014). Script- and cluster-based UCT for StarCraft. In *IEEE Conference on Computatonal Intelligence and Games, CIG.* https://doi.org/10.1109/CIG.2014.6932900.
45. Kacprzyk, J., Zadrozny, S., Nurmi, H., & Bozhenyuk, A. (2021). Towards innovation focused fuzzy decision making by consensus. In *IEEE International Conference on Fuzzy Systems.* https://doi.org/10.1109/FUZZ45933.2021.9494531.
46. Kemke, C., & Walker, E. (2006). Planning with action abstraction and plan decomposition hierarchies. In *2006 IEEE/WIC/ACM International Conference on Intelligent Agent Technology* (pp. 447–451). IEEE. https://doi.org/10.1109/IAT.2006.99.

47. Kheradmandian, G., & Rahmati, M. (2009). Automatic abstraction in reinforcement learning using data mining techniques. *Robotics and Autonomous Systems., 57*, 1119–1128. https://doi.org/10.1016/j.robot.2009.07.002
48. Kipf, T. N., & Welling, M. (2017). Semi-supervised classification with graph convolutional networks. In *5th International Conference on Learning Representations, ICLR 2017 - Conference Track Proceedings* (pp. 1–14).
49. Kocsis, L., & Szepesvári, C. (2006) Bandit based Monte-Carlo planning. In J. Fürnkranz, T. Scheffer, & M. Spiliopoulou (Eds.), *ECML'06 Proceedings of the 17th European conference on Machine Learning* (pp. 282–293). Springer. https://doi.org/10.1007/11871842_29.
50. Konidaris, G. (2019). On the necessity of abstraction. *Current Opinion in Behavioral Sciences., 29*, 1–7. https://doi.org/10.1016/j.cobeha.2018.11.005
51. Konidaris, G., & Barto, A. (2009). Skill discovery in continuous reinforcement learning domains using skill chaining. In *Advances in Neural Information Processing Systems 22 - Proceedings of the 2009 Conference* (pp. 1015–1023).
52. Kroon, M., & Whiteson, S. (2009). Automatic feature selection for model-based reinforcement learning in factored MDPs. In *8th International Conference on Machine Learning and Applications, ICMLA 2009* (pp. 324–330). https://doi.org/10.1109/ICMLA.2009.71.
53. Kruse, R., Borgelt, C., Braune, C., Mostaghim, S., & Steinbrecher, M. (2022). *Computational intelligence*. Springer.
54. Lange, S., & Riedmiller, M. (2010). Deep auto-encoder neural networks in reinforcement learning. *Proceedings of the International Joint Conference on Neural Networks*. https://doi.org/10.1109/IJCNN.2010.5596468
55. Lelis, L. H. S. (2020). Planning algorithms for zero-sum games with exponential action spaces: a unifying perspective. In *Proceedings of the Twenty-Ninth International Joint Conference on Artificial Intelligence* (pp. 4892–4898). https://doi.org/10.24963/ijcai.2020/681.
56. Lelis, L. H. S. (2017). Stratified strategy selection for unit control in real-time strategy games. In *Proceedings of the Twenty-Sixth International Joint Conference on Artificial Intelligence* (pp. 3735–3741). https://doi.org/10.24963/ijcai.2017/522.
57. Levine, J., Congdon, C., Ebner, M., & Kendall, G. (2013). General video game playing. *Dagstuhl Follow-Ups* 1–7. https://doi.org/10.4230/DFU.VOL6.12191.77.
58. Li, Y. (2018). Deep reinforcement learning. In: *ACL 2018 - 56th Annual Meeting of the Association for Computational Linguistics, Proceedings of the Conference Tutorial Abstracts* (pp. 19–21). https://doi.org/10.18653/v1/p18-5007.
59. Liu, K., Fu, Y., Wu, L., Li, X., Aggarwal, C., & Xiong, H. (2021). Automated feature selection: A reinforcement learning perspective. *IEEE Transactions on Knowledge and Data Engineering* 4347 (2021). https://doi.org/10.1109/TKDE.2021.3115477.
60. Majeed, S. J. (2021). Abstractions of general reinforcement learning
61. Mannor, S., Menache, I., Hoze, A., & Klein, U. (2004) Dynamic abstraction in reinforcement learning via clustering. In *Twenty-First International Conference on Machine Learning - ICML '04.* (p. 71). ACM Press. https://doi.org/10.1145/1015330.1015355.
62. Maschler, M., Solan, E., & Zamir, S. (2013) *Game theory*. Cambridge University Press. https://doi.org/10.1017/CBO9780511794216.
63. Mccallum, A. K. (2020). Learning to use selective attention and short-term memory in sequential tasks. *From Animals to Animats 4*. https://doi.org/10.7551/mitpress/3118.003.0039.
64. McGovern, A., & Barto, A. G. (2001). Automatic discovery of subgoals in reinforcement learning using diverse density. In *Proceedings of the 18th International Conference on Machine Learning* (pp. 361–368).
65. Metzen, J. H. (2013). Online skill discovery using graph-based clustering. In *Proceedings of the Tenth European Workshop on Reinforcement Learning* (pp. 77–88).
66. Moradi, P., Shiri, M. E., Rad, A. A., Khadivi, A., & Hasler, M. (2012). Automatic skill acquisition in reinforcement learning using graph centrality measures. *Intelligent Data Analysis, 16*, 113–135. https://doi.org/10.3233/IDA-2011-0513
67. Moraes, R. O., Levi, H. S. (2020). Lelis: asymmetric action abstractions for multi-unit control in adversarial real-time games. In *Proceedings of the AAAI Conference on Artificial Intelligence* (pp. 876–883).

68. Moraes, R. O., Mariño, J. R. H., Mariño, M., & Lelis, L. H. S. (2018). Nested-greedy search for adversarial real-time games. In *Proceedings of the 14th AAAI Conference on Artificial Intelligence and Interactive Digital Entertainment, AIIDE 2018* (pp. 67–73).
69. Olah, C., Mordvintsev, A., & Schubert, L. (2017). Feature visualization. Distill 2. https://doi.org/10.23915/distill.00007.
70. Onisawa, T., & Kacprzyk, J. (Eds.) (1995). *Reliability and safety analyses under fuzziness.* Physica-Verlag. https://doi.org/10.1007/978-3-7908-1898-7.
71. Ontañón, S. (2013). The combinatorial multi-armed bandit problem and its application to real-time strategy games. In *Proceedings of the Ninth AAAI Conference on Artificial Intelligence and Interactive Digital Entertainment* (pp. 58–64).
72. Owsiński, J. W., Stańczak, J., Opara, K., Zadrożny, S., & Kacprzyk, J.: (2021). *Reverse clustering.* Springer International Publishing, Cham. https://doi.org/10.1007/978-3-030-693 59-6.
73. Perez, D., Powley, E. J., Whitehouse, D., Rohlfshagen, P., Samothrakis, S., Cowling, P. I., & Lucas, S. M. (2014). Solving the physical traveling salesman problem: Tree search and macro actions. *IEEE Transactions on Computational Intelligence and AI in Games, 6*, 31–45. https://doi.org/10.1109/TCIAIG.2013.2263884
74. Perez-Liebana, D., Dockhorn, A., Grueso, J. H., & Jeurissen, D. (2020). The design of "stratega": A general strategy games framework
75. Perez-Liebana, D., Guerrero-Romero, C., Dockhorn, A., Xu, L., Hurtado, J., & Jeurissen, D. (2021). Generating diverse and competitive play-styles for strategy games. In *2021 IEEE Conference on Games (CoG)* (pp. 1–8). IEEE. https://doi.org/10.1109/CoG52621.2021.961 9094.
76. Perez-Liebana, D., Liu, J., Khalifa, A., Gaina, R. D., Togelius, J., & Lucas, S. M.: General video game AI: A multi-track framework for evaluating agents, games and content generation algorithms
77. Pickett, M., & Barto, A. G. (2002). PolicyBlocks: An algorithm for creating useful macro-actions in reinforcement learning. In *Proceedings of the 19th International Conference on Machine Learning (ICML)* (pp. 506–513).
78. Powley, E. J., Whitehouse, D., & Cowling, P. I. (2012). Monte Carlo tree search with macro-actions and heuristic route planning for the physical travelling salesman problem. In *2012 IEEE Conference on Computational Intelligence and Games, CIG 2012.* (pp. 234–241). https://doi.org/10.1109/CIG.2012.6374161.
79. Rubinstein, R. Y., & Kroese, D. P. (2016). *Simulation and the Monte Carlo method.* Wiley. https://doi.org/10.1002/9781118631980.
80. Russell, J.S., & Norvig, P. (2003). Artificial intelligence: A modern approach. https://doi.org/10.1017/S0269888900007724.
81. Sebastia, L., Onaindia, E., & Marzal, E. (2006). Decomposition of planning problems. *AI Communications, 19*, 49–81.
82. Şimşek, Ö., & Barto, A. G. (2004). Using relative novelty to identify useful temporal abstractions in reinforcement learning. In *Twenty-First International Conference on Machine Learning - ICML '04.* (p. 95). ACM Press. https://doi.org/10.1145/1015330.1015353.
83. Şimşek, Ö., Wolfe, A. P., & Barto, A. G. (2005). Identifying useful subgoals in reinforcement learning by local graph partitioning. In *ICML 2005 - Proceedings of the 22nd International Conference on Machine Learning* (pp. 817–824). https://doi.org/10.1145/1102351.1102454.
84. Singh, S. P., Jaakkola, T., & Jordan, M. J. (1995). Reinforcement learning with soft state aggregation. *Advances in Neural Information Processing Systems, 7*, 361–368.
85. Sturtevant, N., & Buro, M. (2005). Partial pathfinding using map abstraction and refinement. In *Proceedings of the Twentieth National Conference on Artificial Intelligence and the Seventeenth Innovative Applications of Artificial Intelligence Conference* (pp. 1392–1397).
86. Sturtevant, N., & Buro, M. (2006). Improving collaborative pathfinding using map abstraction. In *Proceedings of the 2nd Artificial Intelligence and Interactive Digital Entertainment Conference, AIIDE 2006* (pp. 80–85).

87. Sutton, R. S. (1988). Learning to predict by the methods of temporal differences. *Machine Learning, 3*, 9–44. https://doi.org/10.1007/BF00115009
88. Sutton, R. S., Precup, D., & Singh, S. (1999). Between MDPs and semi-MDPs: A framework for temporal abstraction in reinforcement learning. *Artificial Intelligence, 112*, 181–211. https://doi.org/10.1016/S0004-3702(99)00052-1
89. Tromp, J., & Farnebäck, G. (2007). Combinatorics of go. In *Computers and games* (pp. 84–99). https://doi.org/10.1007/978-3-540-75538-8_8.
90. Wang, C., Chen, P., Li, Y., Holmgård, C., & Togelius, J. (2021). Portfolio online evolution in StarCraft. In *Proceedings of the AAAI Conference on Artificial Intelligence and Interactive Digital Entertainment* (pp. 114–120).
91. Yang, X., Zhang, G., Lu, J., & Ma, J. (2011). A kernel fuzzy c-means clustering-based fuzzy support vector machine algorithm for classification problems with outliers or noises. *IEEE Transactions on Fuzzy Systems., 19*, 105–115. https://doi.org/10.1109/TFUZZ.2010.2087382
92. Yang, Z., & Ontañón, S. (2019). Guiding Monte Carlo tree search by scripts in real-time strategy games. In *Proceedings of the 15th AAAI Conference on Artificial Intelligence and Interactive Digital Entertainment, AIIDE 2019* (pp. 100–106).

Alexander Dockhorn is Juniorprofessor for Computer Science at the Gottfried Wilhelm Leibniz University Hannover. Following his dissertation at the Otto-von-Guericke University of Magdeburg (OVGU), he worked at the Queen Mary University of London and later came back to the OVGU. During this time, he studied artificial intelligence in general strategy games and began working on abstraction methods for simulation-based search agents in games. He is a member of the Institute of Electrical and Electronics Engineers (IEEE) and currently serves as the chair of the IEEE Games Technical Committee and the Summer School Subcommittee. From 2017 to 2020 he was organizing the Hearthstone AI competition to foster comparability of AI agents in complex card games. See his webpage for a complete list of projects and publications: https://adockhorn.github.io/.

Rudolf Kruse is an Emeritus Professor for Computer Science at the Otto-von-Guericke University of Magdeburg Germany. He obtained his Ph.D. and his Habilitation in Mathematics from the Technical University of Braunschweig in 1980 and 1984 respectively. Following a stay at the Fraunhofer Gesellschaft, he joined the Technical University of Braunschweig as a professor of computer science in 1986. From 1996–2017 he was the leader of the Computational Intelligence Group in Magdeburg. He has co-authored 15 monographs and 25 books as well as more than 400 peer-refereed scientific publications in various areas. He is Fellow of the Institute of Electrical and Electronics Engineers (IEEE), Fellow of the International Fuzzy Systems Association (IFSA), and Fellow of the European Association for Artificial Intelligence (EURAI/ECCAI). His group is successful in various industrial applications, see his webpage www.is.ovgu.de/Team/Rudolf+Kruse.html. His research interests include data science and intelligent systems.

A Tentative Algorithm for Neurological Disorders

Jaime Gil Aluja and Jean Jacques Askenasy

Abstract Increasingly, transversality in research is a source of new findings. In this occasion, two academics from different fields of knowledge have joined their concerns to solve or alleviate a problem that casts a shadow over the lives of so many humans: neurological disorders. With this general objective, the authors have chosen among them, Parkinson's disease (PD). There are two main objectives sought:

- Determining, through the degree or level of intensity of the symptoms perceived, the location and the degree or level of affectation of the neurons.
- Knowing the path or paths followed by the flow of incidence between areas of neurons (contagion), between neurons and symptoms, and between one and the other symptoms, determining in each case, its degree or level.

These two aspects of the work will allow us to know which incidents are most frequently forgotten in non-precise diagnoses. The results obtained through the algorithm developed for this occasion show that, even upon high levels of demand, grades or levels higher than 0.9, 0.8 over 1, contained many oversights. With the presented algorithm, absolutely all oversights are recovered without exception, with the degree or level existing at the beginning.

Keywords Parkinson disease · Fuzzy sets neurological disorders · Diagnosis PD · Pathology PSP · Forgotten effects · Neurons affected · Symptom PD

1 Presentation

Forgetting is part of human nature. And that will continue to increase, due to the rising complexity of the systems in which we find ourselves immersed.

J. Gil Aluja (✉)
Royal Academy of Economic and Financial Sciences, Barcelona, Spain
e-mail: secretaria@racef.es

J. J. Askenasy
Law Tel-Aviv University, Tel-Aviv, Israel

© The Author(s), under exclusive license to Springer Nature Switzerland AG 2023
Y. P. Kondratenko et al. (eds.), *Artificial Intelligence in Control and Decision-making Systems*, Studies in Computational Intelligence 1087,
https://doi.org/10.1007/978-3-031-25759-9_10

The interrelationships between the elements of a system become more and more numerous when, in addition to considering the direct incidences between two of its elements (first-generation relationships), it becomes necessary to consider those that result through a third element that acts intermediary (second generation relationships).

Many are the errors that are made when only direct incidents are considered, forgetting those that take place through another interposed element: the direct path is not always the best, even when it is the shortest.

This is latent idea found in the "Forgotten Effects Theory", whose primary objective is to recover all the forgetfulness to determine, among them, the most important.

To elaborate this theory, initially formulated by professors Kaufmann and Gil-Aluja, Lotfi Zadeh's "Fuzzy Sets Theory" was used.

Subsequent reformulations, with the help of the Principle of Gradual Simultaneity, the incorporation of the concept of the Entropy Beach, as well as new figures capable of a more general formalization, led to what we now call the Theory of Self-Induced Incidents, of a markedly humanistic nature [1].

This line of work has allowed the development of new algorithms aiming to provide solutions to objectives as diverse as the creation of financial products, the harmonization of different generations and cultures in a single project, the distribution of immigrants to suitable tasks in the host country and the very serious problem of assigning the unaccompanied minors (UAMs) to households with compatible cultures between those of the children's countries of origin and those of destination, to name just a few more well-known ones.

We consider that, among the most appropriate tools that we could use to achieve the main sought goal, an early detection of this neurological disorder, the most useful were found in the field of theoretical and technical elements used in advanced research within the Artificial Intelligence.

And so we did, using the existing humanist algorithms as a basis and creating new ones that have later prove to be effective. Why not do it this time?

Albeit the rest is much more ambitious, the objectives to be achieved, would further compensate for the greater efforts devoted to this task.

Furthermore, there is an added value in the current circumstances: this work is intended to contribute to the tribute that scientists from all over the world wish to pay to an academic personality as loved and admired in intellectual circles around the world, such as Dr. Janusz Kacprzyk, Corresponding Academician for Poland of the Royal Academy of Economic and Financial Sciences of Spain, on behalf of which, as its President, I wish to make public my congratulations and that of the members of this high Institution of the Institute of Spain, on the occasion of his 75th anniversary.

Before beginning the narration of our work, we wish to place on record our gratitude to Dr. Yuriy P. Kondratenko and collaborators, for the initiative of organizing this well-deserved tribute through what we, researchers in science, know how to do: work for the benefit of the members of human societies. In our case, to achieve, through health, a better shared well-being.

2 Basic Aspects for the Diagnosis of PD

The Unified Parkinson's Disease (PD) Rating Scale (UPDRS) is an 8-page description of signs classified under 6 headings [2]. In November 2001, Dr. Askenasy published an article describing sleep disorders related to PD and suggested the need for a revision of the UPDRS [3].

To determine a binary 'yes-no' PD risk is extremely difficult for the following reasons:

From a neuroscientific point of view, PD is pathologically heterogeneous, with the most common pathological substrates related to abnormalities in the presynaptic protein α-synuclein [4] or the microtubule binding protein tau [5]. The most common of the Parkinsonian tauopathies is progressive supranuclear palsy (PSP), which is clinically associated to severe postural instability leading to early falls.

The tau pathology of PSP affects both neurons and glia. In idiopathic PD, α-synuclein accumulates in neuronal perikaryal (Lewy bodies) and neuronal processes (Lewy neurites) [6]. Therefore, abnormalities of the presynaptic protein α-synuclein or the microtubule binding protein tau accumulate in neuronal perikaryal (Lewy bodies) and neuronal processes (Lewy neurites) with various speeds.

The disease process is multifocal and involves various central nervous system neurons. The affected neurons may be localized in the amygdala, hippocampus, temporal cortex, cingulate cortex, superior frontal gyrus, motor cortex, caudate nucleus, putamen, globus pallidus, nucleus basalis of Meynert, hypothalamus, thalamus, subthalamus, red nucleus, substantia nigra, oculomotor complex, midbrain, locus coeruleus, pontine tegmentum and nuclei, medullary tegmentum, inferior olivary nucleus, dentate nucleus, and cerebellar white matter. The particular set of neurons affected determines motor and nonmotor clinical presentations affecting both neurons and glia.

From the genetic point of view, variations in PD such as those driven by autosomal dominant mutations in the gene for α-synuclein (a protein regulating synaptic vesicle trafficking), are especially inappropriate for binary 'yes or no' diagnoses. Multiple system atrophy (MSA) is the other major α-synucleinopathy. It is also associated with autonomic dysfunction and in some cases with cerebellar signs. The histopathologic hallmark of MSA is the accumulation of α-synuclein within glial cytoplasmic inclusions (GCI). Uncertainty is an integral aspect of the cellular and molecular development of the disease.

From a clinical point of view, Parkinsonism is a broad field of extrapyramidal disorders encompassing idiopathic PD and Parkinsonian-like Syndromes (PDS). The PDS are of degenerative, vascular, genetic, traumatic, or toxic origin. It can be claimed that the risk of developing Parkinson's is uncertain and the appearance of a tremor at a certain age suggests to many people that it can be the start of a PD. To determine illness risk, it is recommended to send the patient to a clinical neurologist.

From symptomatic point of view, PD varies from exclusively tremorgenic to combinations of the following signs: tremor, rigidity, slowness, amimia (facial expression decline), camptocormia (bent spine syndrome), telegraphic speech, right or left cogwheel rigidity, anterior or posterior push test, on-off phenomenon, absence of alternative upper limb movements, REM behavior disorders, and restless leg syndrome.

Each of these symptoms may range from different amplitudes, intensity, and duration. Tremor can be imperceptible or an obvious tremor affecting all the limbs (upper and lower). The order of symptom appearance is ambiguous. A differential diagnosis implies a decision among Lewy bodies disease, autonomic neural degeneration, multiple system atrophy (MSA aka Shy-Drager Syndrome), progressive nuclear palsy, and MPTP induced PD. Each of these can have identical symptoms.

From diagnostic point of view PD is the result of 3 elements: the knowledge, experience, and intuition of the specialist. The major characteristic of PD is its "diagnostic uncertainty at the start point," which makes a binary 'yes-no' decision inappropriate. Its imprecision provokes animosity between the patient and the specialist and between specialists themselves. The domains of uncertainty include:

(1) The large range of the starting age of PD: 30 to 70
(2) The long pre-symptomatic period since the starting point of the disease: 8–24 years.
(3) The patient has imprecise and subjective complaints.
(4) The physical examination boundary is not exact.
(5) Laboratory results including MRI and Dopa PET may be contradictory.

From the reaction to treatment point of view: the response to treatment may change the diagnosis. The use of paraclinical examinations strengthens the belief of the patient that the neurologist is uncertain. The degree of the physician's uncertainty is often not far from that of the patient. The clinical aspects of Parkinsonism are uncertain [7, 8].

In one word, the prominent feature of PD is 'uncertainty'.

We believe, at the current stage of knowledge, that the answer to such challenges resides in the fuzzy sets theory.

The notion of Fuzzy sets comes from Zadeh [9].

In the thick network of relationships of incidence between the affected neurons and the symptoms we observe, quantities of variations occur but there are also almost inevitable oversights, likely to cause erroneous diagnoses.

The forgotten effects theory provides a solution that recovers all the forgotten relationships with their degree or level of forgivefullness, without error or omission.

The authors of the Forgotten Effects theory are Arnold Kaufmann and Jaime Gil-Aluja [10].

The technical and theoretical elements of the "Fuzzy Sets Theory" and the "Forgotten Effects Theory" allow developing an algorithm, which, thanks to the "Big Data" and the "Digital Revolution", resides in the field of the "Artificial Intelligence".

Usually, the incidences are propagated through a network of sequences which becomes more and more dense and it is therefore very difficult to avoid oversights which very frequently lead to unfavorable secondary consequences for the diagnosis, regardless of the social field. The field of medicine is no exception to this general rule.

Until recently, incidences were valued in the binary domain $\{0, 1\}$, no incidence: 0, total incidence: 1. Currently, we can get closer to human thought, which usually nuances its reasoning. When duality gives way to multivalence, valuations take place in the interval $[0, 1]$.

One can then incorporate the multivalent logics in numerical processes, by employing a numeric-semantic correspondence so that each "valuation" carried out by means of a number in $[0, 1]$, has its correspondence with one or more words of the language usually concerning a concept. In this case, that of incidence. We usually use the endecadary system for this, 11 numbers from 0 to 1. We propose the following correspondence:

0   no effect.
0.1   virtually no impact
0.2   almost no impact
0.3:   very low incidence
0.4   low incidence
0.5   average incidence
0.6   sensitive impact
0.7   large incidence
0.8   high incidence
0.9   very high incidence
1   total incidence

In this way, whenever in our work one of these numbers in $[0, 1]$ appears, our mind will automatically translate it to the words that correspond to it in the previous list, which represent the degree or level of incidence.

Thus, we will have fixed, by a number in $[0, 1]$, the degree or level of incidence of the elements of a set A over the elements of another set B.

3   Establishment of Sets of Incident and Incidental Elements

We will establish, in this work, that the set A, $A = \{a_1, a_2, \ldots, a_8\}$ includes the neurons affecting the correspondent incident on the symptoms of PD. The set of these symptoms then forms the set B, $B = \{b_1, b_2, \ldots, b_{12}\}$.

The concrete name of the 8 types of neurons with lesion in PD, which we have considered are the following:

a_1 Neuron of substantia nigra-pars compacta
a_2 Neuron of the lateral neo striaturn (putamen)
a_3 Retinal amacrine cells
a_4 Noradrenegic neurons the locus coeruleus
a_5 Serotonergic neurons of the dorsal raph nucleus
a_6 Neurons of the acetylcholinergic pons
a_7 Neurons from basolateral nucleus of the amigdala

a_8 Glutaminergic neurons from the parahyppocampal gyrus

On the other side, the name of each of the 12 symptoms of the set of PD retained are:

b_1 Tremor
b_2 Stiffness
b_3 Slownees
b_4 Anemia
b_5 Camptocormia
b_6 Walk without alternating arms
b_7 Walk on a wide base
b_8 Telegraphic words
b_9 On-Off phenomenon
b_{10} Symptoms present in half body
b_{11} Small print writing
b_{12} Olfactory problems

The next step is to establish the degree or level of impact of the affecting neurons on each symptom of PD.

For this we can use the matrix representation. A rectangular matrix of degree 8×12, for example, where will appear in each box the "direct impact" of each neuron affecting each symptom of PD.

4   The Opinion of the Experts

For this, we have brought together a group of specialists in the human brain to count on their expertise. They gave us their opinion which we believe is of the highest quality on the topic at hand.

We did not use the experton algorithm [11] for the valuations, because a clear consensus of our experts seemed sufficient to us.

The set A of affected neurons will be placed as rows, and as columns the symptoms of PD, B. All the information already expressed numerically according to the proposed endecadary semantic system can be presented by the fuzzy matrix $[[\underset{\sim}{M}]]$ next:

$[\underset{\sim}{M}]$	b_1	b_2	b_3	b_4	b_5	b_6	b_7	b_8	b_9	b_{10}	b_{11}	b_{12}
a_1	0.8	0.7	0.8	0.6	0.5	0.7	0.4	0.7	0.7	0.4	0.2	0.9
a_2	0.6	0.5	0.4	0.7	0.7	0.6	0.5	0.3	0.6	0.4	0.6	0.2
a_3	0.9	0.4	0.8	0.3	0.7	0	0.9	0.6	0.7	0.2	0.1	0.8
a_4	0.7	0.5	0.2	0.2	0.2	0.4	0.1	0.3	0.3	0.1	0.1	0.4
a_5	0.6	0.2	0.2	0.4	0.7	0.1	0.5	0.8	0.7	0.6	0.7	0.1
a_6	0.8	0.7	0.6	0.4	0.5	0.3	0.1	0.7	0.3	0.6	0	0.6

(continued)

(continued)

$[\underset{\sim}{M}]$	b_1	b_2	b_3	b_4	b_5	b_6	b_7	b_8	b_9	b_{10}	b_{11}	b_{12}
a_7	0.7	0.7	0.7	0.2	0	0.8	0.2	0.1	0.4	0.1	0	0.8
a_8	0.8	0.6	0.6	0.4	0.3	0.6	0.6	0.2	0.4	0.1	0.4	0.8

This table constitutes a graph presented in matrix form which expresses the "direct incidences" of each neuron affected on each symptom of PD, valued in [0, 1] by means of the endecadary system.

We observe that there is a zero in the boxes $(a_3, b_6)(a_6, b_{11})$ and (a_7, b_{11}), which indicates that the experts considered the total absence of incidence of these affected neurons on these PD symptoms.

The valuation that appears in each box represents the load or the impact potential of each neuron affected on each symptom of PD, generally different. This type of study very often ends with the accumulation of potentials.

However, our objective goes further, because we know that, in addition to the direct incidences, which we have just obtained, there are also others derived from the existence of incidence of the affected neurons on the affected neurons and the impact of PD symptoms on those same PD symptoms.

To add these incidences, which we could call secondary, if we accept that the direct ones are primary, we will continue to detail the process, step by step.

This time, we construct square matrices, $[\underset{\sim}{A}]$ and $[\underset{\sim}{B}]$; the first expresses the incidences of the affected neurons on the affected neurons, and the second the incidences of the symptoms of PD on themselves.

Once the consultations have been carried out with the experts, the following information has been provided, which is presented in matrix form.

$\underset{\sim}{A}$	a_1	a_2	a_3	a_4	a_5	a_6	a_7	a_8
a_1	1	0.8	0.8	0.8	0.4	0.8	1	0.8
a_2	0.6	1	0.6	0.6	0.8	0.8	0.8	0.7
a_3	0.5	0.6	1	0.8	1	0.6	0.4	0.8
a_4	0.6	0.6	0.8	1	1	0.4	0.6	0.8
a_5	0.5	0.8	0.8	1	1	0.8	0.8	0.8
a_6	0.8	0.7	0.6	0.3	0.9	1	0.7	0.8
a_7	0.7	0.7	0.3	0.7	0.7	0.8	1	0.8
a_8	0.8	0.6	0.8	0.9	0.8	0.6	0.7	1

We now prepare to obtain the matrix $[\underset{\sim}{B}]$ which expresses the effects of PD symptoms on themselves.

The result of the direct incidences is therefore reflected on the fuzzy matrix that we reproduce:

$\underset{\sim}{B}$	b_1	b_2	b_3	b_4	b_5	b_6	b_7	b_8	b_9	b_{10}	b_{11}	b_{12}
b_1	1	0.9	1	0	1	0.9	0.8	0	0.9	0.9	0.4	0.8
b_2	0.5	1	0.8	0	0.4	0.9	0.7	0.5	0.7	0.7	0.9	0.7
b_3	0	0	1	0	0	0	0	0	0	0	0	0
b_4	0.3	0.7	1	1	0.7	0.2	0.4	0	0.2	0	0	0.6
b_5	0.2	0.2	0.8	0	1	1	0.8	0.4	0.5	1	0.2	0.3
b_6	0	0	0.2	0	0	1	0.7	0	0.2	0.2	0	0
b_7	0	0	0.2	0	0	0.4	1	0.2	0.2	0.2	0	0
b_8	0	0	0.4	0	0	0	0	1	0	0	0.6	0
b_9	0	0	0.8	0	0.4	0.6	0.8	0.8	1	0.2	0.2	0.2
b_{10}	0	0	0.8	0	0	1	0.8	0.2	0	1	0	0
b_{11}	0	0	0.2	0	0	0	0	0	0	0	1	0
b_{12}	0	0	0	0	0	0	0	0	0	0	0	1

From the information received, we have been able to see in the incidences of PD symptoms between them, that there are many symptoms that do not project incidences on other symptoms. You still must use this matrix [$\underset{\sim}{B}$]. It could be that in some cases, it is enough that one or two of the positive incidences can, through their intermediation, lead us to obtain a good diagnosis.

With this matrix of incidences [$\underset{\sim}{B}$] we have completed the necessary numerical information, already structured by means of the matrices. However, we recall once again that these are not measures (objective values) but valuations (subjective numerical assignments).

This observation allows us to make two important reflections:

The first: when an incidence is valued, this valuation does not correspond to a numerical assignment between the elements that are valued, but to the perception that the experts have of this incidence. Therefore, there is a certain level of subjectivity. And that is why we are going to use different operators from those usually used, as we will see later.

The second: the desire to draw attention to the importance of the symptoms of PD.

A simple look at these 12 symptoms of PD is enough for us to assume the different level of strength of each of these in issuing a correct diagnosis of PD.

5 Technical Elements that Precede the Development of the Algorithm

As we have exposed, the first part of this work is devoted to obtaining the total incidences of each affected neuron on all and every one of the symptoms of PD.

For this, we relied on the algorithm developed in the work which consecrated the Theory of Forgotten Effects [10]. In this work, we started, as we have just done, from 2 frames of reference: A which represents in our case, all the affected neurons and B which represents all the symptoms of PD.

The incidences were expressed using graphs in the form of matrices. We are used, in the field of uncertainty, to express these incidences as follows:

$$\underset{\sim}{M} = A \circ B, \ \underset{\sim}{A} = A \circ A, \ \underset{\sim}{B} = B \circ B$$

As we have seen, the first can be represented by a rectangular fuzzy matrix, the other two by square fuzzy matrices.

We will then see that for our objective, obtaining direct incidences $\underset{\sim}{M}$ that is to say those called first generation incidences, is not sufficient. It is also necessary to consider the incidences of A on A (affected neurons which incident in the affected neurons) and those of B on B (symptoms of PD which incident on the own symptoms of PD.)

The fuzzy matrices $[\underset{\sim}{A}]$ and $[\underset{\sim}{B}]$ are, by construction, square, and a reflexive relation.

Indeed, the incidence of any element of these matrices on itself is total. We must therefore assign the valuation 1 to all the elements of the main diagonal.

Reflexibility makes possible to use another property: associativity.

$$\underset{\sim}{A} \circ \underset{\sim}{M} \circ \underset{\sim}{B} = (\underset{\sim}{A} \circ \underset{\sim}{M}) \circ \underset{\sim}{B}$$

If we consider a reflexive fuzzy matrix like $\underset{\sim}{A}$, when we have $\underset{\sim}{A} \circ \underset{\sim}{M}$ it could not be otherwise, since in $\underset{\sim}{A} \circ \underset{\sim}{M}$ besides the direct incidences, $\underset{\sim}{M}$, we also include those that correspond to the matrix $\underset{\sim}{A}$ which collects the incidences of each affected neuron on each of these.

So, we always get:

$$\underset{\sim}{M} \subseteq \underset{\sim}{A} \circ \underset{\sim}{M}$$

It is the same if $\underset{\sim}{B}$ is a reflective fuzzy matrix.
If we do $\underset{\sim}{M} \circ \underset{\sim}{B}$ we have:

$$\underset{\sim}{M} \subseteq \underset{\sim}{M} \circ \underset{\sim}{B}$$

Ultimately, by combining the two inequalities, we can write:

$$M \subseteq A \circ M\,B$$

We have here a very interesting expression to our approach, because it highlights the fact that when we consider the matrices which express the incidence of the affected neurons on the affected neurons, and those of the symptoms of PD on the symptoms of PD, the incidence levels resulting from the previous composition are greater than or equal to those of the direct incidence matrix. M. Thus, oversights have taken place.

If we call M^* the matrix that gives the first- and second-generation accumulated effects:

$$M^* = A \circ M \circ B$$

We will obviously have:

$$M \subset M^*$$

We can find the forgotten effects by obtaining:

$$D = M^*(-)M$$

The matrix M^* obtained allows an excellent solution to the problem of finding the affected neurons that have a direct or indirect impact with greater intensity on the symptom(s) of PD. It is on these affected neurons on which the greatest efforts must be focused for the effective treatment of the neuronal "assignments" that the corresponding symptom detects.

6   The Phases of the Algorithm

Everything we have exposed before already allows us to develop a fuzzy algorithm, whose phases are stated below:

1. Formation of a set B that includes the symptoms of PD.
2. Establishment of all the affected neurons A capable of influencing in greater or lesser degree or level the symptoms of PD.
3. Elaboration of fuzzy incidence matrices M (incidence of affected neurons on PD symptoms.) A (Incidence of affected neurons on affected neurons) and B (incidence of PD symptoms on PD symptoms).

4. Composition, using the max-min convolution operator, of the matrices $\underset{\sim}{A}$ and $\underset{\sim}{M}$, that is, $\underset{\sim}{A} \circ \underset{\sim}{M}$, taking non-commutativity into account.

5. Composition, using the max-min convolution operator, of matrices $(\underset{\sim}{A} \circ \underset{\sim}{M})$ with $\underset{\sim}{B}$, that is, $\underset{\sim}{M}^* = \underset{\sim}{A} \circ \underset{\sim}{M} \circ \underset{\sim}{B}$.

6. Calculation of the fuzzy matrix $\underset{\sim}{D}$, as the difference between the matrix of first- and second-generation incidences $\underset{\sim}{M}^*$ and that of direct incidences $\underset{\sim}{M}$.

The elements of the matrix $\underset{\sim}{D}$ express the level or degree of all omissions. Their analysis, by taking the greatest valuations, allows us to know the most important "oversights".

It is obvious that it is interesting to discover which are the affected neurons and/or the symptoms of PD which serve as intermediaries, and which had been forgotten.

And it is at this point that we can find the "reason" of the oblivion. This reason sometimes forces us to go back to the original matrices $\underset{\sim}{M}$, $\underset{\sim}{A}$ and $\underset{\sim}{B}$.

We now propose to apply this fuzzy algorithm to this case, using the previous information provided by the experts.

7 Development of the Algorithm Based on Previous Information from Experts

When the algorithm is created and its phases justified, we only have to use the matrices presented before $\underset{\sim}{M}$, $\underset{\sim}{A}$ and $\underset{\sim}{B}$ by following the phases that make up the algorithm.

In our case we have already presented the first three phases that we are going to recover now to be able to complete the process with the following three. These are the ones that will be detailed below:

4^a. Max-min composition $\underset{\sim}{A} \circ \underset{\sim}{M}$

It is well known that the convolution or max-min composition of a square fuzzy matrix $\underset{\sim}{A}$ with a rectangular matrix $\underset{\sim}{M}$ gives a new rectangular matrix $\underset{\sim}{A} \circ \underset{\sim}{M}$. In our case, it is the following:

$\underset{\sim}{A} \circ \underset{\sim}{M}$	b_1	b_2	b_3	b_4	b_5	b_6	b_7	b_8	b_9	b_{10}	b_{11}	b_{12}
a_1	0.8	0.7	0.8	0.7	0.7	0.8	0.8	0.7	0.7	0.6	0.6	0.9
a_2	0.8	0.7	0.7	0.7	0.7	0.8	0.6	0.8	0.7	0.6	0.7	0.8
a_3	0.9	0.6	0.8	0.6	0.7	0.6	0.9	0.8	0.7	0.6	0.7	0.8
a_4	0.8	0.6	0.8	0.6	0.7	0.6	0.8	0.8	0.7	0.6	0.7	0.8
a_5	0.8	0.7	0.8	0.7	0.7	0.8	0.8	0.8	0.7	0.6	0.7	0.8
a_6	0.8	0.7	0.8	0.7	0.7	0.7	0.6	0.8	0.7	0.6	0.7	0.8

(continued)

(continued)

$\underset{\sim}{A} \circ \underset{\sim}{M}$	b_1	b_2	b_3	b_4	b_5	b_6	b_7	b_8	b_9	b_{10}	b_{11}	b_{12}
a_7	0.8	0.7	0.7	0.7	0.7	0.8	0.6	0.7	0.7	0.6	0.7	0.8
a_8	0.8	0.7	0.8	0.6	0.7	0.7	0.8	0.8	0.7	0.6	0.7	0.8

We can already observe in this first part of the convolution that the incidences (valuations that exist in each cell) are equal to or greater than those in the matrix $\underset{\sim}{M}$ of primary incidences. We then have:

$$\underset{\sim}{M} \subset \underset{\sim}{A} \circ \underset{\sim}{M}$$

It could not be otherwise, since in $\underset{\sim}{A} \circ \underset{\sim}{M}$ besides the direct incidences, $\underset{\sim}{M}$, we also include those that correspond to the matrix $\underset{\sim}{A}$ which, it should be remembered, collects the incidences of each affected neuron on each of these.

We are now going to calculate the max-min convolution which also collects the incidences of each symptom of PD on all the symptoms.

5^a Convolution or max-min composition $\underset{\sim}{A} \circ \underset{\sim}{M} \circ B$.

Given the matrix $\underset{\sim}{A} \circ \underset{\sim}{M}$ the convolution $(\underset{\sim}{A} \circ \underset{\sim}{M} \circ B)$ is immediate. The result is:

$\underset{\sim}{A}° \underset{\sim}{M}°B$	b_1	b_2	b_3	b_4	b_5	b_6	b_7	b_8	b_9	b_{10}	b_{11}	b_{12}
a_1	0.8	0.8	0.8	0.7	0.8	0.8	0.8	0.7	0.8	0.8	0.7	0.9
a_2	0.8	0.8	0.8	0.7	0.8	0.8	0.8	0.8	0.8	0.8	0.7	0.8
a_3	0.9	0.9	0.9	0.6	0.9	0.9	0.9	0.8	0.9	0.9	0.7	0.8
a_4	0.8	0.8	0.8	0.6	0.8	0.8	0.8	0.8	0.8	0.8	0.7	0.8
a_5	0.8	0.8	0.8	0.7	0.8	0.8	0.8	0.8	0.8	0.8	0.7	0.8
a_6	0.8	0.8	0.8	0.7	0.8	0.8	0.8	0.8	0.8	0.8	0.7	0.8
a_7	0.8	0.8	0.8	0.7	0.8	0.8	0.8	0.7	0.8	0.8	0.7	0.8
a_8	0.8	0.8	0.8	0.6	0.8	0.8	0.8	0.8	0.8	0.8	0.7	0.8

A glance to the matrix $\underset{\sim}{A} \circ \underset{\sim}{M} \circ \underset{\sim}{B}$ is sufficient to verify the high degree of incidence of each neuron affected on each symptom of PD.

They are all included in the confidence interval [0.6, 0.9] and most in [0.7, 0.9].

One observes a significant difference in some incidences between the original valuation contained in the fuzzy matrix $\underset{\sim}{M}$ and that of the matrix of accumulated incidences $\underset{\sim}{A} \circ \underset{\sim}{M} \circ B$ which we have designated $\underset{\sim}{M}$.

However, despite the quality of the experts, some important "oversights" have occurred, and it is necessary not only to detect them, but also to know the reasons that caused them.

To know the oversights and know the degree or level of each of them, we have several operators at our disposal. For this work, we have considered one of the simplest and most known: subtraction.

6th Obtaining the matrix of forgotten incidences:

Accepting the difference operator as part of the calculation to find the degree or level of forgetfulness of incidences reduces the task to a simple difference of matrices:

$$D = M^*(-)M$$

We will then have:

$$M^* - M = A \circ M \circ B(-)M = D$$

$D=$	$D=M^*(-)M$	b_1	b_2	b_3	b_4	b_5	b_6	b_7	b_8	b_9	b_{10}	b_{11}	b_{12}
	a_1	0	0.1	0	0.1	0.3	0.1	0.4	0	0.1	0.4	0.5	0
	a_2	0.2	0.3	0.4	0	0.1	0.2	0.3	0.5	0.2	0.4	0.1	0.6
	a_3	0	0.5	0.1	0.3	0.2	0.9	0	0.2	0.2	0.7	0.6	0
	a_4	0.1	0.3	0.6	0.4	0.6	0.4	0.7	0.5	0.5	0.7	0.6	0.4
	a_5	0.2	0.6	0.5	0.3	0.1	0.7	0.3	0	0.1	0.2	0	0.7
	a_6	0	0.1	0.2	0.3	0.3	0.5	0.7	0.1	0.5	0.2	0.7	0.2
	a_7	0.1	0.1	0.1	0.5	0.8	0	0.6	0.6	0.4	0.7	0.7	0
	a_8	0	0.2	0.2	0.2	0.5	0.2	0.2	0.6	0.4	0.7	0.3	0

We already perceived in the 4th phase of the algorithm, having found the semi composition $A \circ M$, that the accumulation of incidences did not excessively increase the value of the fuzzy matrix. On the other hand, we were beginning to guess a few omissions.

With obtaining the Forgotten Incidence Matrix, all, all "forgets" present themselves with their degree or level of forgetfulness. Each element of the matrix D shows the corresponding level of forgetting.

The information contained in the oversight matrix allows us to focus our attention on the highest valuation. In our work, the largest omission corresponds to the incidence (a_3, b_6).

As always (we will never cease to repeat it), here again we emphasize the important role of the concept of degree or level. Since the beginning we have adopted the endecadary system, we suggest continuing with this one. It's convenient and we are essentially required to use it.

However, we should perhaps justify why we spoke of one or more thresholds, to fix the minimum level that marks the importance of forgetting.

As we have commented, not all elements of the PD symptom set have the same "strength" in diagnosis. Moreover, the intensity, degree, or level of this force changes over time and from patient to patient.

In this work, we have opted, to simplify the exposure without deviating from the desired goal, to use a main level, $\alpha \geq 0.8$, for all the symptoms of PD, which gives us two major omissions: (a_7, b_5) with an incidence 0.8 and (a_3, b_6) whose oblivion is 0.9.

As a complement, we also offer a quick look at omissions with a level $\alpha = 0.7$.

We will separate, then, the incidences whose valuation of forgetting are the most important and we will present some brief reflections arising from these results.

8   Analysis of Forgotten Incidences

Sometimes there is a special interest in a specific incidence relationship: we have no problem studying it deeply, even when the level of forgetfulness is smaller than the established threshold. The flexibility of the algorithm allows it.

We will explore and present the path the clinical analyst can follow to find the highest level of total incidence of affected neurons on a symptom of PD, and also highlight the biggest oversights that may have taken place.

In this study, the greatest omissions took place, as we have pointed out, in the incidence a_3 on b_6 in a level $\alpha = 0.9$ and a_7 on b_5 in a level $\alpha = 0.8$. The experts had initially given, for the first incidence, a valuation of 0, "no incidence", and on the other hand through other elements of the set of affected neurons and symptoms of PD, an overall incidence was found to be 0.9 "very strong incidence". For the second, the incidence fixed for the experts was first valued at 0, and, considering the secondary incidences, the total incidence is 0.8.

For an analysis of this concrete incidence, we usually follow the phases marked by the algorithm. We first consider the matrix $\underset{\sim}{A}$ which connects the affected neurons which incident on the affected neurons: there are only two incidences which are equal to or greater than 0.9 when starting from a_4.

It is necessary, according to the fourth phase of the algorithm, to carry out the convolution or max-min composition $\underset{\sim}{A} \circ \underset{\sim}{M}$. On our following sagittal representation, we ignore all the elements of the set A as elements to have an incidence, except those whose valuations are equal to or greater than 0.9. There are only two: a_1 and a_3.

For the incidence of a_1, on all the configurator elements, b_j, $j = 1, 2, \ldots, 12$, the convolutions or max-min compositions that concern us are the incidences that come out precisely from a_1 and a_3.

When we start from a_1 we only retain the incidence accumulated in $\underset{\sim}{A} \circ \underset{\sim}{M}$ with a force equal to or greater than 0.9. This is the symptom of PD: b_{12}.

And in the same way, for the incidence of a_3 on b_j, $j = 1, 2, \ldots, 12$, the valuations resulting from the convolution $\underset{\sim}{A} \circ \underset{\sim}{M}$ which are equal to or greater than 0.9, correspond to the symptoms of PD b_1 and b_7.

It only remains finally, to expose the valuations resulting from the convolution between $(\underset{\sim}{A} \circ \underset{\sim}{M})$ and $\underset{\sim}{B}$, that is to say $\underset{\sim}{A} \circ \underset{\sim}{M} \circ \underset{\sim}{B}$.

For more visual clarity, we limit ourselves, first, to incidences in which their valuation is equal to or greater than 0.9.

The following network expresses what we have just pointed out.

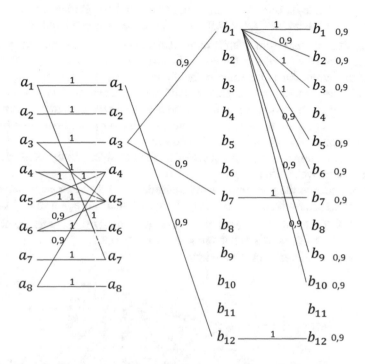

Our readers will have noticed that the anterior sagittal graph contains only the arcs that we have considered the most essential for the analysis.

This network, obviously simplified, does not explicitly consider either the affected neurons or the symptoms of PD which act as "intermediaries", as a consequence of the convolutions.

From this sagittal graph it is possible to find the total incidences contained in the fuzzy matrix $[\underset{\sim}{M}^*]$, whose valuation is equal to or greater than 0.9. We present them below:

$(a_1,a_1,b_{12}, b_{12}) \rightarrow 0.9$ $(a_3,a_3,b_1, b_6) \rightarrow 0.9$
$(a_3,a_3,b_1, b_1) \rightarrow 0.9$ $(a_3,a_3,b_1, b_9) \rightarrow 0.9$
$(a_3,a_3,b_1, b_2) \rightarrow 0.9$ $(a_3,a_3,b_1, b_{10}) \rightarrow 0.9$
$(a_3,a_3,b_1, b_3) \rightarrow 0.9$ $(a_3,a_3,b_7, b_7) \rightarrow 0.9$
$(a_3,a_3,b_1, b_5) \rightarrow 0.9$

A first look at these incidences leads us to select from among them, those that flow only between a set of affected neurons, A, and a single PD symptom element, B. This is the case of:

(a_1,a_1,b_{12}, b_{12})
(a_3,a_3,b_1, b_1)

(a_3, a_3, b_7, b_7)

where, the valuation of the incidence in the initial matrix, $[\underset{\sim}{M}]$, is always the same as in the matrix of accumulated incidences, $[\underset{\sim}{M}^*]$. In these three occurrences it is always 0.9. There is, then, no "forgotten effect" and the elements (a_1, b_{12}) (a_3, b_1) and (a_3, b_7) of the matrix $[\underset{\sim}{D}] = [\underset{\sim}{M}^*] - [\underset{\sim}{M}]$ will be assigned a null valuation.

Now, if this is clear from a purely numerical perspective, the same does not happen from the point of view of the diagnosis of the disease, since, although the degree or level of the symptoms is the same in the direct flow as in the accumulated, 0.9, the period of time in which this "degree or level" takes place in the first is instantaneous and in the second at various times due to the need to flow the incidence from a_i to a_k, $i, k = 1, 2, \ldots, 8$, and from b_j to b_1, in our cases a_1 to a_1 and b_{12} to b_{12}; from a_3 to a_3 and b_1 to b_1 and from a_3 to a_3 and b_7 to b_7.

Hence the interest, for the purposes of clinical analysis, in the construction and study of the above sagittal graph.

We now turn to the presentation and preliminary study of the accumulated valuations, which are found in the matrix $[\underset{\sim}{M}^*]$ with valuations equal to or greater than 0.8, $\alpha \geq 0.8$. To do this, we present the following sagittal graph under the conditions established in the case of $\alpha \geq 0.9$, but now with $\alpha \geq 0.8$.

The graph obtained is the following:

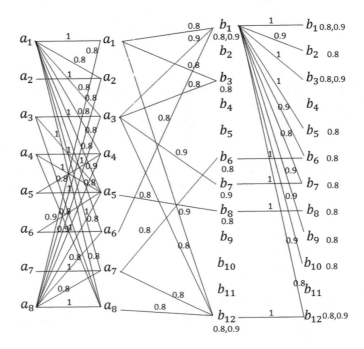

This graph shows that, for certain vertices, the incidence deposit has two capacities (degrees or levels), 0.8 and 0.9, depending on the flow of the channels that converge in it (arcs arriving at the vertex).

9 Intermediation Role of Some Affected Neurons and Certain Symptoms of PD

Our previous computation using the max-min convolution operator contains all affected neurons and all PD symptoms that play the role of intermediaries in the global circulation of incidence flows through the network of incidence channels.

As it has already been pointed out, the incident flow only crosses affected neurons or indirect PD symptoms and, therefore, follows a longer path, if the indirect channels allow a flow equal to or greater than the direct path.

In the study that we are carrying out, it is already feasible for us to establish all, all, the possible ways to link, at the desired levels, the affected neurons with the symptoms of PD. We have chosen high incidence levels, 0.9 and 0.8, on the understanding that the process is valid for any level.

We believe that, for the purposes of the methodology of this work and given the high importance of oblivions, we are going to present firstly those incidence relationships whose oblivion is greater than 0.9 and 0.8. These are the incidence of Retinal amacrine cells on Walk without alternating arms, with a oblivion level of 0.9, and the incidence of Neurons from basolateral nucleus of the amygdala on symptoms present in half body, with a oblivion of 0.8.

Incidence of retinal amacrine cells on walk without alterning arms

Initially estimated valuation: 0

Accumulated incident-incidence valuation: 0.9

Valuation difference (forgotten effect): 0.9

Relevant key interposed incidents: Retinal amacrine cells

Tremor
We expose the corresponding paths through the following subgraphs (Figs. 1 and 2).

Incidence of neurons from basolateral nucleus of the amygdala on camptocormia

Case 1.

Initially estimated valuation: 0

Accumulated incident-incidence valuation: 0.8

Valuation difference (forgotten effect): 0.8

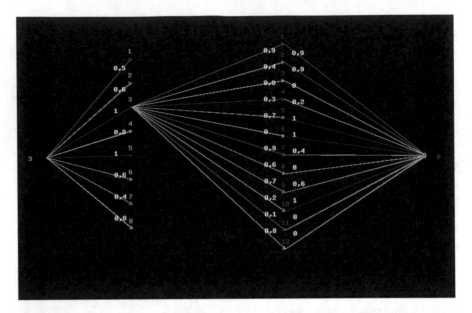

Fig. 1 Sagittal graph of incidence of retinal amacrine cells on walk without alterning arms

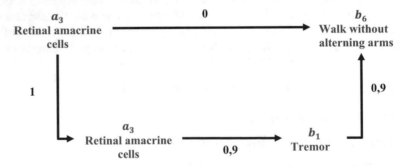

Fig. 2 Incidence of retinal amacrine cells on walk without alterning arms

Interposed Key Valuations: Neurons de pons acetylcholnergic

Tremor (Figs. 3 and 4).

Case 2.

Initially estimated valuation: 0

Accumulated incident-incidence valuation: 0.8

Valuation difference (forgotten effect): 0.8

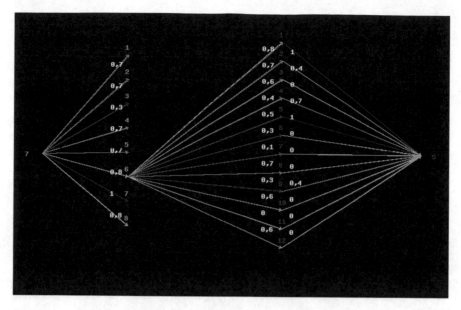

Fig. 3 Sagittal graph of incidence of neurons from basolateral nucleus of the amygdala on camptocormia (Case 1)

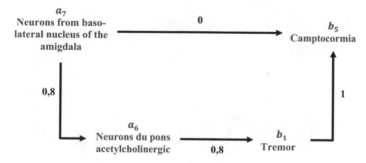

Fig. 4 Incidence of neurons from basolateral nucleus of the amygdala on camptocormia (Case 1)

Relevant interposed key assessment: Glutaminergic neurons from parahyp-pocampal gyrus

Tremor (Figs. 5 and 6).

So far, we have focused our discussion on a high requirement for forgetting, $\alpha \geq 0.8$. It is legitimate to ask if, by minimally lowering the level, the number of incidents with important oblivion increases significantly.

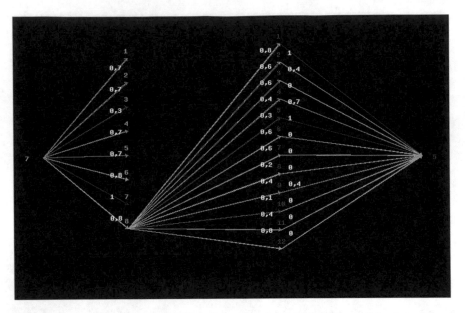

Fig. 5 Sagittal graph of incidence of neurons from basolateral nucleus of the amygdala on camptocormia (Case 2)

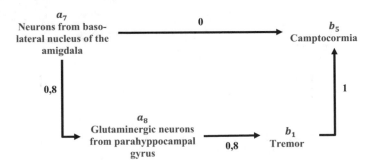

Fig. 6 Incidence of neurons from basolateral nucleus of the amygdala on camptocormia (Case 2)

We have looked for the answer and the result shows that by reducing one tenth, up to $\alpha = 0.7$, to the previously described forgotten incidences, ten more other incidences must be added with the consequent paths. These are the following:

(a_3, b_{10}), (a_4, b_7), (a_4, b_{10}), (a_5, b_6), (a_5, b_{12}), (a_6, b_7), (a_6, b_{11}), (a_7, b_{10}), (a_7, b_{11}), (a_8, b_{10})

10 Conclusions

More than 40 years ago, the incorporation of Lotfi Zadeh's fuzzy concept into economic research works allowed Kaufmann and Gil-Aluja to develop a set of nine pioneering works in the new sense that Economic Science was taking, laying the foundations of what later it would be called the School of Humanist Economics of Barcelona. Among these works we highlight for our purposes one of them, Models for the investigation of forgotten effects on which these authors formulated the "Forgotten Effects Theory" [10].

This theory is timeless and of great applicability today, also in speculative works, and especially in the political, social, economic and financial management fields.

We ventured into the field of neurology thanks to the confluence of concerns of two scientists of such distant nationalities as the Israeli and the Spanish, both convinced that human thought still has many limitations, as it is not yet able to establish the reasoned and imaginative formalization of a long chain of connections that stretch far and wide in increasingly complex brain networks.

In this regard, we have taken information in this work and created an algorithm, in which we start from a set of 8 affected neurons and 12 symptoms of PD. It has been observed that from a relatively small number of direct relationships between affected neurons, and PD symptoms, the incidences immediately derive to a network of very high complexity.

The direct incidence matrix is therefore made up of $8 \times 12 = 96$ elements. However, the mental connections to cover all the relationships necessary to issue complete and at the same time personalized diagnoses are numerous, very numerous so as not to have oblivions.

If we consider the incidences of a single element of the affected neurons on a single element of the set of PD symptoms of this work, the mental connections would be as numerous as the arcs of the following sagittal graph:

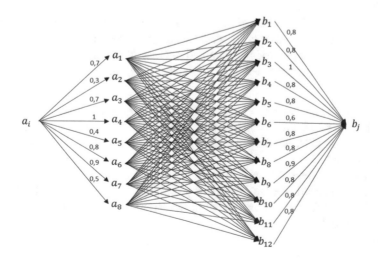

$$i = 1, 2, \ldots, 8$$
$$j = 1, 2, \ldots, 12$$

Beware; to consider all the incidence connections necessary to know at what level or degree each affected neuron acts on each affected neuron, each affected neuron on each symptom of PD and each symptom of PD on each symptom of PD, throughout the chain of incidences, we would have to multiply the arcs of this graph by 96.

It is not strange, then, that Transhumanism unites with "Big Data" and the "Digital Revolution", to turn the activity of those who are forced to issue diagnoses into livable, on which depend perhaps, human lives.

We have moved forward in time: with a humanistic algorithm like the one we are proposing, through which we are able (already today and by ourselves) to know absolutely all the possible mental connections that exist between the affected neurons and the symptoms of PD, thus, to find its degree or level of incidence without error or omission.

In the same way that we act in this way for PD, we can do it to solve other problems that remain in the diagnosis of other degenerative diseases.

The structure of the algorithm has been conceived in such a way that it has the highly recommended qualities by our teacher Arnorld Kaufmann. We mean flexibility and adaptability.

We have no doubt that its use can take place, not only in the medical space in which we have located it, but also in many others, we would say in all of them.

If somebody would ask us for any proof, there goes a memory from more than 15 years ago...

This is an interview that the great celebrated journalist, Lluis Amiguet [12], published in the newspaper "La Vanguardia", on July 8, 2006, with one of the most prestigious medical doctors in Spain in the twentieth century. We refer to Academic Hon. Dr. Ciril Rozman.

We do not wish to, nor can we, change a single letter of it, so we reproduce exactly the questions and answers to which we attribute an exceptional interest for our work.

After the illustrious interviewee pointed out that "the human being is a system in which everything is interrelated", the interviewer continues:

- For example......
  "I have treated many cases where the origin of the illness was far away from where the disorder arose. I remember that a specialist sent me a patient with a huge liver and asked me to locate the tumor. Well, it wasn't a tumor."
- What was it?
  "I did a more general examination and discovered that his heart was very fast and so I detected a problem of hyperthyroidism, which accelerated his heart rate and therefore forced the liver to work harder, and this overexertion caused the inflammation".
- Good example.

"Another patient who was vomiting was sent to me as sick to his stomach. I examined him and discovered that he had a very large prostate that obstructed his urinary tract and caused dysfunction in the kidneys, which caused accumulation of urea in the blood and intoxicate the gastric mucosa, that's why he was vomiting".

Human immersed in a system, everything depends on everything, in the same way as in all areas of life in society. We never tire of reminding ourselves that communication and management instruments must be flexible and adaptive to reach a high degree or level of effectiveness.

We expect a lot for the future of humanistic algorithms, created and developed in the field of Artificial Intelligence.

In the same way that the mobile phone has been irreversibly introduced into human lives, starting the digitization of our common existence, we would like that, from this modest work, awareness of the immense possibilities of Artificial Intelligence because of the fusion of Artificial Reasoning and Artificial Imagination.

This call is aimed at reducing, alleviating, or preventing the suffering of those who are waiting today, but also for those who will be waiting tomorrow, for a cure through an early and safe diagnosis. It is the eternal struggle for life and against the irrationality of death.

The future of the authors of this essay is as limited as their hopes of seeing their dream of a better world come true.

References

1. Gil Aluja, J. (2020). Personal contributions for a new theory of self-induced incidences. Harnessing tangible and intangible assets in the context of European integration and globalization challenges ahead. In *Proceedings of ESPERA 2019* (vol. 1).
2. Fahn, S., Mardsen, D., Goldstein, M., & Calne, D. (Ed.) (1987). *Recent developments in Parkinson's disease* (vol. 2, pp. 153–163, 293–304). Macmillan Health Care Information.
3. Askenasy, J. J. (2001). Approaching disturbed sleep in late Parkinson's disease: First step toward a proposal for a revised UPDRS. *Parkinsonism and Related Disorders, 8*(2), 123–131.
4. Burré, J. (2015). The synaptic function of α-synuclein. *Journal of Parkinson's Disease, 5*(4), 699–713.
5. Lathuilière, A., Valdés, P., Papin, S., Cacquevel, M., Maclachlan, C., Knott, G. W., & Schneider, B. L. (2017). Motifs in the tau protein that control binding to microtubules and aggregation determine pathological effects. *Scientific reports, 7*(1), 1–18.
6. Braak, H., Sandmann-Keil, D., Gai, W., & Braak, E. (1999). Extensive axonal Lewy neurites in Parkinson's disease: A novel pathological feature revealed by α-synuclein immunocytochemistry. *Neuroscience letters, 265*(1), 67–69.
7. Askenasy, J. J., Weitzman, E. D., & Melvin D. Y. (1987). Are periodic movements in sleep a basal ganglia dysfunction? *Journal of Neural Transmission, 70*(3), 337–347.
8. Askenasy, J. J., & Yahr, M. D. (1990). Parkinsonian tremor loses its alternating aspect during non-REM sleep and is inhibited by REM sleep. *Journal of Neurology, Neurosurgery and Psychiatry, 53*(9), 749–753.
9. Żadeh, L. A. (1965). Fuzzy sets. *Information and Control, 8*(3), 338–353.
10. Kaufmann, A., & Gil Aluja, J. (1988). *Modelos para la investigación de efectos olvidados*. Ed. Milladoiro.

11. Kaufmann, A., & Gil Aluja, J. (1993). *Técnicas especiales para la gestión de expertos.* Milladoiro, Vigo.
12. Amiguet, Ll. (2006). Solo es un buen médico quien es buena persona. *La Vanguardia.*

On the Use of Quasi-Sigmoids in Function Approximation Problems with Neural Networks

Francesco Carlo Morabito, Maurizio Campolo, and Cosimo Ieracitano

Abstract The paper aims to introduce and illustrate a novel kind of nonlinearity to be used as an activation function for the neurons of a Neural Network (NN). It is shown how the proposed function can be regarded as a generalization of the standard sigmoid. For this reason, it is referred to as quasi-sigmoid. The features of a backpropagation algorithm based on *quasi-sigmoids* are illustrated and commented. The properties of quasi-sigmoidal networks are compared to basic sigmoidal models on benchmark test cases as well as on sample function approximation problems in electromagnetics. The performance of quasi-sigmoidal networks as function approximators are shown to be generally superior to sigmoidal one by reason of the flexibility introduced and/or in terms of degrees of freedom of the model. As a by-product, the proposed activation function allows to carry out data-driven detections of nonlinearity in the data.

Keywords Deep learning · Neural networks · Quasi-Sigmoid · Inverse problems

1 Introduction

The determination of the weights of a Neural Network (NN) based on a suitable optimization/cost criterion is commonly carried out by means of the backpropagation algorithm (*backprop* hereafter) and its numerous variants. A relevant role in this procedure is played by the nonlinearity of the neurons (i.e., the activation function) and by the derivative of the activation function. Typically, the sigmoid (or squashing function) is used, also in the version of the hyperbolic tangent. Although commonly adopted, since it stems directly from the biological neural counterpart, the sigmoid is not the only, or necessarily the most convenient choice. By means of different functions, a better matching to the problem under study can be achieved, possibly reducing the computational burden of searching for an optimal set of weights of an

F. C. Morabito (✉) · M. Campolo · C. Ieracitano
AI_Lab, DICEAM Department, University "Mediterranea" of Reggio Calabria, Via Dell'Università, 89127 Reggio, Calabria, Italy
e-mail: morabito@unirc.it

© The Author(s), under exclusive license to Springer Nature Switzerland AG 2023
Y. P. Kondratenko et al. (eds.), *Artificial Intelligence in Control and Decision-making Systems*, Studies in Computational Intelligence 1087,
https://doi.org/10.1007/978-3-031-25759-9_11

input–output mapping. For example, in recent schemes based on Deep Learning (DL), the piecewise function known as ReLU (Rectified Linear Unit) is used to both reduce the number of free model's parameters to learn and to regularize the mapping. In particular, function approximation problems can take advantage of a variety of functions commonly used in numerical analysis, e.g., polynomials, Gaussian functions, wavelets, Chebyshev polynomials, and so on [13–15].

The aim of this paper is to introduce a novel nonlinear function, here referred to as quasi-sigmoid, which can be regarded as a generalization of the standard sigmoid. Backprop can easily incorporate quasi-sigmoids for the activation of the neurons; the potential advantages can then be simply assessed for the case at hand. Moreover, analytical considerations based on the features of quasi-sigmoids could lead to some interesting developments. For example, it is possible to give some insights on a successful technique proposed in [1] to accelerate backprop's convergence.

The remainder of the paper is organized as follows: Sects. 2 and 3 are devoted to a short review of backprop and to point out some considerations about the learning process in NNs; Sect. 4 shows that an NN of quasi-sigmoids could be substituted by an equivalent sigmoidal NN of proper topology. In the Sects. 5 and 6 a training procedure for the characteristic parameter of the quasi-sigmoids is derived, and the proposed approach is compared to the technique based on the introduction of a variable sigmoidal gain [2]. Section 7 presents some experimental results on the advantageous features of quasi-sigmoids: three test cases are presented, the first one concerns a simple function approximation problem and the other two regard the solution of weakly nonlinear identification problems in low-frequency electromagnetics. Finally, concluding comments are given in Sect. 8.

2 Static Backpropagation Algorithm in Function Approximation Problems

A two-layer feedforward NN can be described by the following well known equations:

$$
\begin{aligned}
y_l &= f_0(net y_l) = f_0\left(\sum_{k=1}^{N_h} w_{lk} h_k - b_l\right) l = 1, N_0 \\
h_j &= f_h(net h_j) = f_h\left(\sum_{i=1}^{N_i} v_{ji} x_i - c_j\right) j = 1, N_h
\end{aligned}
\tag{1}
$$

where f_h, f_0 are the activation functions of the hidden and the output layer, respectively. In particular, f_h is generally chosen as a sigmoidal function, i.e., $f_h(net h) = (1 + e^{-kneth})^{-1}$ while f_0 is often selected to be a linear function in the context of function approximation problems. From now on, the sigmoidal functions of any layer will be generically indicated with $f_s(x)$. A quasi-sigmoid is defined by adding a weighted in–out connection to each sigmoidal node and it will be denoted as $f_{qs}(x)$, where:

$$f_{qs}(x) = f_s(x) + k(x) \tag{2}$$

The weights' corrections for w depend on the derivative of the activation function, on the output error as well as on some typical design parameters (e.g., the learning rate, the momentum coefficient, the decay coefficient, etc.). The corrections on the hidden layer depend on the scaled *deltas* of the preceding layers, on the above mentioned parameters and on the activation function derivative of the hidden nodes. In the sigmoidal case, the $f_s'(x)$ introduces a suppression factor in the correction term which depends on the activation level of the neuron [3]. This is believed to introduce a stabilizing effect on the convergence properties of the method because the corrections are larger when the output is close to 0.5. Accordingly, the weights which are connected to the units in their midrange are the most changed. This is reasonable in a classification context. However, when $y_l = 1(0)$ and the target value is $d_l = 0(1)$, we have $y_l(1 - y_l) = 0$, and $\Delta w_{lk} = 0$. As a consequence, a maximum output error corresponds to no change in the weights. The same conclusion applies for the hyperbolic tangent function. By using a linear output the Δw_{lk} corrections just depend on the output errors and then the effect disappears. Since the corrections Δv_{ji} depend also on f_h', they are mostly influenced by this effect. The suppression factor impact is generally minimized by using heuristics, for example limiting the output range within (0.2, 0.8). The NN's outputs are then forced to stay away from the saturation regions. A different technique was proposed in [1]: the derivatives of the sigmoidal functions are modified by adding a small positive offset to them, prior to scaling the local error. The offset is noted to give some benefits; however, rules are not given for determining an optimal offset value. In addition, just one offset value has been used for all the weights of a layer. In contrast, quasi-sigmoids allow to automatically incorporate a different offset for each neuron and to determine its optimal value by implementing the learning technique proposed in Sect. 6. The offset allows to split the correction term in two parts: the first one is proportional to the sigmoidal derivative while the second one, proportional to the output error, dominates the correction in the case of a large error. Indeed, we have:

$$f'_{qs}(x) = f_s(x)[1 - f_s(x)] + k \tag{3}$$

The correction term is now given by:

$$\Delta w_{lk} = (y_l - d_l)h_k[y_l(1 - y_l) + k] = \delta_l h_k[\delta\left(f'_s\right)_l + k] \tag{4}$$

In this case, when $y_l = 0(1)$ and $d_l = 1(0)$, $|\Delta w_{lk}| = h_k, k \neq 0$. By a proper choice of k, one can benefit from the offset term without affecting too much the learning when $y_l \cong d_l$. Figure 1 shows the quasi sigmoid and its derivative, for different choices of the characteristic parameter k.

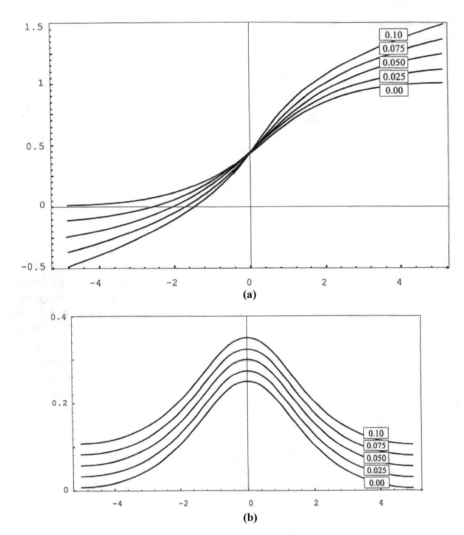

Fig. 1 The quasi-sigmoid (**a**) and its derivative (**b**) for different choices of the characteristic parameter k. It differs from the standard sigmoid by the added term linear k x, whose free parameter can be learned as the other weights of the representation. In the limit of $k = 0$, the standard sigmoidal nonlinearity is obtained

3 Considerations on the Learning Process in Multilayer NN

The strategy of decision making in classification and in function approximation problems is generally different. In the former class of problems, during learning the NN attempts to build an internal representation of the task in terms of features. Once assumed that a feature is present in the database, and that the NN has captured that feature in terms of its internal representation, the presence or absence of the feature in the incoming pattern is associated to the firing of one or more hidden nodes. A complex decision is carried out by combining the effects of more hidden nodes. In classification problems, the target of the processing is a binary-like concept. The goal of the minimization procedure is to firmly commit each node towards an ON/OFF condition. Once this condition has been reached, the suppression factor goes to zero, and the learning stops. In a pattern recognition problem, the approach is similar. In general, a robust representation of such problems requires a high variance of the weights incoming to a node and solutions where the average magnitude of the sigmoidal derivative is small. By analyzing the standard weight correction term in backprop, i.e.,

$$\Delta w = w(n+1) - w(n) = -\eta \frac{\delta E}{\delta w} = h_i f'_s(x) \tag{5}$$

one realizes that solutions in which a hidden node is firing ($h_i \neq 0$), and the output node has a nonzero derivative are penalized. This means that the output attempts to avoid the steep region of the sigmoid. The effect is not relevant in the early phases of learning where the output error dominates the correction and a proper choice of the initial weights seeding forces the net inputs to each node to assume a near zero value. However, once reduced the error, the learning proceeds balancing the effect of the two terms. For an input-hidden weight, there are two suppression factors and solutions in which both the involved nonlinearities are saturating will be favoured. This destabilizes solutions that hold the hidden nodes on the slope of the sigmoids. The final trained NN representation shows good properties of robustness: a little change of an input produces a little change in the output, as the derivatives $f'_s(x)$ are near zero.

In a function approximation problem, a successful mapping generally requires that the hidden nodes stay on the steep region of the sigmoid. In this case, it can be convenient to introduce a variable gain parameter which allows to flexibly deform the sigmoid like a spline, so matching the function to be approximated [2]. In such cases, backprop tends to minimize the influence of the "analog" properties of the representation, by also finding a solution to relevant inverse problems in biomedical engineering applications [5]. In the case of weakly nonlinear mappings as the ones commonly encountered in electromagnetic problems, at least in some regions of the multidimensional space of the solutions [6], it is advisable to use an NN model with a layer of weights which skips the hidden nonlinear layer, by directly connecting the inputs and the linear outputs. The global mapping is here constrained to have a significant linear component, which can also be determined off-line by various linear techniques. Actually, this linear trend implements a strong a priori knowledge about the mapping to be interpolated. This information may simplify the training, leaving to the nonlinear part of the NN the task of refining the approximation. In the course of the training the hidden nodes of such a NN are constrained to just yield a nonlinear correction term. Their outputs are forced to be very small, i.e., to sample the knee of the sigmoid, in a region where the transformation is mostly nonlinear. Indeed, by using standard backprop, the inputs show typically a larger variance with respect to the activations of the hidden nodes (as a consequence of the initial choice of relatively small weights). This implies that the in–out connections will be favored during learning, and this ultimately will favor the introduction of a dominant linear component on the mapping.

On the contrary, the proposed activation function has the potential of self-discovering whether the linear regressor can be useful in determining an accurate global fit or not. This will be shown by the regression results of a simple benchmark test proposed in Sect. 7, where the accuracy of a "high-way link" NN is one order of magnitude smaller than the proposed model.

The distributions of the outputs of a typical sigmoidal hidden node for a NN with and without direct linear connections and for a NN of quasi-sigmoids in a test function approximation problem are depicted in Fig. 2. Table 1 shows the mean values and the standard deviations, estimated on the training dataset, of the activities of two typical hidden nodes, for different NN models trained on the same test problem (discussed in Sect. 7). The quasi-sigmoidal nodes have $k = 0.09$. The quasi-sigmoidal NN tends to have a behavior in between the other two models, and in the limit of $k \to 1$ the hidden neurons show very different mean values and standard deviations of their outputs.

An interesting effect of introducing quasi-sigmoids derives from the following experimental consideration. A sigmoidal hidden layer NN yields a non-orthogonal expansion of a function. On the other hand, a NN with a linear hidden layer can

Fig. 2 Distributions of the typical hidden node activities for a quasi-sigmoidal NN (**a**), and for a NN with and without direct linear link (**b**). The activities for the hidden nodes in a quasi-sigmoidal NN are randomly distributed around 0.5 with larger variance with respect standard sigmoidal NN

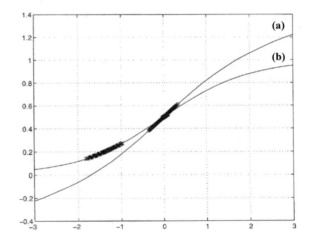

Table 1 Mean values and Standard Deviations of the activities of two hidden nodes for three different NN models in a test function approximation problem

NN type	Mean value	Standard Deviation
Sigmoidal	0.5125, 0.5203	0.0075, 0.0063
Quasi Sigmoidal	0.5487, 0.4321	0.0253, 0.0185
Sigmoidal with in–out links	0.2084, 0.1944	0.0344, 0.0307

work as an asymmetric principal component extractor [4, 7]. This means that the NN extracts the best linear combinations of the input data with respect to the minimization of the reconstruction squared error on the outputs. By using a hidden layer of quasi-sigmoids, the final effect is to force the learning to implicitly carry out a similar process.

By means of a statistical analysis of the hidden nodes' outputs, we have verified that the introduction of the linear term has sort of a whitening effect: the correlation coefficients among the hidden node outputs are strongly reduced with respect to the sigmoidal case. This is in contrast to the case of a NN with direct in–out connections in which the hidden nodes result in practice often duplicate. The topology with direct links reproduces quite well the linear part of the mapping but its refinement is difficult and tedious, requiring a post-processing pruning procedure. The hidden node outputs are indeed highly correlated. From a learning perspective there is also a question concerning the temporal scheduling of the training. The convenience of firstly (or separately) training the linear part is indeed problem-dependent. In the quasi-sigmoidal approach, the linear mapping is embodied in the hidden activation functions and thus there are no a priori choices to be carried out. Furthermore, the use of such functions implies a strong reduction of the degrees of freedom of the model, which impacts positively on the mapping's generalization properties and may act as a regularizer that overcome ReLu in this kind of problems.

4 Sigmoidal Equivalent of a Quasi-Sigmoidal NN

The first question we pose ourselves in order to define the properties of quasi-sigmoidal functions is their unboundness which may in principle represent a problem. In this Section, an informal interpretation of quasi-sigmoidal networks in terms of standard sigmoidal ones will be given.

A sigmoid (squashing) function is defined as a bounded function $f_s : \Re \to [0, 1]$, such that $\lim_{x \to -\infty_-} s(x) = 0$ and $\lim_{x \to +\infty_-} s(x) = 1$. The well known hyperbolic tangent function can be obtained from the basic sigmoid by a simple transformation, i.e.: $f_{th}(x) = -1 + 2 f_s(x)$. The quasi-sigmoid here proposed can easily be generalized to include as a special case both sigmoid and hyperbolic tangent functions, i.e.:

$$f_{qs}(x) = a_o + a_1(x) + a_2 f_2(x) \tag{6}$$

Starting from this formulation, for suitable values of a_o, a_1, a_2, the different sigmoidal activation functions can be obtained. Based on this more general formulation, a learning procedure could be introduced by considering the coefficients as free parameters. Actually, this is carried out in next Section.

It is commonly accepted that the universal approximation capabilities hold for bounded activation functions. The quasi sigmoid is an unbounded but continuous, nonconstant, monotonically increasing, differentiable, smooth function. The quasi sigmoid also satisfies a Lipschitz condition. In this sense, a multilayer feedforward NN of quasi-sigmoidal functions at the hidden layer and a linear output can be supposed to be capable of arbitrarily accurate approximating to any (vector) real-valued function over a compact set. Although this is then not requested, in principle, by backprop, and it does not seem a strict requirement in the proofs given in [8], we would like to stress that an NN of quasi-sigmoids is functionally equivalent to a properly designed sigmoidal NN.

In particular, when considering a NN with no squashing at the output layer, by using compact notations, we have:

$$\underline{f}_{qs} = \underline{b} + \underline{\underline{wh}}_{qs}\left(\underline{c}, \underline{\underline{v}}, \underline{x}\right) \tag{7}$$

where:

$$\underline{h}_{qs}\left(\underline{c}, \underline{\underline{v}}, \underline{x}\right) = k\left(\underline{\underline{v}}\underline{x} - \underline{c}\right) + \underline{h}_s\left(\underline{\underline{v}}\underline{x} - \underline{c}\right) \tag{8}$$

Hence:

$$\underline{y}_{qs} = \underline{b}^* + \underline{\underline{\gamma}}^*\underline{x} + \underline{\underline{wh}}_s\left(\underline{\underline{v}}\underline{x} - \underline{c}\right) \tag{9}$$

where $\underline{b}^* = k\underline{\underline{wc}}$, and $\gamma^* = k\underline{\underline{wc}}$.

For a NN of sigmoidal hidden layer, linear output, and a direct in–out link, we also have:

$$\underline{y}_s = \underline{b} + \underline{\underline{\lambda}}\underline{x} + \underline{\underline{wh}}_s\left(\underline{\underline{v}}\underline{x} - \underline{c}\right) \tag{10}$$

which is formally equivalent to (9).

In the case of quasi-sigmoidal outputs, we have:

$$\begin{aligned}
\underline{y}_{qs} &= f_{qs}\left(\underline{b} + \underline{\underline{wh}}_{qs}\left(\underline{\underline{v}}\underline{x} - \underline{c}\right)\right) == \underline{b}^* + \underline{\underline{\lambda}}\underline{x} + m\underline{\underline{wh}}_s\left(\underline{\underline{v}}\underline{x} - \underline{c}\right) \\
&+ f_s\left[\underline{b} + k\underline{\underline{w}}\left(\underline{\underline{v}}\underline{x} - \underline{c}\right) + \underline{\underline{wh}}_s\left(\underline{\underline{v}}\underline{x} - \underline{c}\right)\right]
\end{aligned} \tag{11}$$

where:

$$f_{qs}(x) = m(x) + f_s(x) \tag{12}$$

$$h_{qs}(x) = k(x) + h_s(x) \tag{13}$$

and $\underline{b}^* = m\underline{b} - mk\underline{\underline{wc}}$, and $\lambda = mk\underline{\underline{wv}}$.

The same mapping can easily be obtained by considering a sigmoidal NN, like the one described by the following expression:

$$\underline{y}_s = \underline{b} + \underline{\underline{\lambda}}\underline{x} + \underline{\phi}\underline{h}_s\left(\underline{\underline{v}}\underline{x} - \underline{c}\right)$$
$$+ f_s\left[\underline{d} + \underline{\underline{g}}\underline{x} + \underline{\underline{w}}\left(\underline{h}_s\left(\underline{\underline{v}}\underline{x} - \underline{c}\right)\right)\right] \tag{14}$$

In conclusion, the validity of the following statements has been shown:

- in the case of a NN with linear output, there is an equivalence between a NN with quasi-sigmoids at the hidden layer and a NN with sigmoids at the hidden layer and a direct layer of connections between the inputs and the outputs;
- in the case of an NN with sigmoidal outputs, there is an equivalence between a NN with hidden and output layers of quasi-sigmoids and a NN with two sigmoidal hidden layers and a linear output in which each layer is fully connected to the successive layers.

Figure 3 pictorially represents the above described equivalences.

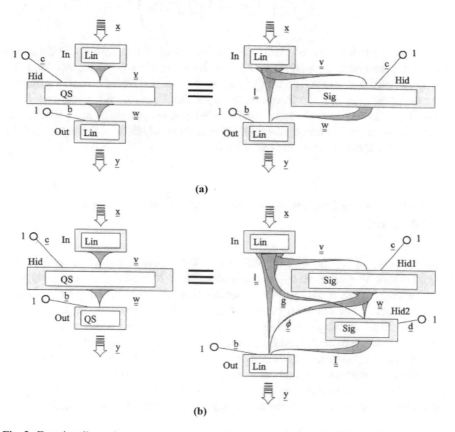

Fig. 3 Functionally equivalent models for sigmoidal and quasi-sigmoidal NN's. **a** linear output; **b** quasi-sigmoidal output

5 A Gradient Descent Procedure on the Steepness Parameter of Quasi-Sigmoids

One can improve the flexibility of quasi-sigmoidal mappings by introducing a gradient descent procedure on the coefficients of the linear part of the quasi-sigmoids, from now on referred to as $k'_h s$. By starting from small random values of $k'_h s$ (between 0.01 and 0.05), one can compute the derivative of the error function with respect to k_h.

Experiments show that learning the $k'_h s$ speeds up the training and stabilizes the whole procedure. In particular, in the limit case, the quasi-sigmoids could reduce to standard sigmoids ($k_h = 0$). It can be noted from the following relationship, valid in the case of linear output NN:

$$\frac{\partial E}{\partial k_h} = \left(\sum_{i=1}^{N_0} \delta_i w_{ih} \right) \left(\sum_{j=1}^{N_i} v_{hj} x_j \right) \tag{15}$$

that the gradient descent procedure on k_h has the effect of constraining the net inputs to the hidden nodes $\sum v_{hj} x_j$ to be not too large. In (15), δ_i represents the error computed on the output layer and $\sum_i \delta_i w_{ih}$ is the so called scaled error. This implies that the quasi-sigmoids are forced to work far from the saturation regions. Thus, there is a tendency to exploit the flexibility of quasi-sigmoids in the linear range, by avoiding, in the meantime, of exploring regions in which the unboundness of the functions could originate stability problems.

6 Quasi-Sigmoids and Gain Parameters in Standard Sigmoids

By expanding in Taylor series around net $= 0$ a sigmoid, a quasi-sigmoid and a sigmoid with varying gain, α, we notice that, close to the origin, a quasi-sigmoid with varying linear slope is somewhat equivalent to a standard sigmoid with a varying gain parameter. The common beneficial effect is to add an extra flexibility to the representation, allowing reduction of the training time and a better shaping of the function to be approximated. Although the use of a variable gain parameter can speed up learning, the functional expression of the resulting model remains unchanged, that is, the gain parameter can be embodied in the weight matrix, without changing neither the topology, nor the functional behavior of the NN. On the contrary, as it has been shown in the previous section, the quasi-sigmoids have also an impact on the model.

7 Experimental Results in Function Approximation Problems

In this Section, the performance of sigmoidal NN with and without direct input–output connections are compared to that of quasi-sigmoidal NN on three test function approximation problems.

The first problem concerns the approximation of a simple sinusoidal mapping. This simulation was selected with the aim of stressing the different behavior of quasi-sigmoids with respect to NN with direct in–out link. This is indeed a problem in which the NN with direct input–output connections is expected to show the worst performance. The different 1-3-1 NN models were trained from the same initial weight configuration ten times by using 150 data pairs (x_i, y_i). Table 2 shows the averaged RMS estimation error achieved on a test dataset of 100 patterns. The quasi-sigmoidal nodes were given the same k coefficient $(= 0.02)$. Table 2 shows that the sigmoidal mapping $(k = 0)$ yields the best performance. Hence, it was decided to start a training procedure on the k coefficients, by using the already trained quasi-sigmoidal NN, with a technique reminiscent of transfer learning. The procedure shows that by correcting just one k_h coefficient from 0.020 to 0.022 the quasi-sigmoidal NN improves its performance over that of the sigmoidal NN. By training simultaneously the k_h coefficients and the weights, starting from the values of $k_h = 0.020$, there was a significant improvement in the performance. The final values of k_h' s were 0.014, 0.012, 0.021. The performance of quasi-sigmoidal networks on some other benchmark approximation problems are reported in [11].

The details of the other two problems not strictly pertaining this work, as well as the results obtained, are discussed elsewhere [9, 10]. Here, we merely point out some aspects of interest for our discussion.

The problem concerns the reconstruction of the boundary of a plasma column in the vacuum chamber of a tokamak reactor for nuclear fusion applications. From the reading of a set of shielded magnetic measurements taken as close as possible to the plasma, but outside the vacuum chamber, it is required to infer the position and shape of the plasma column.

The input of the identification procedure is thus given by the set of available measurements. The aim of the identification process is to estimate the plasma configuration in terms of a set of geometric and shape parameters. In principle, the determination of the plasma shape would require the integration of a partial differential equation known as Grad-Shafranov (GS) equation with the boundary conditions

Table 2 RMS error in approximating a sinusoidal mapping via various different NN mappings

Sigmoidal NN	0.0071
Sigmoidal NN plus direct in–out link	0.0703
Quasi Sigmoidal NN (fixed k_h)	0.0089
Quasi Sigmoidal NN (variable k_h)	0.0055
Parameter Spread	0.309

obtained from the magnetic measurements. The solution of the GS equation in numerical codes is found from the condition of best fit with the measurement data of the plasma configuration calculated. To solve the problem faster, methods have been developed based on the statistical analysis of properly generated data bases simulating the expected input–output relationships. A large number of training pairs near those expected experimentally, listing the measured (input) and determined (output) plasma parameters, have been numerically generated. The NN is used to determine off-line the non-linear functional relationship between input and output parameters. The resulting model is then used on-line to track a real plasma discharge [12], From an analytical viewpoint, the corresponding problem (Cauchy problem for elliptic equation) is strongly ill-posed showing instability of the solution with respect to small error in specifying the boundary conditions. In this sense, it is of practical relevance to obtain also a little improvement of the performance.

The plasma boundary is typically specified by means of a finite number of state parameters. If the plasma has a circular cross-section, three parameters completely describe its shape. In the case of elongated boundary, more parameters are needed. In this paper, also for this more demanding case we consider just three parameters. The involved geometric parameters are pictorially depicted in Fig. 4, which represents an ideal cross-section of the fusion reactor.

The circular plasma case, for which an analytical description of the mapping is available, was solved by generating 300 training patterns of 10 inputs and 3 outputs each. The inputs are computed values of the magnetic flux in selected locations,

Fig. 4 Pictorial representation of the parameters to be regressed for the described magneto-hydrodynamic problem

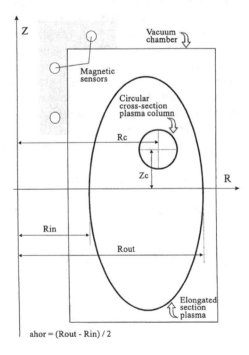

ahor = (Rout - Rin) / 2

Table 3 Physical RMS error and degrees of freedom in a functional approximation problem in nuclear fusion via various different NN mappings

	Rc (mm)	Zc (mm)	∧	DoF
Sigmoidal NN	5.1	27.8	13.9	87
Sigmoidal NN + direct in–out link	3.4	9.8	8.3	117
Quasi Sigmoidal NN (fixed k_h)	4.4	14.8	11.2	87
Quasi Sigmoidal NN (variable k_h)	4.0	10.2	9.8	93

while the outputs are the coordinates of the geometric center of the circular column (R_e and Z_c) and a third variable which describes the shape of the plasma within the chamber, ∧. By using standard pruning techniques, it was decided to use 6 hidden nodes. As always, the starting configuration was the same for the different models tested, and the experiment was repeated ten times with different initial randomly chosen parameters. The performances of different models is reported in Table 3.

The results are computed by taking the mean value of ten different NNs to filter the impact of initial seeding. In this case, the mapping is weakly nonlinear, thus it is rather expected that the NN with direct input–output linear connections is superior with respect to both the sigmoidal model and the quasi-sigmoidal model with fixed k_h' s.

However, if the k_h' s coefficients are considered as free parameters, the learning phase highlights the linear dependency by increasing the k_h' s values. In this case, the performance of the proposed model approaches that of the best model by using a reduced number of degrees of freedom, thus improving the sparsity of the representation (of about 20%).

In the case of elongated cross-section plasmas, the parameters to be interpolated are the R coordinate of the innermost and of the outermost boundary points (R_{in}, R_{out}), and the horizontal radius (a_{hor}) of the plasma column. The database is composed of 1000 in patterns each of 34 inputs and 3 outputs. The inputs are yet magnetic fluxes and fields plus a current measurement. The patterns (plasma equilibria) have been generated by using a highly validated numerical code, available at the *Max-Planck Institut fuer Plasmaphysics* in Garching bei Muenchen, Germany [9]. The performances of three different NN models (sigmoidal, sigmoidal + direct linear term, quasi-sigmoidal) in terms of physical RMS error on an out-of-sample database of 250 cases are reported in Table 4. The NN's training has been carried out by starting from the same initial weights seeding. The coefficients of the linear term of the quasi-sigmoids have been held fixed to the value of 0.03. All the NN's have 12 hidden nodes. In the case of the model with direct connection between the input and the output layers, the learning of the linear part has been carried out via standard backprop. A proper schedule has been introduced for the learning rate and the momentum coefficient, holding the same strategy for the three models. In particular, the learning rate was reduced from 0.05 to 0.005 in five successive steps. The momentum term

Table 4 Performance of three different NN models (sigmoidal, sigmoidal + direct linear term, quasi-sigmoidal) in terms of physical RMS error on an out-of-sample database of 250 cases

	R_{in} (mm)	R_{out} (mm)	a_{hor} (mm)	DoF
Sigmoidal NN	6.8	5.0	2.9	459
Sigmoidal NN + direct in–out link	4.7	2.5	2.2	561
Quasi Sigmoidal NN (fixed k_h)	3.9	1.9	1.8	459
Quasi Sigmoidal NN (variable k_h)	2.6	1.7	1.4	471

was also changed from 0.4 to 0.85 in five steps. Table 4 shows that there is a dramatic improvement of the performance with respect to the sigmoidal model (about three times better) and of about 50% with respect to the model with in–out connections (which has a much larger number of degrees of freedom).

The use of an unbounded activation function has never given rise to instabilities during training. Another effect of the quasi-sigmoids seems to be an enhanced extrapolation capability of the whole model. This appears to be a quite reasonable issue: if one considers that the sigmoids are saturating functions, when an input exceeds its training set maximum, the net input to a hidden node can force saturation, i.e., its output becomes insensitive to incremental changes of the input. On the contrary, in the quasi-sigmoidal case, the output will be nearly linearly extrapolated.

8 Concluding Remarks

The reduction of the weighting effect of the sigmoid derivative on the weight correction term of the backpropagation algorithm may yield some benefits in function approximation problems. The use of an activation function which explicitly provides such flexibility has been proposed and commented. Being the quasi-sigmoid functions unbounded, some doubts upon its universal approximation capabilities can be cast. The results given here do not provide a rigorous proof of the universal approximation capabilities of the quasi-sigmoidal NN's. The aim of this paper is rather to show constructively how such a model can play a role in practical function approximation problems. At the same time, the equivalence shown to standard sigmoidal models can be viewed as a simple a posteriori justification of the counterintuitive stability properties of the proposed activation function. On the other hand, the use of an equivalent sigmoidal model carrying a direct link between the input and the output nodes is not convenient both in terms of number of degrees of freedom and of computational complexity of the learning. The quasi-sigmoidal model has a self-tuning capability, as it determines, by itself, the weights of the linear parts of the underlying mappings. The determination of the optimal NN topology cannot be made in advance, thus, the decision about the use of the direct link is a design choice which could be disadvantageous in the particular problem. Here, the NN is left to determine during learning whether the mapping possesses (or not) a linear component.

Acknowledgements This theoretical contribution is presented in honor of Professor Janusz Kacprzyk, who has been working for many decades on finding suitable engineering representations for inverse problems mainly based on fuzzy logic and the related membership functions. The engineering community working on computational intelligence is grateful to him for spreading knowledge and friendship.

References

1. Fahlman, S. E. (1989). Faster-learning variations on back-propagation: An empirical study. In D. Touretzky, G. Hinton, & T. Sejnowski (Eds.), *Proceedings of 1988 Connectionist Models Summer School* (pp. 38–51). San Marco, CA: Morgan Kaufmann.
2. Kruschke, J. K., & Movellan, J. R. (1991). Benefits of gain: Speeded learning and minimal hidden layers in back-propagation networks. *IEEE Transactions, on Systems, Man, and Cybemetics, 21*(1), 273–280.
3. Humpert, B. K. (1994). Improving back propagation with a new error function. *Neural Networks, 7*, 1191–1192.
4. Kung, S. Y. (1993). *Digital neural networks*. Prentice Hall.
5. Azzerboni, B., Finocchio, G., Ipsale, M., La Foresta, F., & Morabito, F. C. (2002). A New approach to the detection of muscle activation by ICA and wavelet transform. *Lecture Notes in Computer Science, Springer-Verlag, Berlin-Heidelberg, 2846*, 109–116.
6. Morabito, F. C. (2000). Independent component analysis and feature extraction for NDT data. *Materials Evaluation, 58*(1), 85–92.
7. Baldi, P. F., & Hornik, K. (1995). Learning in linear neural networks: A survey. *IEEE Transactions on Neural Networks, 6*(4), 837–858.
8. Hornik, K., Stinchcombe, M., & White, H. (1989). Multilayer feedforward networks are universal approximators. *Neural Networks, 2*, 359–366.
9. Coccorese, E., Martone, R., & Morabito, F. C. (1994). Identification of non-circular plasma equilibria using neural network approach. *Nuclear Fusion, 34*(10), 1349–1363.
10. Campolo, M., La Foresta, F., Labate, D., Morabito, F. C., Lay-Ekuakille, A., & Vergallo, P. (2011). ECG-derived respiratory signal using empirical mode decomposition. In *Medical Measurements and Applications Proceedings (MeMeA), 2011 IEEE International Workshop on*, IEEE, Piscataway (NJ) (Vol. 1, pp. 399–403).
11. Morabito, F. C. (1996) Function modeling in sigmoidal and quasi-sigmoidal backpropagation. In *Proceedings of the World Congress on Neural Networks,* WCNN'96, San Diego, CA (pp. 229–233). New York: Lawrence Erlbaum Associates.
12. Morabito, F. C. (1995, November). Equilibrium parameters recovery for experimental data in Asdex-Upgrade elongated plasmas. In *Proceedings of the 1995 IEEE International Conference on Neural Networks,* Perth, Western Australia (Vol. 2, pp. 937–941). New York, NY, USA: IEEE Press.
13. Dubey, S. R., Kumar Singh, S., & Chaudhuri, B. B. A comprehensive survey and performance analysis of activation functions in Deep Learning. arXiv:2109.14545.
14. Molina, A., Schramowski, P., & Kersting, K. P. (2020). Activation units: End-to-end learning of flexible activation functions in deep networks. In *International conference on learning representations*.
15. Macedo, D., Zanchettin, C., Oliveira, A. L., & Ludermir, T. (2019). Enhancing batch normalized convolutional networks using displaced rectifier linear units: A systematic comparative study. *Expert Systems with Applications, 124*, 271–328.

Human-Centric Question-Answering System with Linguistic Terms

Nhuan D. To(iD), Marek Z. Reformat(iD), and Ronald R. Yager(iD)

Abstract Current Question-Answering Systems can automatically answer questions posed by humans in a natural language. However, most of them can only answer questions that do not contain imprecise concepts and lead to short answers. This paper introduces a Human-Centric Question-Answering system capable of answering questions containing user-defined, personalized linguistic terms. The system works with information represented in the form of knowledge graphs. We describe the system and present its main components, emphasizing a few extensions that make the system distinctive. We illustrate the execution of the system with a few examples.

Keywords Question-answering systems · Linguistic terms · Knowledge graphs · Human-centric · Natural language processing

1 Introduction

Question-Answering Systems (QASs) involve multiple tasks such as information retrieval, natural language processing, and machine learning techniques. They are required to construct systems that automatically answer questions posed by humans in a natural language. Many such systems have been developing so far.

N. D. To (✉)
Nam Dinh University of Technology Education, Nam Dinh, Vietnam
e-mail: tdnhuan@nute.edu.vn

M. Z. Reformat
University of Alberta, Edmonton, Canada T6G 1H9
e-mail: reformat@ualberta.ca

University of Social Sciences, 90-113 Łódź, Poland

R. R. Yager
Iona College, New Rochelle, NY, USA
e-mail: yager@panix.com

King Abdelaziz University, Jeddah, Saudi Arabia

© The Author(s), under exclusive license to Springer Nature Switzerland AG 2023   239
Y. P. Kondratenko et al. (eds.), *Artificial Intelligence in Control and Decision-making Systems*, Studies in Computational Intelligence 1087,
https://doi.org/10.1007/978-3-031-25759-9_12

The QAS is composed of three sections: question processing, processing of retrieved documents, and answer processing. Once a user asks a question, the system analyzes it and transforms it into a query. The system queries a knowledge base. The base could be a structured database, for example, WDAqua-core0 [5], or an unstructured set of documents as in the case of the Wikipedia DrQA [3]. Finally, the obtained answer is examined, and the most relevant information is presented to the user.

Yet, there are two aspects of questions that the current QAS does not address—presence of imprecise terms, such as 'large,' 'new,' 'close' that occur in questions quite often, and their very user-dependent meaning.

Here, both issues are being addressed. A Human-Centric Question-Answering system is introduced and described. The essential capabilities of the system can be summarized as follows:

- the system accepts questions with linguistic terms—it can recognize linguistic terms in a question and contains procedures for collecting data necessary to construct a meaningful answer to the asked question;
- the system is equipped with a graphical interface that enables users to define their own, personalized definitions of linguistic terms.

The system is described, and its main components are presented. Finally, the execution of the system is illustrated with a few examples.

2 Question-Answering System *LingTeQA*: Overview

The proposed Human-Centric Question-Answering System—called *Linguistic Term Question-Answering system (LingTeQA)*—is equipped with several components enabling it to provide the users with a more human-friendly experience, Fig. 1.

The system—*LingTeQA*—allows for using vague concepts, asking questions that required additional processing, and providing more summary-like answers. Several algorithms and methods have been developed necessary for doing the following tasks: (1) template generation from pairs ⟨*question—SPARQL*[1]*query*⟩; (2) question to query translation based on generated templates; (3) data collection from Knowledge Graphs (KGs) by executing generated queries; (4) construction of definitions of user-based linguistic terms and quantifiers (if any) so questions that contain such terms can be answered accordingly to their understanding of the terms; (5) data summarization in forms of linguistic summaries and aggregated values (if suitable).[2]

Section 3 introduces a basic form of representing questions—trees—that is a vital idea utilized in the process of building a repository of templates for the system. The construction of the repository is described in Sect. 4. The system's utilization, i.e.,

---

[1] SPARQL is a semantic query language for querying information represented in the form of RDF triples. RDF—Resource Description Framework—is a method for representing graph data.

[2] A summarization process is a topic of other publications.

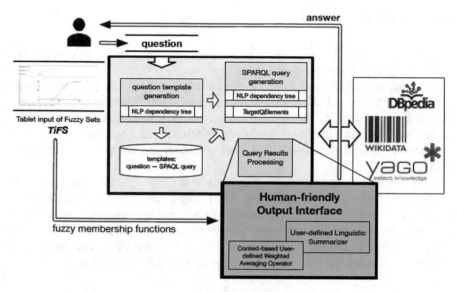

Fig. 1 *LingTeQA*: A human-centric question answering system

answering question, is explained in Sect. 5. Section 6 focuses on answering questions with linguistic terms: how to detect terms, how to collect data, and how to analyze to provide answers. The related work is included in Sect. 8.

3 Question Representation

Questions asked in natural language are represented using a hierarchical structure called a *dependency tree*. Such a tree is a set of connected labeled nodes, each reachable via a unique path from a distinguished root node.

3.1 Phrasal Dependency Tree—Definition

A *phrasal dependency tree* of a sentence is a directed graph. The sentence phrases are nodes, and grammatical relations are edge labels. A phrase, also called a *constituent*, is a set of words that act together as a unit. Every phrase has a head that determines its category. If the head is a noun, the phrase is a *Noun Phrase*, abbreviated NP, or a proper noun—PN. If the head is a verb, the phrase is a *Verb Phrase*—VP.

3.2   Tree Generation

Once the user enters her question, the following steps are performed to generate a phrasal dependency tree representing the question.

The sentence is processed using the dependency parser, *spaCy* [9]. The obtained typed dependencies are *RDF triples*, i.e., pairs of words from the question with a relation between them. For example, given the question *'What is the time zone of Salt Lake City?'* the parser generates typed dependencies whose graphical representation is a tree as illustrated in Fig. 2. The *part-of-speech* (POS) tags are presented in parenthesis.

The dependency tree, also called a word-level tree, is further modified using specialized heuristics. As a result, it becomes a phrase-level tree. In particular, words that involve a multiword-expression relation recognized as *compounds* are combined to form phrases. Part-of-speech tags of phrases are also renamed to indicate common *Noun Phrase* (NP), *Proper Noun phrase* (PN), *Verb Phrase* (VP). The phrase-level dependency tree for the above exemplary question is illustrated in Fig. 3.

4   *LingTeQA*: Template Repository Construction

The *template generator* is responsible for constructing a pair ⟨*Question_ template— Query_template*⟩, denoted by ⟨Qst, Qrt⟩, from a given a pair ⟨*natural language question—SPARQL query*⟩ named ⟨Qs, Qr⟩. This process is done in the context of a specific Knowledge Graph (K) identified by the user. The template generator is composed of a question template generator and a query template generator.

Fig. 2 Word-level dependency tree of the question *'What is the time zone of Salt Lake City?'*

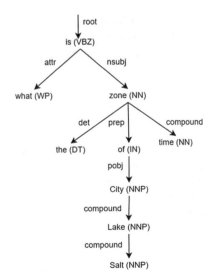

4.1 Question Template Generator

Construction of a *question template* (Qst) based on a given question (Qr) is performed by traversing a phrasal dependency tree of the question and identifying the part-of-speech tags and the *accessing paths* of every visited node. The accessing path of a node in the tree is a sequence built of edge labels starting from the root following the connections to reach the node. Algorithm 1 shows the process of generating a question template.

Algorithm 1 Question template generation

1: **procedure** GENERATING A QUESTION TEMPLATE(Qs : *a user question*)
2:    $wordLevelDependencyTree \leftarrow spaCyDependencyParser(Qs)$
3:    $phrasalLevelDependencyTree \leftarrow treeTransformation(wordLevelDependencyTre)$
4:    $questionTemplate \leftarrow preOrderTreeTraversal(phrasalLevelDependencyTree)$
5:    $return\ questionTemplate$
6: **end procedure**

An example of constructed question template (Qst) for the question *'What is the time zone of Salt Lake City?'*, whose phrasal-level dependency tree is depicted in Fig. 3, is shown in Fig. 4.

4.2 Query Template Generator

Given a pair ⟨*natural language question—SPARQL query*⟩ and a target Knowledge Graph (K),[3] the query template generator constructs a *SPARQL query template* (Qrt)

Fig. 3 Phrase-level dependency tree of the question *'What is the time zone of Salt Lake City?'*

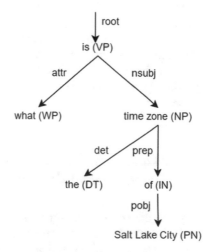

---

[3] A Knowledge Graph identified by the user as a data source for querying.

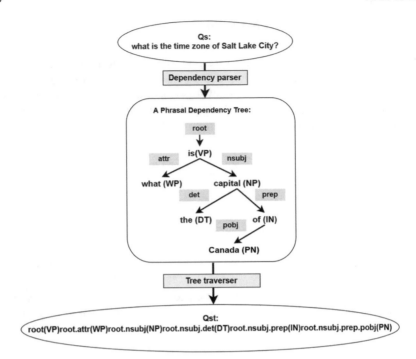

Fig. 4 Question template generation process—a question template (Qst) generated based on the question: *'What is the time zone of Salt Lake City?'*

from the SPARQL query (Qr). It is done by replacing the elements of the SPARQL query (Qs) with the items and phrases from the knowledge graph (K) mapped to the phrases of the question (Qs). Algorithm 2 presents the query template generation process.

The complete process of generating a pair ⟨Question_ template—Query_template⟩, i.e., ⟨Qst, Qrt⟩ performed on the question *'What is the time zone of Salt Lake City?'* is shown in Fig. 5. The process of mapping phrases form a natural language question into Knowledge Graph's semantic items (URIs) is described in the next subsection.

4.3 Mapping Natural Language Expressions into Knowledge Graph's Semantic Items

Mapping the *natural language expressions/phrases* found in the user's question into target Knowledge Graph's semantic items is performed by matching such graph's items as *classes*, *entities*, and *properties* identified with URIs to the expressions that contain the URI labels.

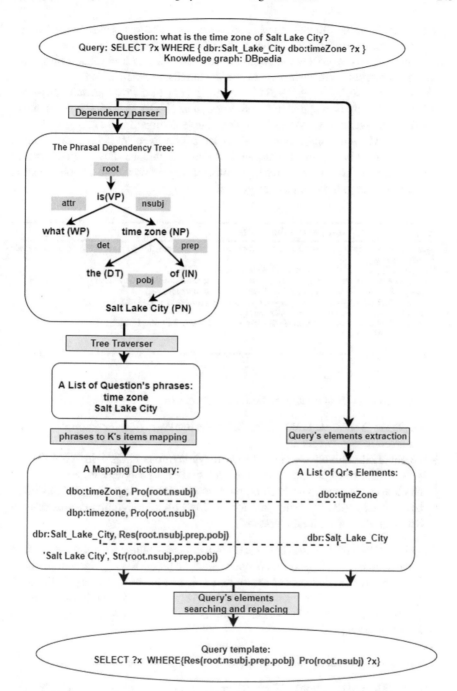

Fig. 5 Query template generation process—a query template (Qrt) generated based on the question: '*What is the time zone of Salt Lake City?*'

However, the variability of natural language phrases and terms used in the Knowledge Graph means differences between the phrases and the terms. Usually, external sources such as WordNet and some lexicons are used to find related phrases—synonyms, different word surfaces—to narrow the discrepancies.

For example, well-known QASs—AquaLog [13] and Power-Aqua [14]—use WordNet [6] for finding synonyms and antonyms. NEQA [1], on the other hand, uses manually created lexicons such as predicate lexicon and class lexicon.

In *LingTeQA*, items and property labels from WordNet, Wikidata, and a small manually created lexicon are used to obtain additional phrases. The original phrases and the additional phrases are then mapped into the target Knowledge Graph semantic items with respect to their lexical categories.

Algorithm 2 Query template generation

1: **procedure** GENERATING A QUERY TEMPLATE(Qs : *a user question*; Qr : *a SPARQL query*; K : *aknowledgegraph*)
2:    $wordLevelDependencyTree \leftarrow spaCyDependencyParser(Qs)$
3:    $phrasalLevelDependencyTree \leftarrow treeTransformation(wordLevelDependencyTree)$
4:    $questionPhrases \leftarrow phraseExtraction(phrasalLevelDependencyTree)$
5:    $mappingDictionary \leftarrow phrases - To - KG - elements(questionPhrases, K)$
6:    $queryElements \leftarrow queryElementExtraction(Qr)$
7:    $queryTemplate \leftarrow mappingAndReplacing(Qr, queryElements, mappingDictionary)$
8:    $return queryTemplate$
9: **end procedure**

- A verb phrase 'VP' is treated in the following steps:

  – the phrase is converted into a noun phrase using WordNet; for example, 'wrote' is converted into 'writer';
  – the obtained noun phrase is used to find synonymous phrases in WordNet; for instance, the word 'writer' is used to find 'author' as a synonym;
  – the phrases are used to find synonymous expressions using 'alt' labels in Wikidata; for example, by using 'author' to look up synonymous expressions in Wikidata, the system finds 'creator';

  the original verb phrase and additional noun phrases obtained via the three above mentioned steps are mapped into properties of the target Knowledge Graph using a label matching procedure; for example, the label matching between DBpedia's properties and obtained phrases is used by *LingTeQA* to identify the following set of properties *'dbp:writer', 'dbo:writer', 'dbp:author', 'dbo:author', 'dbp:creator', and 'dbo:creator'*.

- A common noun phrase 'NP' is processed in the following way:

  – it is used to find its canonical form (lemma) using WordNet, for example, 'writer' is a lemma of 'writers';

- the obtained noun phrase is mapped into the target Knowledge Graph class, for example, the class *'dbo:Writer* is obtained when mapping 'writer' into a DBpedia's class, further the obtained noun 'writer' is used to extract synonymous phrases using WordNet and Wikidata;

as mentioned above the system extracts 'author' and 'creator' as synonyms of 'writer'; the obtained phrases are mapped into DBpedia's properties such as *'dbp:writer', 'dbo:writer', 'dbp:author', 'dbo:author', 'dbp:creator', and 'dbo:creator'*.

- A proper noun phrase 'PN' is processed as follows:

  - it is mapped into the target Knowledge Graph entity (individual) using 'entity lookup' APIs and querying the Knowledge Graph using SPARQL. *LingTeQA* uses WikiData's API, DBpedia's API to obtain a list of entities. In case an empty result list is obtained, the system uses the API of the other Knowledge Graph and then it uses the property *owl:sameAs* to find entities in the other Knowledge Graph. Next, the system selects among obtained entities ones whose label is best matched with the proper noun;

  for example, the system retrieves *wd:Q16, wd:Q1121436, wd:Q2569593,* ... using Wikidata's API with the keyword 'Canada'. The system selects *wd:Q16* because its label best matches the keyword. Finally, the *owl:sameAs* property is used to procure *'dbr:Canada* as the targeted resource in DBpedia. The proper noun is also mapped into a string constant.

- an adjective 'AJ' is processed accordingly:

  - if the phrase belongs to a nationality/language lexicon, a corresponding country's name is retrieved, and the retrieved proper noun is processed similarly to a proper noun mentioned above; if it does not, WordNet is used to find a corresponding noun;

  for example, WordNet returns a noun 'height' when the adjective 'high' is queried. WordNet is also used to find attributes of the adjective. For instance, attributes of 'high' are 'level', 'degree', 'grade'. Obtained noun phrases are then processed similarly to common noun phrases. A comparative adjective or adverb is also mapped into comparative signs using a manual lexicon. For instance, 'higher/more' is translated into the sign '>'.

- a cardinal phrase, identified by a 'CD', is converted into a number using our own defined function.

Some expressions do not correspond to any vocabulary element. Examples are quantifiers like 'the most', comparative expressions like 'more than', cardinals, and superlatives. Instead, these expressions correspond to aggregation operations in

SPARQL, such as 'FILTER,' 'ORDER,' and 'LIMIT.' A fixed, dataset-independent meaning lexicon is used for mapping them into the query elements.

A placeholder in a query template (Qrt) is created to encode the mapping process when the question's phrase is mapped into a query element, such as a KG resource, a constant, or a comparative sign. A placeholder is a string of a function-like format. Its name is a three-letter string that indicates the type of the target element. For instance, 'Cla' means class, 'Res' means resource, 'Pro' means property. Its argument is a string that specifies the position of the phrase is the phrasal dependency tree determined by an accessing path. For example, 'Res(root.nsubj.prep.pobj)' is a placeholder indicating that a phrase attached to the node determined by the accessing path 'root.nsubj.prep.pobj' is mapped into a resource of the target Knowledge Graph during the process of instantiation of the query template.

5 *LingTeQA*: Answering Questions

During answering questions, *LingTeQA* processes a question asked by the user. In the beginning, the system analyzes it using the procedure applied during the template generation process and creates a question template. This template is used as a key to retrieve a corresponding SPARQL template from the template repository. The retrieved SPARQL template is populated with specific information extracted from the asked question and semantic items obtained from a mapping process (Sect. 4.3). Linguistic term(s) are detected. In the case when, there is no linguistic term, the system invokes a procedure for answering regular questions, otherwise, it invokes a process for answering questions with linguistic terms (Sect. 6.4). Algorithm 3 provides an overview of the answering question process.

Algorithm 3 Answering question process

1: **procedure** ANSWERING A QUESTION(Qs : *a user question*; T : *template repository*; wn : *WordNet*; KG : *a knowledge graph*)
2:     $phrasalLevelDependencyTree \leftarrow dependencyTreeGeneration(Qs)$
3:     $questionTemplate \leftarrow questionTemplateGeneration(phrasalLevelDependencyTree)$
4:     $queryTemplate \leftarrow queryTemplateRetrieval(questionTemplate, T)$
5:     $linguisticTermDictionary \leftarrow linguisticTermDetectingAndMapping(Qs, wn, KG)$
6:     **if** $linguisticTermDictionary \neq \emptyset$ **then**
7:         $answeringQuestionWithoutLTs(phrasalLevelDependencyTree, queryTemplate, KG)$
8:     **else**
9:         $answeringQuestionWithLTs(phrasalLevelDependencyTree,$
10: $queryTemplate, linguisticTermDictionary, KG)$
11:     **end if**
12: **end procedure**

To illustrate the question answering procedure of *LingTeQA*, let us take an exemplary question '*What is the capital of Canada?*'. Figure 6 shows the process.

Fig. 6 *LingTeQA*'s process
of answering questions
without linguistic terms

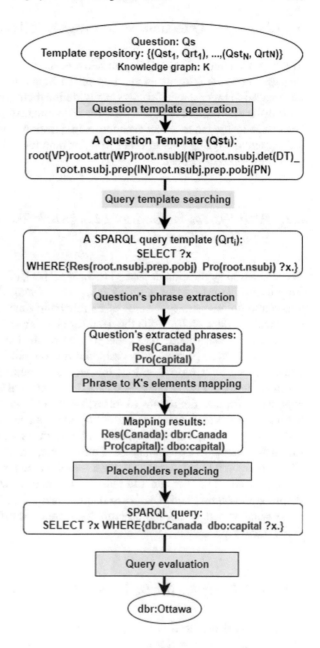

6 *LingTeQA*: Questions with Linguistic Terms

The presented above *LingTeQA* system has been extended to handle questions with *linguistic terms*. A specialized module, called *Tablet input of Fuzzy Sets* (TiFS) has been added in order to enable users easily define their membership functions representing their perception of imprecise concepts—linguistic terms. A special algorithm has been constructed to detect such terms in questions. Additional processes of collecting and analyzing data in order to determine the degrees of satisfaction of the terms have been developed.

6.1 iPad *System for Defining Linguistic Terms and Quantifiers*

An iPad-based application described here, *TiFS* [21], addresses a need for a system simplifying the process of defining fuzzy sets. Using their fingers, the users are allowed to define—draw—shapes of membership functions. The simplicity of the graphical interface of *TiFS*, and the need for entering only basic information makes the process of constructing fuzzy sets very convenient and straightforward.

In order to manage previously defined terms on multiple domains, the *TiFS* provides the user with several options located at the bottom of the screen. There are (from left to right, Fig. 7): NEW—to clean the content of input fields and prepare the application for new definitions of terms; DELETE—to remove an already existing definition of the term that is identified by the domain name and the term's name; SAVE—to store the entered definition of a term into the application storage, the stored data includes: a domain name, a term name, range and values of the term, and its shape entered by a user; RETRIEVE—to load the previously defined term from storage based on the domain and term names, and show the membership function shape on the screen; DISPLAY—tto display fuzzy function membership values of the previously defined and stored term, the term to be displayed is identified by the domain and term names.

6.2 Web Interface for Defining Linguistic Terms

LingTeQA [19] has been also implemented as a Web application. It is accessible at www.lingteqa.site, and it contains a user-interface, based on *TiFS*, for entering membership functions defining linguistic terms.

Whenever the system answers a question containing a linguistic term, it collects data from a specified (target) Knowledge Graph by executing a constructed query. It prepares a coordinate plane for users to enter their understanding of the term. The system determines an acceptable range: it finds out the minimum and the maximum

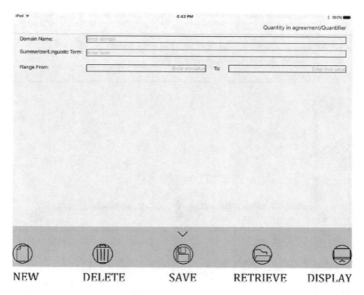

Fig. 7 Interface of *TiFS* for defining linguistic terms—iPad version

of the obtained results and uses them to scale the x-axis properly. A screenshot of the interface when answering the question *'give me large cities in Poland by population'* based on the data collected from DBpedia is shown in Fig. 8.

The selection of the option **See collected data** allows to inspect that actual data, while selection of **See histogram plot** shows the distribution of data. Screenshots of the collected data and its histogram plot are given in Figs. 9 and 10, respectively.

The application allows to draw or redraw (the option **redraw**) a shape of membership function on the provided coordinate plane. The function is submitted with the option **sendData**.

6.3 Function Fitting

Although *TiFS* provides users with an easy-to-use interface to draw shapes of membership functions, their drawings will never be perfect. Therefore, function fitting (curve fitting) is a needed technique to adjust values of the function parameters to best describe a set of data determined by the user-drawn shape.

There are commonly used categories of membership functions. They are triangular, trapezoidal, Γ-shape, S-shape, Gaussian, and Exponential-like functions. All of them are defined in the universe of real numbers. In the system, a parametric fitting approach is adopted. A user-drawn membership function and corresponding function obtained after the fitting process are illustrated in Fig. 11. Hereafter, the term 'a membership function defined by a user' means its system-fitted version.

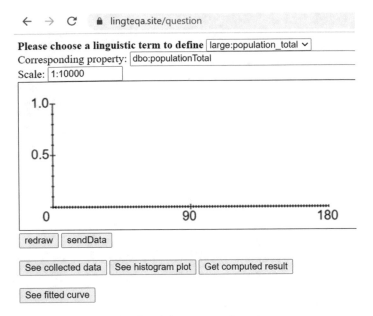

Fig. 8 Interface of *TiFS* for defining linguistic terms—web version

Fig. 9 Polish cities with largest population—top five

	city	population_total
76	http://dbpedia.org/resource/Warsaw	1793579.0
48	http://dbpedia.org/resource/Łódź	679941.0
9	http://dbpedia.org/resource/Wrocław	643782.0
68	http://dbpedia.org/resource/Poznań	534813.0
13	http://dbpedia.org/resource/Gdańsk	470907.0

Fig. 10 Polish cities with largest population—histogram

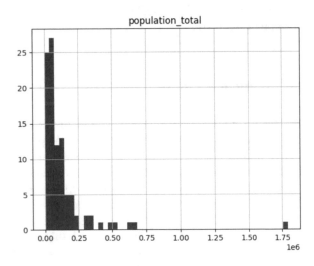

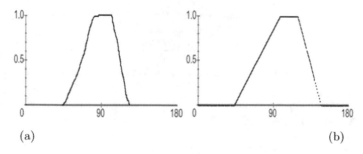

Fig. 11 *TiFS*—a membership function: **a** user-drawn; **b** system-fitted

6.4 Detecting Linguistic Terms in Questions

Humans can easily detect and capture the meanings of linguistic concepts. However, for a system/process it is not easy to determine *velocity* as the descriptor of *high speed* while *height* or *tallness* as the descriptor of *high building*. This section described the procedure that allows the *LingTeQA* system to (1) detect linguistic terms in a question, and (2) identify a descriptor (*property* of a triple in the target Knowledge Graph) for the detected linguistic term.

Although a concept of *linguistic term* is used quite often, there is no formal definition of it. In *LingTeQA*, a method based on WordNet [15] is applied for identifying linguistic terms.

WordNet's nouns, verbs, adjectives, and adverbs are grouped into sets of cognitive synonyms (*synsets*), each expressing a distinct concept. Synsets are distinguished by character codes indicating their types. Each synset contains a definition explaining the word's meaning and its attributes. An attribute of a noun is an adjective representing the noun's characterization. For example, the noun *weight* has two attributes—adjectives *light* and *heavy*. A list of adjectives and adverbs from WordNet has been constructed for the system. The list contains adjectives and adverbs that have non empty synset attribute sets or which definitions contain at least one of the following words: *'amount'*, *'quantify'*, *'quantifier'*, *'imprecise'*, *'approximate'*, *'size'*, *'range'*, *'scope'*, *'degree'*, *'quantity'*, *'quantities'*. This list constitutes a list of the system's recognizable linguistic terms.

Via analyzing the question, the system treats a constituent as a linguistic term if this term or its synonym or antonym belongs to the previously collected linguistic terms. Once a linguistic term is detected, the system identifies a base variable for the term. It is accomplished by determining a noun that the linguistic term modifies.

The process of detecting linguistic terms is presented as Algorithm 4, while the process of answering a question with linguistic terms as a diagram is given in Fig. 12.

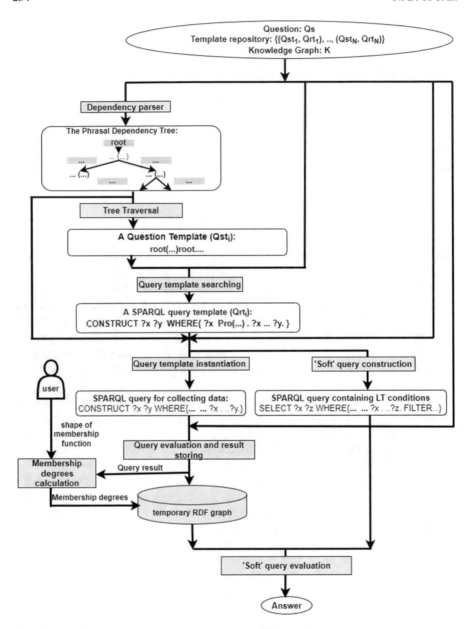

Fig. 12 *LingTeQA* process of answering questions with linguistic terms

---

Algorithm 4 Identify and map linguistic terms to KG's properties

---

1: **procedure** DETECT LINGUISTIC TERMS(Q : *a user question*; *wn* : *WordNet*; *KG* : *a knowledge graph*)
2:     $POSofQ \leftarrow POS\_tagger(Q)$
3:     $adjPhrases \leftarrow ExtractAdjPhrase(POSofQ)$
4:     $linguisticTerms \leftarrow \emptyset$
5:     **while** $i <= len(adjPhrases)$ **do**
6:         $adj, noun \leftarrow splitAdjPhrase(adjPhrases[i])$
7:         $attributes \leftarrow getAttributes(adj, wn)$
8:         **if** $attributes \neq \emptyset$ **then**
9:             $pro \leftarrow noun2pro(noun, KG)$
10:            **if** $pro \neq \emptyset$ **then**
11:                $dic.add(adjPhrases[i], pro)$
12:            **else**
13:                $cla \leftarrow noun2class(noun, KG)$
14:                **if** $cla \neq \emptyset$ **then**
15:                    $instances \leftarrow getInstances(cla, KG)$
16:                    $superlative \leftarrow getSuperlative(adj, wn)$
17:                    $nounsInKG \leftarrow getMostNoun(instances, KG, noun, superlative)$
18:                    $pro \leftarrow noun2pro(nounsInKG, KG)$
19:                    **if** $pro \neq \emptyset$ **then**
20:                        **for** $(p \in pro)$ **do**
21:                            $dic.add(adjPhrases[i], p)$
22:                        **end for**
23:                    **end if**
24:                **end if**
25:            **end if**
26:        **end if**
27:        $i \leftarrow i + 1$
28:    **end while**
29:    *return dic*
30: **end procedure**

---

7 *LingTeQA*: Answering Questions with Linguistic Terms

The method applied to answer a question that contains a linguistic term is illustrated here. The considered question-request is *'Give me large cities in Canada by population'*.

The detected linguistic term in the question is *large*. The phrasal dependency tree of the question is shown in Fig. 13.

Fig. 13 Phrasal dependency tree for the example question with the term *large*

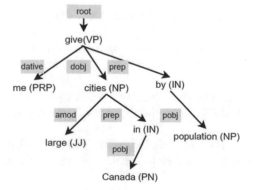

Table 1 Canadian cities and their population based on DBpedia

City	Population total
http://dbpedia.org/resource/Montreal	1704694
http://dbpedia.org/resource/Ottawa	934243
http://dbpedia.org/resource/Winnipeg	705244
http://dbpedia.org/resource/Vancouver	631486
http://dbpedia.org/resource/Quebec_City	531902

The template (Qst) of the question constructed automatically is:

root(VP)root.dative(PRP)root.dobj(NP)root.prep(IN)root.dobj.amod(JJ)
root.dobj.prep(IN)root.prep.pobj(NP)root.dobj.prep.pobj(PN)

Based on it, *LingTeQA* identifies the corresponding SPARQL query template (Qrt) in the template repository. This template is:

SELECT DISTINCT ?a ?b
WHERE{?a Pro(type) Cla(root.dobj). ?a Pro(root.prep.pobj) ?b.
?a Pro(root.dobj.prep) Res(root.dobj.prep.pobj).}

LingTeQA instantiates this query template. As a result, the SPARQL query is constructed. The query is executed against the specified Knowledge Graph. The data—cities and their population—are extracted, in this example, from DBpedia. The constructed query (Qr$_1$) is:

SELECT DISTINCT ?city ?populationTotal
WHERE{?city rdf:type dbo: City. ?city dbo:populationTotal ?populationTotal.
?city dbo: country dbr: Canada.}

The system also constructs a 'soft' query (Qr2) containing linguistic filter condition:

SELECT DISTINCT ?city (?populationTotal as ?MembershipGrade)
WHERE{?a rdf:type dbo:City. ?a dbo:populationTotal ?populationTotal.
FILTER(LANG(?populationTotal) =' large').}
where prefixes used in the queries are defined as follows:

PREFIX dbo: <https://dbpedia.org/ontology/>
PREFIX rdfs: <http://www.w3.org/2000/01/rdf-schema#>

By evaluating the query Qr1 against DBpedia, *LingTeQA* receives a list of cities along with their population. First five cities in the query's result are given in Table 1.

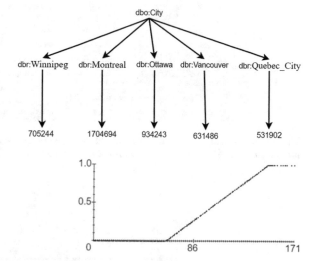

Fig. 14 Temporary RDF graph storing data for answering the question with the linguistic-term *large*

Fig. 15 Membership function drawn by the user describing linguistic term *large* (range: 0–171*10k)

The collected data are stored in the form of an RDF graph (G) whose graphical representation is given in Fig. 14.

The collected data is analyzed, and the system determines a range for the universe of discourse for defining the membership function representing the term *large*. A coordinate is presented to the user to draw a 'shape' of membership function describing her interpretation of the linguistic term *large*. Figure 15 shows a user-drawn fitted 'shape' depicting the term *large*.

LingTeQA calculates, based on the fitted fuzzy set, membership grades corresponding to the obtained values of populations of cities. It also constructs triples with objects representing calculated degrees of membership for the term *large*. The triples are added to the collected cities' nodes. We visualize the updated RDF graph in Fig. 16.

By evaluating the constructed 'soft' query Qr2, *LingTeQA* provides a list of the top five cities with the highest membership grades for the term *large*. As it is known,

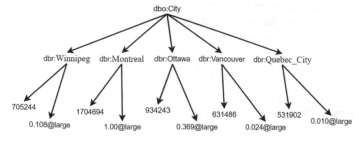

Fig. 16 RDF graph of collected data: cities with their populations, and degrees to which the populations satisfy the linguistic term *large*

← → C 🔒 lingteqa.site/runsquery

	city	MembershipGrade
31	http://dbpedia.org/resource/Montreal	1.000
286	http://dbpedia.org/resource/Ottawa	0.369
45	http://dbpedia.org/resource/Winnipeg	0.108
171	http://dbpedia.org/resource/Vancouver	0.024
195	http://dbpedia.org/resource/Richmond,_British_Columbia	0.011

Fig. 17 Answer to the question *'give me large cities in Canada by population'* with the user-provided membership function representing the term *large* shown in Fig. 15

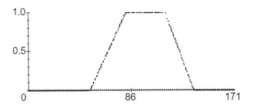

Fig. 18 'shape' of membership function drawn by user describing linguistic term *large* (10k people)

← → C 🔒 lingteqa.site/runsquery

	city	MembershipGrade
286	http://dbpedia.org/resource/Ottawa	1.000
45	http://dbpedia.org/resource/Winnipeg	0.655
171	http://dbpedia.org/resource/Vancouver	0.409
241	http://dbpedia.org/resource/Quebec_City	0.076
0	http://dbpedia.org/resource/Barkmere,_Quebec	0.000

Fig. 19 Answer to the question *'give me large cities in Canada by population'* with on the user-provided membership function representing the term *large* shown in Fig. 18

the grades depend on the personal perception of the term *large* provided by the user. A screenshot of the question result shown to the user is shown in Fig. 17.

However, if the user draws an alternative shape of the membership function representing the term *large* as illustrated in Fig. 18 then the user is presented with a different result, Fig. 19.

8   Related Work

An essential aspect of the utilization of fuzzy sets and fuzzy set-based technologies in multiple areas of industrial and commercial applications is related to their construction processes. Since introducing the concept of fuzzy sets, there have been multiple examples of methods and techniques addressing the issue of building the most suitable fuzzy sets, which means determining shapes of membership functions.

8.1 Construction of Fuzzy Sets

In general, there are several techniques for constructing fuzzy sets. They can be classified into user-driven and data-driven techniques.

The user-driven approaches can be further divided into direct methods and indirect methods.

In a direct method, an expert is expected to assign to each given element $x \in X$ a membership grade $A(x)$ that, in her opinion, best captures the meaning of the linguistic term represented by the fuzzy set A. There are two ways to do this: (1) via defining the membership function in terms of a justifiable mathematical formula; or (2) via exemplifying the function for selected elements of X. The latter approach requires answering questions such as "what is the degree of membership of x in A" or "what is the degree of compatibility of x with L_A" [12].

In an indirect method, an expert is asked questions of the form 'what are elements of X belong to fuzzy set A at a degree not lower than α?' where α is a certain level (threshold) of membership grades in [0, 1]. The expert identifies the elements from the corresponding α-cut of A. Using the process for several selected values of α, a fuzzy set is finally constructed from the α-cuts. Another method in this group involves a group of experts. Each of them is expected to answer questions: 'does x belong to the concept represented by a fuzzy set A?' for each element $x \in X$. The membership grade $A(x)$ is then calculated as the ratio of the number of 'yes' answers to the number of asked experts. The simplicity of these methods is their advantage. However, they could exhibit a lack of continuity because the membership grades are separately computed for elements of the universe of discourse [16].

The priority method introduced by Saaty [18] forms another interesting alternative. In this case, an expert is requested to compare elements in X in pairs according to their relative weights of belonging to A with a scale of 5, 7, or 9 levels. The expert will assign a high value of the available scale to the entry of a so-called reciprocal matrix at the position of row i-th and column j-th if x_i is strongly preferred to x_j when being considered in the context of the fuzzy set whose membership function we would like to estimate. The value of 1 at (i, j) position indicates that x_i and x_j are equally preferred. The eigenvector associated with the largest eigenvalue of the reciprocal matrix is then the needed fuzzy set. This method of constructing a fuzzy set helps the expert focus on only two elements once at a time, thus reducing uncertainty and hesitation while leading to a higher level of consistency [16].

In the data-driven approaches, fuzzy sets can be formed based on numeric data through their clustering. Fuzzy C-Means (FCM) is one of the commonly used mechanisms of fuzzy clustering. FCM clustering is completed through a sequence of iterations starting from a randomly initialized partition matrix and updating clusters' prototypes and the partition matrix until a certain termination criterion has been satisfied. The final partition matrix indicates the way of allocating the data to corresponding clusters. An entry u_{ik} is the membership degree of data x_k in the ith cluster [16]. There are also several data-driven methods in the literature. They differ in the complexity of construction processes. Among a variety of methods, membership functions are constructed using statistical, and probability-based algorithms, different clustering algorithms, entropy, and evolutionary computation [4, 8, 20]. In particular, the authors of [20] describe an unsupervised technique that uses self-organizing maps to generate fuzzy membership function. Another unsupervised method is proposed in [4]. The authors propose a method based on bandwidth mean-shift and robust statistics to construct membership functions. They use it to build triangular and trapezoidal functions representing underlying data. Also, the technique can determine the number of such functions. Chen et al. [4] proposes a gradient pre-shaped fuzzy C-means (GradPFCM) algorithm to generate better transparent membership functions. GradPFCM will preserve the predefined transparent shapes of membership functions during the gradient descent-based optimization of the clustering algorithm.

Fuzzy sets constructed using the methods mentioned above are almost invariably normalized, convex, and distinct. While those properties are undoubtedly valuable for many applications, they limit the selection of general shapes that may be used for representing membership functions for linguistic terms [7].

8.2 Fuzzy Queries and Relational Databases

To the best of our knowledge, *LingTeQA* is the first system that can answer questions containing linguistic terms over RDF knowledge graphs. However, fuzzy query languages extending the standard query language (SQL) and tools for fuzzy querying to relational databases were already proposed. Here we name a few noticeable ones.

FQUERY [10, 11] is a family of fuzzy-logic-based querying systems. It was developed and implemented by Kacprzyk and collaborators as a Microsoft Access "add-on" to extend its capabilities of handling fuzzy terms. FQUERY uses a set of predefined fuzzy terms maintained and developed by users. The fuzzy values are defined as fuzzy sets on $[-10, 10]$ interval, whereas the fuzzy linguistic quantifiers are defined as fuzzy sets on $[0, 10]$ interval instead of the original unit one. The membership functions of fuzzy values are trapezoidal. The system has been successfully applied in querying databases and data summarization with linguistic terms.

Summary SQL [17] is a fuzzy query language introduced by Rasmussen and Yager. In Summary SQL, an attribute in a database (DB) is associated with a collection of fuzzy concepts defined via membership functions as fuzzy subsets over the attribute domain. Summary SQL allows for using linguistic terms in a query.

For example, "select all persons where the height is tall" is such a fuzzy query. The query's result is a ranked fuzzy subset over the elements from the DB, including objects and their degree of satisfaction with the question. However, the authors did not mention how to enter membership functions defining fuzzy concepts.

Bosc and Pivert introduced a fuzzy set-based extension of the query language SQL called SQLf [2]. SQLf recognizes such extensions as selection, join, and projection with fuzzy conditions. They can also be applied to nesting operators and set-oriented operators. In addition, the language allows for the partitioning of relations involving groups based on fuzzy quantifiers.

Recently, Zadrożny and Kacprzyk proposed a novel use of fuzzy terms in analytic functions that is a part of the SQL syntax [22]. In particular, they introduced a unified approach to modeling a fuzzy version of the GROUP BY clause with linguistic terms. The terms can be used in grouping expressions and standard aggregate functions. Their proposed concept of a flexible analytic clause of the SQL's SELECT instructions is promising and has a potential application in the linguistic summarization of data.

9 Conclusion

Constructing SPARQL queries to extract information from RDF datasets is a challenging task. As far as we know, the existing state-of-the-art Question-Answering systems can answer simple questions, yet questions with imprecise linguistic terms are not considered.

A system capable of answering questions with linguistic terms represented by user-defined membership functions—called *Linguistic Term Question-Answering system (LingTeQA)*—is introduced. This system automatically performs such tasks as: (1) constructing a query to collect 'intermediate' data; (2) providing a user-friendly interface for users to draw membership functions, enabling personalization of linguistic terms [21]; and (3) transforming obtained data into answers expected by a user.

The user-friendly interface allows users to draw shapes of membership functions associated with fuzzy terms representing terms and concepts. A simple process of inputting shapes as definitions of membership functions allows the users to modify or change their definitions of concepts.

The interface is integrated with *LingTeQA* and provides the users with a useful and easy way of interacting with the question-answering process and obtaining answers to questions that contain personalized definitions of linguistic terms describing imprecise concepts.

References

1. Abujabal, A., Saha Roy, R., Yahya, M., & Weikum, G. (2018). Never-ending learning for open-domain question answering over knowledge bases. *Proceedings of the World Wide Web Conference, WWW, 2018*, 1053–1062. https://doi.org/10.1145/3178876.3186004
2. Bosc, P., & Pivert, O. (1995). SQLf: A relational database language for fuzzy querying. *IEEE Transactions on Fuzzy Systems, 3*(1), 1–17. https://doi.org/10.1109/91.366566
3. Chen, D., Fisch, A., Weston, J., & Bordes, A. (2017) Reading Wikipedia to answer open-domain questions. *Proceedings of ACL 2017—55th Annual Meeting of the Association for Computational Linguistics, 1*, 1870–1879 (2017). https://doi.org/10.18653/v1/P17-1171
4. Chen, L., & Philip Chen, C.L. (2008) Gradient pre-shaped fuzzy C-means algorithm (GradPFCM) for transparent membership function generation. In *Proceedings of IEEE International Conference on Fuzzy Systems* (pp. 428–433). https://doi.org/10.1109/FUZZY.2008.4630403.
5. Diefenbach, D., Singh, K., & Maret, P. (2017). WDAqua-core0: A question answering component for the research community. *Communications in Computer and Information Science, 769*, 84–89. https://doi.org/10.1007/978-3-319-69146-6_8
6. Fellbaum, C. (1998). *WordNet?: an electronic lexical database*. MIT Press.
7. Garibaldi, J. M., & John, R. I. (2003). Choosing membership functions of linguistic terms. *Proceedings of IEEE International Conference on Fuzzy Systems, 1*, 578–583. https://doi.org/10.1109/fuzz.2003.1209428
8. Hasuike, T., Katagiri, H., Tsubaki, H., & Tsuda, H. (2012). Constructing membership function based on fuzzy shannon entropy and human's interval estimation. In Proceedings of IEEE International Conference on Fuzzy Systems (pp. 10–15). https://doi.org/10.1109/FUZZ-IEEE.2012.6251199
9. Honnibal, M., & Johnson, M. (2015). An improved non-monotonic transition system for dependency parsing. In *Proceedings of EMNLP 2015: Conference on Empirical Methods in Natural Language Processing* (pp. 1373–1378). https://doi.org/10.18653/v1/d15-1162
10. Kacprzyk, J., & Zadrożny, S. (1995). Fquery for access: Fuzzy querying for a windows-based DBMS. In *Fuzziness in database management systems* (pp. 415–433). https://doi.org/10.1007/978-3-7908-1897-0_18
11. Kacprzyk, J., & Zadrozny, S. (1999). On interactive linguistic summarization of databases via a fuzzy-logic-based querying add-on to microsoft access®. *LNCS, 1625*, 462–472. https://doi.org/10.1007/3-540-48774-3_52
12. Klir, G.J., & Yuan, B. (1995). *Fuzzy sets and fuzzy logic: Theory and applications*. Prentice Hall PTR
13. Lopez, V., & Motta, E. (2004). Ontology-driven question answering in AquaLog. *LNCS, 3136*, 89–102. https://doi.org/10.1007/978-3-540-27779-8_8
14. Lopez, V., Uren, V., Sabou, M., & Motta, E. (2011). Is question answering fit for the semantic web?: A survey. *Semantic Web, 2*(2), 125–155. https://doi.org/10.3233/SW-2011-0041
15. Miler, G. (1995). WordNet: A lexical database for English. *Communications of the ACM, 38*(11), 39–41.
16. Pedrycz, W., & Gomide, F. (2007). *Fuzzy systems engineering: toward human-centric computing*. Wiley
17. Rasmussen, D. (1997). *Summary SQL—A Fuzzy Tool For Data Mining, 1*(98), 49–58.
18. Saaty, T. L. (1988). What is the Analytic Hierarchy Process?
19. To, N.D., & Reformat, M. (2020). Question-answering system with linguistic terms over RDF knowledge graphs. In *Proceedings of IEEE International Conference on Systems, Man, and Cybernetics (SMC)* (pp. 4236–4243). https://doi.org/10.1109/SMC42975.2020.9282949
20. Wang, R., & Mei, K. (2010). Analysis of fuzzy membership function generation with unsupervised learning using self-organizing feature map. *Proceedings of CIS2010: International Conference on Computational Intelligence and Security, (1)*, 515–518. https://doi.org/10.1109/CIS.2010.118

21. Yager, R. R., Reformat, M. Z., & To, N. D. (2019). Drawing on the iPad to input fuzzy sets with an application to linguistic data science. *Information Sciences, 479*, 277–291.
22. Zadrozny, S., & Kacprzyk, J. (2020). Fuzzy analytical queries: A new approach to flexible fuzzy queries. In *Proceeding of IEEE International Conference on Fuzzy Systems*. https://doi.org/10.1109/FUZZ48607.2020.9177556

Computational Intelligence in Control and Decision Support Processes

OWA Operators in Pensions

Anton Figuerola-Wischke, Anna M. Gil-Lafuente, and José M. Merigó

Abstract The public pension system crisis, arising mainly from the changing demographic, has hit different countries worldwide. For governments and citizens, it is very important to have reliable information regarding pensions in order to make decisions with a maximum degree of effectiveness and to ensure a decent income in retirement. This study presents a new method for optimizing forecasts of the average pension by using the ordered weighted averaging (OWA) operator, the induced ordered weighted averaging (IOWA) operator, the generalized ordered weighted averaging (GOWA) operator, the induced generalized ordered weighted averaging (IGOWA) operator, and particular forms of the probabilistic ordered weighted averaging (POWA) operator and the quasi-arithmetic ordered weighted averaging (Quasi-OWA) operator. It also accounts for inflation or deflation, providing a more realistic assessment of the average pension. The main advantage of this approach is the possibility to include the attitudinal character of experts or decision-makers into the calculation. The study also presents an illustrative example of how to forecast the real average pension for all autonomous communities of Spain by using this new approach.

Keywords Decision-making · Forecasting · Aggregation operator · OWA operator · Average pension · COVID-19

A. Figuerola-Wischke (✉) · A. M. Gil-Lafuente
Department of Business, Faculty of Economics and Business, University of Barcelona, 690
Diagonal Ave., 08034 Barcelona, Spain
e-mail: anton.figuerola@ub.edu

A. M. Gil-Lafuente
e-mail: amgil@ub.edu

J. M. Merigó
School of Information, Systems, and Modelling, Faculty of Engineering and Information
Technology, University of Technology Sydney, 81 Broadway St., Sydney, NSW 2007, Australia
e-mail: jose.merigo@uts.edu.au

1   Introduction

As stated in Chap. 3 of the European Pillar of Social Rights [9], "everyone in old age has the right to resources that ensure living in dignity" (p. 58). But can governments ensure it? The continuous increase in life expectancy and the low rates of fertility implies a declining ratio of workers to pensioners. Consequently, countries find it very difficult to guarantee the long-term financial sustainability of their pension systems. Japan is the most affected country by these demographic changes: in 2019 there were 47 older dependents per 100 in the working-age population [28]. Despite reforms in some countries, for example Spain is increasing gradually the statutory retirement age from 65 to 67 by the year 2027 [7], they are not enough to solve the problem. Also, many of these reforms lead to a reduction in the level of individual public pension income. Additional factors that have a significant influence on the sustainability of public pension systems are those related to the economic growth and employment, among others [26].

Especially in the current situation, it is important for governments to have reliable and accurate pension forecasts in realistic terms in order to conduct the best pension policy decision-making. Also, it is important for citizens, so that they can plan properly their retirement and therefore reduce the risk of poverty when they retire.

The aim of this paper is to optimize average pensions forecasts. To accomplish this, a new method called the ordered weighted averaging real average pension (OWARAP) is presented. This new method is built under the ordered weighted averaging (OWA) operator [30] and it includes the consumer price index (CPI) to adjust average pensions forecasts for inflation. The OWA operator introduced by Yager is an increasing popular aggregation method [1, 6, 15] that aggregates the information underestimating or overestimating it according to the attitudinal character of the decision-maker. The advantage of using the OWA operator is the possibility to add in a more flexible way the opinion that the decision-maker has about future scenarios. Thus, this operator is extremely helpful when dealing with uncertain environments.

This paper also considers other extensions of the OWA operator to optimize pensions forecasts. These extensions are the induced ordered weighted averaging (IOWA) operator [36], the generalized ordered weighted averaging (GOWA) operator [34], the induced generalized ordered weighted averaging (IGOWA) operator [24], and families of the probabilistic ordered weighted averaging (POWA) operator [23]. The main characteristic of the IOWA operator is that the reordering of the arguments is carried out by another variable that Yager and Filev called order-inducing variable. The main characteristic of the GOWA operator is the addition of a parameter controlling the power to which the argument values are raised. The IGOWA operator was introduced by Merigó and Gil-Lafuente and combines the main characteristics of the IOWA and the GOWA operator, so it uses generalized means and order-inducing variables. The main feature of the POWA operator is the possibility of unifying the attitudinal character and the probabilistic information under the same formulation.

Throughout the literature we can find a large variety of forecasting models that incorporate the family of OWA operators. Among them: Merigó et al. [25] apply the OWA and the unified aggregation operator (UAO) for sales forecasting. Cheng et al. [2] propose a forecasting model that incorporates OWA and adaptive network-based fuzzy inference systems (ANFIS) and which is utilized for predicting stock prices. Huang and Cheng [17] developed an OWA based time series model to predict air quality. Flores-Sosa et al. [12] estimates exchange rates by unifying the IOWA operator with linear regression in a single formulation.

The remaining of this paper is organized as follows. Section 2 briefly reviews basic preliminaries regarding aggregation operators. Section 3 introduces the OWARAP operator. Section 4 studies some new extensions of the OWARAP operator. Section 5 proposes an algorithm to calculate the real average pension, develops an illustrative example, and compares the new approach with traditional methods. Section 6 discusses the applicability of aggregation operators to address the COVID-19 impacts on pensions. Section 7 summarizes the finding and conclusions of the study.

2 Preliminaries

This section briefly defines the concept of aggregation and aggregation operator and reviews the OWA operator, the IOWA operator, the GOWA operator, and the IGOWA operator.

Aggregation is the process of combining several numerical values into a single representative value, and an aggregation operator (also called aggregation function) performs this operation [14]. Aggregation operators have been applied to a wide range of fields such as social choice and voting [19], portfolio selection [20, 21], inflation calculations [8, 22], retirement planning [11], and many others [18]. The most common aggregation operators are the arithmetic average (simple average), the weighted average, and the OWA.

The OWA operator from Yager [30] provides a parameterized class of mean type aggregation operators that lies between the minimum and the maximum. The OWA operator can be defined as follows.

Definition 1 An *OWA* operator of dimension n is a function $F : R^n \rightarrow R$ that has associated a weighting vector W of dimension n $W = (w_1, \ldots, w_n)$ with $w_j \in [0, 1]$ and $\sum_{j=1}^{n} w_j = 1$, in which:

$$OWA(a_1, \ldots, a_n) = \sum_{j=1}^{n} w_j b_j, \tag{1}$$

where b_j is the jth largest element of the arguments $a_1, \ldots, a_n$, that is $(b_1, \ldots, b_n)$ is $(a_1, \ldots, a_n)$ reordered from largest to smallest.

An interesting characteristic of this type of operator is that it includes the classical methods for decision-making into a single formulation. This can be achieved through choosing different manifestations of the weighting vector W. The optimistic criterion or maximax criterion selects the most favorable result of each alternative, that is $w_1 = 1$ and $w_j = 0$ for $\forall \, j \neq 1$. The pessimistic criterion or maximin criterion selects the most unfavorable result of each alternative, that is $w_j = 0$ and $w_n = 1$ for $\forall \, j \neq n$. The Laplace criterion is obtained when $w_j = 1/n$ for $\forall \, j$, so it is assumed that all alternatives have equal probability to occur. The Hurwicz criterion is found when $w_1 = \alpha$, $w_n = 1 - \alpha$, and $w_j = 0$ for $\forall \, j \neq 1, n$, so it takes into account both the best and the worst alternative.

Note that if the reordering process is carried out from smallest to largest, we get the ascending ordered weighted averaging (AOWA) operator [31].

One appealing extension of the OWA operator is the IOWA operator developed by Yager and Filev [36]. In this operator, the step of reordering is carried out using order-inducing variables. This allows to consider other factors in the reordering process and not only to the degree of optimism. It can be defined as follows.

Definition 2 An *IOWA* operator of dimension n is a function $F : R^n \times R^n \to R$ that has associated a weighting vector W of dimension n $W = (w_1, \ldots, w_n)$ with $w_j \in [0, 1]$ and $\sum_{j=1}^{n} w_j = 1$, in which:

$$IOWA(\langle u_1, a_1 \rangle, \ldots, \langle u_n, a_n \rangle) = \sum_{j=1}^{n} w_j b_j, \tag{2}$$

where b_j is the a_i value of the *IOWA* pair $\langle u_i, a_i \rangle$ having the jth largest u_i value. u_i is referred as the order-inducing variable and a_i as the argument variable.

Another interesting extension is the GOWA operator. The GOWA operator was introduced by Yager [34] and it combines the OWA operator with generalized means [5]. This operator is defined as follows.

Definition 3 A *GOWA* operator of dimension n is a function $F : R^n \to R$ that has associated a weighting vector W of dimension n $W = (w_1, \ldots, w_n)$ with $w_j \in [0, 1]$ and $\sum_{j=1}^{n} w_j = 1$, in which:

$$GOWA(a_1, \ldots, a_n) = \left(\sum_{j=1}^{n} w_j b_j^\lambda \right)^{1/\lambda}, \tag{3}$$

where λ is a parameter such that $\lambda \in (-\infty, +\infty)$ and b_j is the jth largest of the argument variable a_i.

Note that if $\lambda = -1$ we obtain the ordered weighted harmonic averaging (OWHA) operator [34], if $\lambda = 0$ the ordered weighted geometric (OWG) operator [3, 4], if $\lambda = 1$ the OWA operator, and if $\lambda = 2$ the ordered weighted quadratic averaging (OWQA) operator [34].

Merigó and Gil-Lafuente [24] introduced the IGOWA operator. This operator uses the main characteristics of the IOWA operator and the GOWA operator. The IGOWA operator is defined as follows.

Definition 4 An *IGOWA* operator of dimension n is a function $F : R^n \times R^n \to R$ that has associated a weighting vector W of dimension n $W = (w_1, \ldots, w_n)$ with $w_j \in [0, 1]$ and $\sum_{j=1}^{n} w_j = 1$, in which:

$$IGOWA(\langle u_1, a_1 \rangle, \ldots, \langle u_n, a_n \rangle) = \left(\sum_{j=1}^{n} w_j b_j^\lambda \right)^{1/\lambda}, \tag{4}$$

where b_j is the a_i value of the *IGOWA* pair $\langle u_i, a_i \rangle$ having the jth largest u_i value. u_i is referred as the order-inducing variable, a_i as the argument variable, and λ is a parameter such that $\lambda \in (-\infty, +\infty)$.

All the above-mentioned operators satisfy the conditions of monotonicity, commutativity, boundedness, and idempotency. For a detailed proof consult [24, 30, 34, 36].

3 The Ordered Weighted Averaging Real Average Pension

In this section the OWARAP operator, the induced ordered weighted averaging real average pension (IOWARAP) operator, and the generalizations of these two are presented.

3.1 The OWARAP Operator

The OWARAP operator is a new aggregation function that measures the future average pension adjusted for price changes and which is built under the family of the OWA operator. This operator provides a parametrized family of aggregation operators between the minimum and the maximum real average pension.

An important feature of the OWARAP operator is its possibility of unifying different opinions of a set of experts or decision-makers into a collective result without losing any information. Another interesting advantage is the capability to consider a wide range of situations and alternatives, thus providing a better understanding of the problem.

Most countries publish information regarding pensions in current prices. However, people can obtain a true picture of the average pension by using the OWARAP operator. In this sense, citizens, governments, and companies can know if there will be a purchasing power reduction in the retirement income, thus improving the quality and effectiveness of decision-making. The OWARAP operator can be defined as follows.

Definition 5 An *OWARAP* operator of dimension n is a function $F : R^n \to R$ that has associated a weighting vector W of dimension n $W = (w_1, \ldots, w_n)$ with $w_j \in [0, 1]$ and $\sum_{j=1}^{n} w_j = 1$, in which:

$$OWARAP(p_1, \ldots, p_n) = \left(\frac{100}{CPI}\right) \sum_{j=1}^{n} w_j P_j, \tag{5}$$

where P_j is the jth largest element of a set of nominal average pensions $p_1, \ldots, p_n$ (referred as arguments) and CPI is the consumer price index.

Example 1 Assume the following collection of nominal pensions ($p_1 = 920$, $p_2 = 890$, $p_3 = 950$, $p_4 = 980$) and weighting vector $W = (0.40, 0.25, 0.25, 0.10)$. If we assume that $CPI = 105$, then the aggregation process is solved as follows:

$$\left(\frac{100}{105}\right) \times (0.40 \times 980 + 0.25 \times 950 + 0.25 \times 920 + 0.10 \times 890) = 903.33.$$

The OWARAP operator is a mean operator that satisfies the properties of monotonicity, commutativity, and boundedness. These properties are expressed in the following theorems:

Theorem 1 Monotonicity. Let F be the OWARAP operator. If $p_i \geq \widehat{p}_i$ for all i, then, $F(p_1, \ldots, p_n) \geq F(\widehat{p}_1, \ldots, \widehat{p}_n)$.

Theorem 2 Commutativity (symmetry). In the sense that the initial indexing of the arguments does not meter. So, if F is the OWARAP operator, then, $F(p_1, \ldots, p_n) = F(\widehat{p}_1, \ldots, \widehat{p}_n)$, where $(\widehat{p}_1, \ldots, \widehat{p}_n)$ is any permutation of $(p_1, \ldots, p_n)$.

Theorem 3 Boundedness. Since the aggregation is delimited. Let F be the OWARAP operator. Then, $\left(\frac{100}{CPI}\right) Min\{p_i\} \leq F(p_1, \ldots, p_n) \leq \left(\frac{100}{CPI}\right) Max\{p_i\}$.

Another noteworthy aspect is the measures for characterizing the weighting vector and the type of aggregation it performs. This work focuses on four characterizing features introduced by Yager: the *alpha* value of W (degree of or-ness measure) [30], the dispersion measure [30], the balance operator [32], and the divergence of W [33].

The first measure (degree of or-ness) refers to the attitudinal character associated with a weighting vector and it is denoted as $\alpha(W)$ or also as $AC(W)$ [35]. It can be defined as follows:

$$\alpha(W) = \sum_{j=1}^{n} w_j \left(\frac{n-j}{n-1}\right). \tag{6}$$

As we can see, $\alpha(W) \in [0, 1]$. The closer $\alpha(W)$ is to 1, the higher the level of preference for larger values in the aggregation.

The second measure is the measure of dispersion or entropy and it is denoted as $H(W)$ or also as $Disp(W)$ [35]. Its definition is as follows:

$$H(W) = -\sum_{j=1}^{n} w_j \ln(w_j). \tag{7}$$

Table 1 Particular cases of measures for characterizing a weighting vector

Measure	Criterion		
	Pessimistic	Laplace	Optimistic
$\alpha(W)$	0	0.5	1
$H(W)$	0	$\ln(n)$	0
$Bal(W)$	-1	0	1
$Div(W)$	0	$\frac{n+1}{12(n-1)}$	0

It can be shown that $H(W)$ has a value between 0 and the natural logarithm of n. That is $H(W) \in [0, \ln(n)]$.

The third, is the balance operator $Bal(W)$, which measures the degree of favoritism towards higher values (optimistic values) or lower values (pessimistic values). Its formula is as follows:

$$Bal(W) = \sum_{j=1}^{n} w_j \left(\frac{n+1-2j}{n-1} \right). \tag{8}$$

The balance operator can range from -1 to 1, that is $Bal(W) \in [-1, 1]$. For values of $Bal(W)$ close to -1 the aggregation emphasizes the lower values. For values of $Bal(W)$ close to 1 the aggregation emphasizes the higher values.

The fourth, is the measure of divergence $Div(W)$. It measures the divergence of the weights against the degree of or-ness measure. It can be defined by using the following expression:

$$Div(W) = \sum_{j=1}^{n} w_j \left(\frac{n-j}{n-1} - \alpha(W) \right)^2. \tag{9}$$

In Table 1, we can see when the special cases of the pessimistic, Laplace, and optimistic criterion are met according to the measure outcome.

3.2 The IOWARAP Operator

A new extension of the OWARAP operator is developed by using order-inducing variables, called the IOWARAP operator. It can be defined as follows.

Definition 6 An *IOWARAP* operator of dimension n is a function $F : R^n \times R^n \to R$ that has associated a weighting vector W of dimension n $W = (w_1, \ldots, w_n)$ with $w_j \in [0, 1]$ and $\sum_{j=1}^{n} w_j = 1$, in which:

$$IOWARAP(\langle u_1, p_1 \rangle, \ldots, \langle u_n, p_n \rangle) = \left(\frac{100}{CPI} \right) \sum_{j=1}^{n} w_j P_j, \tag{10}$$

where P_j is the p_i value of the *IOWARAP* pair $\langle u_i, p_i \rangle$ having the jth largest u_i value. u_i is referred as the order-inducing variable and p_i as the nominal average pension variable. CPI is the consumer price index.

Hence, the main difference between the OWARAP operator and the IOWARAP operator resides in the process of reordering the set of values $p = (p_1, \ldots, p_n)$. In the case of the OWARAP, the reordering is made based on the magnitude of the values to be aggregated. By contrast, the reordering step of the IOWARAP depends upon the values of their associated order-inducing variables.

Example 2 Assume we have the following four IOWARAP pairs $\langle u_i, p_i \rangle$: $\langle 7, 920 \rangle$, $\langle 3, 890 \rangle$, $\langle 9, 950 \rangle$, $\langle 5, 980 \rangle$. Assume that the weighting vector is $W = (0.40, 0.25, 0.25, 0.10)$ and $CPI = 105$. We get the following calculation result:

$$\left(\frac{100}{105} \right) \times (0.40 \times 950 + 0.25 \times 920 + 0.25 \times 980 + 0.10 \times 890) = 899.05.$$

3.3   Generalized Aggregation Operators

By using generalized means in the OWARAP operator we obtain the generalized ordered weighted averaging real average pension (GOWARAP) operator. The analyst can obtain a wide range of particular cases of the GOWARAP operator by using different values of the parameter *lambda*. Its definition would be the following.

Definition 7 A *GOWARAP* operator of dimension n is a function $F : R^n \to R$ that has associated a weighting vector W of dimension n $W = (w_1, \ldots, w_n)$ with $w_j \in [0, 1]$ and $\sum_{j=1}^{n} w_j = 1$, in which:

$$GOWARAP(p_1, \ldots, p_n) = \left(\frac{100}{CPI} \right) \left(\sum_{j=1}^{n} w_j P_j^{\lambda} \right)^{1/\lambda}, \tag{11}$$

where λ is a parameter such that $\lambda \in (-\infty, +\infty)$, P_j is the jth largest of the nominal average pension variable p_i, and CPI is the consumer price index.

By analyzing the parameter λ, we can see that when $\lambda = -1$ the ordered weighted harmonic averaging real average pension (OWHARAP) operator is obtained. If $\lambda = 0$ we form the ordered weighted geometric real average pension (OWGRAP) operator. With $\lambda = 1$ we obtain the OWARAP operator. When $\lambda = 2$ we form the ordered weighted quadratic averaging real average pension (OWQARAP) operator.

Note that when $\lambda = -\infty$ and $w_n \neq 0$ we get the smallest argument adjusted for inflation as the aggregated value, that is $(100/CPI)P_n$. However, if $\lambda = -\infty$ and $w_1 = 1$ (that is $w_j = 0$ for all $j \neq 1$) we get the largest argument of the collection p_i adjusted for inflation, which is $(100/CPI)P_1$. In the case where $\lambda = +\infty$ and $w_1 \neq 0$ we get the largest argument adjusted for inflation as the value of the resulting aggregation, that is $(100/CPI)P_1$. Otherwise, when $\lambda = +\infty$ and $w_n = 1$ (that is $w_j = 0$ for $\forall\ j \neq n$), we obtain the smallest argument adjusted for inflation, which is $(100/CPI)P_n$.

Other families of the GOWARAP operator could be developed by choosing different values of the parameter λ and weighting vector W.

Example 3 Consider the same collection of arguments, weighting vector, and CPI as in Example 1. If we assume $\lambda = -1$, then, the aggregation result is:

$$\left(\frac{100}{105}\right) \times \left(0.40 \times 980^{-1} + 0.25 \times 950^{-1} + 0.25 \times 920^{-1} + 0.10 \times 890^{-1}\right)^{1/-1}$$

$$= 902.37.$$

Using generalized means and order-inducing variables we obtain the induced generalized ordered weighted averaging real average pension (IGOWARAP) operator. Its definition is as follows.

Definition 8 An *IGOWARAP* operator of dimension n is a function $F : R^n \times R^n \to R$ that has associated a weighting vector W of dimension n $W = (w_1, \ldots, w_n)$ with $w_j \in [0, 1]$ and $\sum_{j=1}^n w_j = 1$, in which:

$$IGOWARAP(\langle u_1, p_1 \rangle, \ldots, \langle u_n, p_n \rangle) = \left(\frac{100}{CPI}\right)\left(\sum_{j=1}^n w_j P_j^\lambda\right)^{1/\lambda}, \quad (12)$$

where P_j is the p_i value of the *IGOWARAP* pair $\langle u_i, p_i \rangle$ having the jth largest u_i value. u_i is referred as the order-inducing variable, p_i as the nominal average pension variable, λ is a parameter such that $\lambda \in (-\infty, +\infty)$, and CPI is the consumer price index.

Example 4 If we take the IOWARAP pairs from Example 2 and a *lambda* equal to -1, then, we get the following aggregation result:

$$\left(\frac{100}{105}\right) \times \left(0.40 \times 950^{-1} + 0.25 \times 920^{-1} + 0.25 \times 980^{-1} + 0.10 \times 890^{-1}\right)^{1/-1}$$

$$= 898.26.$$

4   Some Other Extensions of the OWARAP Operator

In this section the use of probability in the OWARAP operator is analyzed. To do so, the POWA operator and the probabilistic induced ordered weighted averaging (PIOWA) operator are used. Also, the particular generalization of the quasi-arithmetic ordered weighted averaging (Quasi-OWA) operator is studied.

The probabilistic aggregation (PA) operator is an aggregation function where the aggregation process is done according to the probability associated to each argument. A PA operator is defined as follows.

Definition 9 A *PA* operator of dimension n is a function $F : R^n \rightarrow R$ such that:

$$P A(a_1, \ldots, a_n) = \sum_{i=1}^{n} v_i a_i, \tag{13}$$

where a_i is the ith argument variable and each argument a_i has associated probability v_i with $\sum_{i=1}^{n} v_i = 1$ and $v_i \in [0, 1]$.

The POWA operator introduced by Merigó [23], is an aggregation function that unifies the probability and the OWA operator (attitudinal character) in the same formulation and according to the degree of importance of these two concepts in the aggregation process. Therefore, it provides a unified framework between decision-making problems under uncertainty and under risk. The POWA operator can be defined as follows.

Definition 10 A *POWA* operator of dimension n is a function $F : R^n \rightarrow R$ that has associated a weighting vector W of dimension n $W = (w_1, \ldots, w_n)$ with $w_j \in [0, 1]$ and $\sum_{j=1}^{n} w_j = 1$, in which:

$$P O W A(a_1, \ldots, a_n) = \sum_{j=1}^{n} \hat{v}_j b_j = \beta \sum_{j=1}^{n} w_j b_j + (1 - \beta) \sum_{i=1}^{n} v_i a_i, \tag{14}$$

where b_j is the jth largest of the a_i, each argument a_i has associated probability v_i with $\sum_{i=1}^{n} v_i = 1$ and $v_i \in [0, 1]$, $\hat{v}_j = \beta w_j + (1 - \beta) v_j$ with $\beta \in [0, 1]$, and v_j is the probability v_i ordered according to b_j, that is, according to the jth largest of a_i.

Note that if the parameter β is equal to 1, we obtain the normal OWA operator, and if β is equal to 0, we get the PA operator. Then, by taking into consideration Eqs. 1 and 13, the POWA operator can be formulated alternatively as $P O W A = \beta(O W A) + (1 - \beta) P A$.

If the reordering process is carried out whit order-inducing variables, rather than based on the magnitude of the arguments, we get the PIOWA operator. This operator can be defined as follows.

Definition 11 A *PIOWA* operator of dimension n is a function $F : R^n \times R^n \rightarrow R$ that has associated a weighting vector W of dimension n $W = (w_1, \ldots, w_n)$ with $w_j \in [0, 1]$ and $\sum_{j=1}^{n} w_j = 1$, in which:

$$PIOWA(\langle u_1, a_1 \rangle, \ldots, \langle u_n, a_n \rangle) = \sum_{j=1}^{n} \hat{v}_j b_j$$

$$= \beta \sum_{j=1}^{n} w_j b_j + (1 - \beta) \sum_{i=1}^{n} v_i a_i,$$

(15)

where b_j is the a_i value of the *PIOWA* pair $\langle u_i, a_i \rangle$ having the jth largest u_i value. u_i is referred as the order-inducing variable and a_i has associated probability v_i with $\sum_{i=1}^{n} v_i = 1$ and $v_i \in [0, 1]$, $\hat{v}_j = \beta w_j + (1 - \beta) v_j$ with $\beta \in [0, 1]$, and v_j is the probability v_i ordered according to b_j, that is, according to the jth largest of u_i.

An interesting generalization of the OWA operator is the Quasi-OWA operator, presented by Fodor et al. [13]. By using quasi-arithmetic means, the Quasi-OWA operator provides a more general formulation, including a wide range of particular cases that are not considered in the GOWA operator. The Quasi-OWA operator can be defined as follows.

Definition 12 A *Quasi-OWA* operator of dimension n is a function $F : R^n \to R$ that has associated a weighting vector W of dimension n $W = (w_1, \ldots, w_n)$ with $w_j \in [0, 1]$ and $\sum_{j=1}^{n} w_j = 1$, in which:

$$Quasi - OWA(a_1, \ldots, a_n) = g^{-1}\left(\sum_{j=1}^{n} w_j g(b_j)\right),$$

(16)

where b_j is the jth largest of the a_i and $g(b)$ is a strictly continuous monotonic function.

Note that if $g(b) = b$ we get the OWA operator and if $g(b) = b^\lambda$ the GOWA operator.

By adding probabilities, the Quasi-OWA operator can be extended to a quasi-arithmetic probabilistic ordered weighted averaging (Quasi-POWA) operator, which can be defined as follows.

Definition 13 A *Quasi-POWA* operator of dimension n is a function $F : R^n \to R$ that has associated a weighting vector W of dimension n $W = (w_1, \ldots, w_n)$ with $w_j \in [0, 1]$ and $\sum_{j=1}^{n} w_j = 1$, in which:

$$Quasi - POWA(a_1, \ldots, a_n) = g^{-1}\left(\sum_{j=1}^{n} \hat{v}_j g(b_j)\right),$$

(17)

where b_j is the jth largest of the a_i, each argument a_i has associated probability v_i with $\sum_{i=1}^{n} v_i = 1$ and $v_i \in [0, 1]$, $\hat{v}_j = \beta w_j + (1 - \beta) v_j$ with $\beta \in [0, 1]$, and v_j is the probability v_i ordered according to b_j, that is, according to the jth largest of a_i, and $g(b)$ is a strictly continuous monotonic function.

Another interesting operator is the quasi-arithmetic probabilistic induced ordered weighted averaging (Quasi-PIOWA). It is an extension of the Quasi-POWA operator that uses order-inducing variables in the reordering step. It is defined as follows.

Definition 14 A *Quasi-PIOWA* operator of dimension n is a function $F : R^n \times R^n \to R$ that has associated a weighting vector W of dimension n $W = (w_1, \ldots, w_n)$ with $w_j \in [0, 1]$ and $\sum_{j=1}^{n} w_j = 1$, in which:

$$Quasi - PIOWA(\langle u_1, a_1 \rangle, \ldots, \langle u_n, a_n \rangle) = g^{-1}\left(\sum_{j=1}^{n} \hat{v}_j g(b_j)\right), \quad (18)$$

where b_j is the a_i value of the *Quasi-PIOWA* pair $\langle u_i, a_i \rangle$ having the jth largest u_i value. u_i is referred as the order-inducing variable and a_i has associated probability v_i with $\sum_{i=1}^{n} v_i = 1$ and $v_i \in [0, 1]$, $\hat{v}_j = \beta w_j + (1 - \beta)v_j$ with $\beta \in [0, 1]$, and v_j is the probability v_i ordered according to b_j, that is, according to the jth largest of u_i, and $g(b)$ is a strictly continuous monotonic function.

The probabilistic ordered weighted averaging real average pension (POWARAP) operator is an extension of the OWARAP operator that includes the probability in the aggregation process. To do so, it is assigned a probability of occurrence to the different values of the average pension of each scenario. So, this operator provides a parametrized family of aggregation operators between the probabilistic minimum and probabilistic maximum average pension. The POWARAP operator can be defined as follows.

Definition 15 A *POWARAP* operator of dimension n is a function $F : R^n \to R$ that has associated a weighting vector W of dimension n $W = (w_1, \ldots, w_n)$ with $w_j \in [0, 1]$ and $\sum_{j=1}^{n} w_j = 1$, in which:

$$POWARAP(p_1, \ldots, p_n) = \left(\frac{100}{CPI}\right)\left(\sum_{j=1}^{n} \hat{v}_j P_j\right)$$

$$= \left(\frac{100}{CPI}\right)\left(\beta \sum_{j=1}^{n} w_j P_j + (1 - \beta) \sum_{i=1}^{n} v_i p_i\right), \quad (19)$$

where P_j is the jth largest value of a set of nominal average pensions p_i. Each p_i has associated probability v_i with $\sum_{i=1}^{n} v_i = 1$ and $v_i \in [0, 1]$, $\hat{v}_j = \beta w_j + (1 - \beta)v_j$ with $\beta \in [0, 1]$, and v_j is the probability v_i ordered according to P_j, that is, according to the jth largest of p_i. CPI is the consumer price index.

Example 5 Suppose we have the following vector of arguments ($p_1 = 920, p_2 = 890, p_3 = 950, p_4 = 980$), weighting vector $W = (0.40, 0.25, 0.25, 0.10)$, and probabilistic vector $V = (0.4, 0.10, 0.40, 0.10)$. If we assume that the weighting vector W has a degree of importance of 30% and a CPI of 105 points, then, the POWARAP aggregation can be calculated as follows:

$$\hat{v}_1 = 0.30 \times 0.40 + 0.70 \times 0.10 = 0.19.$$
$$\hat{v}_2 = 0.30 \times 0.25 + 0.70 \times 0.40 = 0.355.$$
$$\hat{v}_3 = 0.30 \times 0.25 + 0.70 \times 0.40 = 0.355.$$
$$\hat{v}_4 = 0.30 \times 0.10 + 0.70 \times 0.10 = 0.10.$$

$$POWARAP = \left(\frac{100}{105}\right)$$
$$\times (0.19 \times 980 + 0.355 \times 950 + 0.355 \times 920 + 0.10 \times 890)$$
$$= 894.33.$$

Alternatively, it can be calculated as follows:

$$POWARAP = \left(\frac{100}{105}\right)$$
$$(0.30 \times (0.40 \times 980 + 0.25 \times 950 + 0.25 \times 920 + 0.10 \times 890)$$
$$+0.70 \times (0.40 \times 920 + 0.10 \times 890 + 0.40 \times 950 + 0.10 \times 980))$$
$$= 894.33.$$

If we use order-inducing variables in the reordering step we form the probabilistic induced ordered weighted averaging real average pension (PIOWARAP) extension, which can be defined as follows.

Definition 16 A *PIOWARAP* operator of dimension n is a function $F : R^n \times R^n \rightarrow R$ that has associated a weighting vector W of dimension n $W = (w_1, \ldots, w_n)$ with $w_j \in [0, 1]$ and $\sum_{j=1}^{n} w_j = 1$, in which:

$$PIOWARAP(\langle u_1, p_1 \rangle, \ldots, \langle u_n, p_n \rangle) = \left(\frac{100}{CPI}\right)\left(\sum_{j=1}^{n} \hat{v}_j P_j\right)$$

$$= \left(\frac{100}{CPI}\right)\left(\beta \sum_{j=1}^{n} w_j P_j + (1 - \beta) \sum_{i=1}^{n} v_i p_i\right), \tag{20}$$

where P_j is the p_i value of the *PIOWARAP* pair $\langle u_i, p_i \rangle$ having the jth largest u_i value. u_i is referred as the order-inducing variable and p_i as the nominal average pension variable. p_i has associated probability v_i with $\sum_{i=1}^{n} v_i = 1$ and $v_i \in [0, 1]$, $\hat{v}_j = \beta w_j + (1 - \beta) v_j$ with $\beta \in [0, 1]$, and v_j is the probability v_i ordered according to P_j, that is, according to the jth largest of u_i. CPI is the consumer price index.

The quasi-arithmetic probabilistic ordered weighted averaging real average pension (Quasi-POWARAP) operator is an extension of the POWARAP operator that uses quasi-arithmetic means. It is defined as follows.

Definition 17 A *Quasi-POWARAP* operator of dimension n is a function $F : R^n \rightarrow R$ that has associated a weighting vector W of dimension n $W = (w_1, \ldots, w_n)$ with

$w_j \in [0, 1]$ and $\sum_{j=1}^{n} w_j = 1$, in which:

$$Quasi - POWARAP(p_1, \ldots, p_n) = \left(\frac{100}{CPI}\right)\left(g^{-1}\left(\sum_{j=1}^{n} \hat{v}_j g(P_j)\right)\right), \quad (21)$$

where P_j is the jth largest value of a set of nominal average pensions p_i. Each p_i has associated probability v_i with $\sum_{i=1}^{n} v_i = 1$ and $v_i \in [0, 1]$, $\hat{v}_j = \beta w_j + (1 - \beta)v_j$ with $\beta \in [0, 1]$, and v_j is the probability v_i ordered according to P_j, that is, according to the jth largest of p_i, and $g(b)$ is a strictly continuous monotonic function. CPI is the consumer price index.

The quasi-arithmetic probabilistic induced ordered weighted averaging real average pension (Quasi-PIOWARAP) operator is an aggregation operator that unifies the PIOWARAP operator with the Quasi-OWA operator. This operator is defined as follows.

Definition 18 A *Quasi-PIOWARAP* operator of dimension n is a function $F : R^n \times R^n \to R$ that has associated a weighting vector W of dimension n $W = (w_1, \ldots, w_n)$ with $w_j \in [0, 1]$ and $\sum_{j=1}^{n} w_j = 1$, in which:

$$Quasi - PIOWARAP(\langle u_1, p_1 \rangle, \ldots, \langle u_n, p_n \rangle)$$
$$= \left(\frac{100}{CPI}\right)\left(g^{-1}\left(\sum_{j=1}^{n} \hat{v}_j g(P_j)\right)\right), \quad (22)$$

where P_j is the p_i value of the *Quasi-PIOWARAP* pair $\langle u_i, p_i \rangle$ having the jth largest u_i value. u_i is referred as the order-inducing variable and p_i as the nominal average pension variable. p_i has associated probability v_i with $\sum_{i=1}^{n} v_i = 1$ and $v_i \in [0, 1]$, $\hat{v}_j = \beta w_j + (1 - \beta)v_j$ with $\beta \in [0, 1]$, and v_j is the probability v_i ordered according to P_j, that is, according to the jth largest of u_i, and $g(b)$ is a strictly continuous monotonic function. CPI is the consumer price index.

5 Real Average Pensions Forecasting with OWARAP Operators

5.1 Proposed Methodology

In contrast to traditional forecasting methodologies, the OWARAP operator and its extensions provide greater flexibility allowing to underestimate or overestimate the information according to the opinion and knowledge of one or more experts (multi-expert). Thus, they enable to optimize pension forecasts in complex environments and consequently help governments to improve its assessments of policy decisions. This new formulation and measurement can be applied to any country and region.

In order to forecast the real average pension through the use of the OWARAP operator and its extensions, it is proposed the following algorithm:

- **Step 1**. Data collection. The first step consists in gathering data regarding the average pension in current prices.
- **Step 2**. Scenario-based forecasting. This step consists in generate forecasts of the average pension in current prices for the different scenarios that may occur in the future $S = (S_1, \ldots, S_n)$. These forecasts shall be referred as the argument variables $p = (p_1, \ldots, p_n)$.
- **Step 3**. Reordering of the arguments. That is building the vector $P_d = (P_1, \ldots, P_n)$ for each expert d. For the OWARAP, GOWARAP, POWARAP, and Quasi-POWARAP operators, the reordering process is based on the values of the arguments. By contrast, for the IOWARAP, IGOWARAP, PIOWARAP, and Quasi-PIOWARAP operators, the reordering step is carried out using the order-inducing variables vector $U_d = (u_1, \ldots, u_n)$ which the experts must have determined previously.
- **Step 4**. Definition of the weighting weights and aggregation parameters. In this step each expert shall determine its weighting vector $W_d = (w_1, \ldots, w_n)$ according to their attitudinal character. Additionally, for the probabilistic operators, experts shall define the probability vector $V_d = (v_1, \ldots, v_n)$ and the parameter β.
- **Step 5**. Sub-aggregation. Forecasts calculated in Step 2 and reordered in Step 3 are individually aggregated for each expert based on their weighting vector and probability vector determined in Step 4. This is done using the OWA, IOWA, GOWA, IGOWA, POWA, PIOWA, Quasi-POWA, and Quasi-PIOWA operators.
- **Step 6**. Final aggregation. All the results obtained in Step 5 are aggregated into a single outcome. To do so, the most suitable aggregation function shall be used according to the degree of importance that is given to the assessments made by each expert.
- **Step 7**. Inflation adjustment. Finally, the effect of price inflation or deflation is removed from the results obtained in Step 6, thereby getting the OWARAP and its extensions. This is done by dividing the data by the corresponding CPI and multiplying by 100.

5.2   Illustrative Example

In the following section an illustrative multi-person example is presented. This example will calculate the OWARAP and the extensions IOWARAP, GOWARAP, IGOWARAP, POWARAP, and PIOWARAP for all the regions of Spain for the year 2023. Also, it will consider only contributory retirement pensions. The following explains the main steps necessary to calculate the future average pension in real terms using the OWARAP operator and its extensions.

- **Step 1**. First, we collect historical monthly data regarding the average pension and the CPI. Our data was gathered from the National Social Security Institute (INSS is its acronym in Spanish) and the National Statistics Institute (INE is its acronym in Spanish). The most recent data available at date of preparation of this study is December 2020.
- **Step 2**. Once we have collected all the data, we generate the estimations of the current average pension for three different scenarios. The first scenario S_1 contemplates economic growth, the second scenario S_2 assumes the economy will remain unchanged, and the third scenario S_3 considers an economic downturn. Table 2 shows the values for each situation. For comparative purposes, in the same table we add the results of applying some of the classical decision approaches, which are the optimistic criterion, the pessimistic criterion, and the Laplace criterion, referred as OC, PC, and LC, respectively. Also, with the OWA and AOWA operators.
- **Step 3**. Afterward, we proceed to reorder the pension estimates obtained in the previous step. In order to determine the inducing vector a group of three experts is selected. To do this, the experts assess each scenario on a scale of 1 to 10. The different vectors are as follows:

Table 2 Spain's nominal average pensions 2023 scenario forecasts by region and aggregated results. Values in euro

Region	S1	S2	S3	OC	PC	LC	OWA	AOWA
Andalusia	1,173	1,118	1,062	1,173	1,062	1,118	1,103	1,132
Aragon	1,369	1,304	1,239	1,369	1,239	1,304	1,287	1,321
Balearic Islands	1,207	1,150	1,092	1,207	1,092	1,150	1,134	1,165
Basque Country	1,597	1,521	1,445	1,597	1,445	1,521	1,501	1,541
Canary Islands	1,206	1,148	1,091	1,206	1,091	1,148	1,133	1,163
Cantabria	1,392	1,326	1,260	1,392	1,260	1,326	1,308	1,343
Castile and Leon	1,295	1,234	1,172	1,295	1,172	1,234	1,217	1,250
Castile La Mancha	1,211	1,153	1,095	1,211	1,095	1,153	1,138	1,168
Catalonia	1,327	1,264	1,201	1,327	1,201	1,264	1,247	1,281
Ceuta	1,407	1,340	1,273	1,407	1,273	1,340	1,322	1,358
Community of Navarre	1,478	1,407	1,337	1,478	1,337	1,407	1,388	1,426
Community of Madrid	1,518	1,445	1,373	1,518	1,373	1,445	1,426	1,465
Extremadura	1,079	1,027	976	1,079	976	1,027	1,014	1,041
Galicia	1,102	1,050	997	1,102	997	1,050	1,036	1,064
La Rioja	1,252	1,192	1,133	1,252	1,133	1,192	1,176	1,208
Melilla	1,366	1,301	1,236	1,366	1,236	1,301	1,283	1,318
Principality of Asturias	1,563	1,489	1,414	1,563	1,414	1,489	1,469	1,509
Region of Murcia	1,162	1,106	1,051	1,162	1,051	1,106	1,092	1,121
Valencian Community	1,195	1,138	1,081	1,195	1,081	1,138	1,123	1,153

- Expert 1. $U_1 = (6, 9, 7)$.
- Expert 2. $U_2 = (5, 7, 9)$.
- Expert 3. $U_3 = (5, 8, 7)$.

- **Step 4**. We continue with the construction of the weighting vector W and probability vector V. Note that the experts are considering subjective probabilities. On these, we place a level of importance of 35% and 65%, respectively. That is a parameter β equal to 0.35. The weighting and probability vectors are as follows:

  - Expert 1. $W_1 = (0.20, 0.40, 0.40)$ and $V_1 = (0.30, 0.50, 0.20)$.
  - Expert 2. $W_2 = (0.15, 0.30, 0.55)$ and $V_2 = (0.30, 0.35, 0.35)$.
  - Expert 3. $W_3 = (0.25, 0.30, 0.45)$ and $V_3 = (0.30, 0.40, 0.30)$.

- **Step 5**. Through the OWA operator, the IOWA operator, the OWHA operator, the OWQA operator, the induced ordered weighted harmonic averaging (IOWHA) operator, the induced ordered weighted quadratic averaging (OWQA) operator, the POWA operator, and the PIOWA operator we add the estimations of the different scenarios according to the opinion and knowledge of the experts (Tables 3, 4, and 5).
- **Step 6**. Next, we fuse the aggregated results of the experts into a single outcome (see Table 6). To do so, we use the arithmetic mean because we consider the assessments of each expert equally important. However, if we believe that the judgements of each expert are not equally relevant, we could employ the OWA operator, the IOWA operator, and many more.
- **Step 7**. Finally, we perform the inflation adjustment to obtain the OWARAP, IOWARAP, OWHARAP, OWQARAP, induced ordered weighted harmonic averaging real average pension (IOWHARAP), induced ordered weighted quadratic averaging real average pension (IOWQARAP), POWARAP, and PIOWARAP operators. The CPIs employed and the final aggregated results can be seen in Tables 7 and 8, respectively.

In Tables 3, 4, and 5 we can see different scenarios of the estimated nominal average pension of Spain's regions based on the attitudinal character of each expert. Note that depending on the weights and order-inducing variables that the expert chooses for the aggregation, the average pension can change considerably. Table 6 unifies the results obtained by the group of experts into a single outcome, which afterwards is expressed in real prices in Table 8. With all this information, the decision-maker gain a better understanding of the pension development and can make better decision.

As may be seen, different results are obtained depending on the type of aggregation operator used. For example, in the Community of Madrid the real average pension for the year 2023 ranges from 1,321 euro to 1,354 euro. We can also see that the lowest estimations are obtained with the OWHARAP operator and the highest with the IOWQARAP operator. Moreover, the operators that include order-inducing variables in the reordering step produce greater forecasts. This can be summarized as follows (from less to more):

Table 3 Expert 1 aggregated results of the 2023 nominal average pensions of Spain's regions. Values in euro

Region	OWA_1	$IOWA_1$	$OWHA_1$	$OWQA_1$	$IOWHA_1$	$IOWQA_1$	$POWA_1$	$PIOWA_1$
Andalusia	1,106	1,118	1,105	1,107	1,115	1,119	1,117	1,121
Aragon	1,291	1,304	1,289	1,292	1,301	1,305	1,304	1,308
Balearic Islands	1,138	1,150	1,136	1,139	1,147	1,151	1,149	1,153
Basque Country	1,506	1,521	1,504	1,507	1,518	1,523	1,521	1,526
Canary Islands	1,137	1,148	1,135	1,137	1,146	1,149	1,148	1,152
Cantabria	1,313	1,326	1,311	1,313	1,323	1,327	1,325	1,330
Castile and Leon	1,221	1,234	1,220	1,222	1,231	1,235	1,233	1,238
Castile La Mancha	1,142	1,153	1,140	1,142	1,151	1,154	1,153	1,157
Catalonia	1,252	1,264	1,250	1,253	1,262	1,266	1,264	1,268
Ceuta	1,327	1,340	1,325	1,328	1,338	1,342	1,340	1,345
Community of Navarre	1,393	1,407	1,391	1,394	1,404	1,409	1,407	1,412
Community of Madrid	1,431	1,445	1,429	1,432	1,443	1,447	1,445	1,450
Extremadura	1,017	1,027	1,016	1,018	1,025	1,028	1,027	1,031
Galicia	1,039	1,050	1,038	1,040	1,048	1,051	1,050	1,053
La Rioja	1,180	1,192	1,179	1,181	1,190	1,193	1,192	1,196
Melilla	1,288	1,301	1,286	1,288	1,298	1,302	1,300	1,305
Principality of Asturias	1,474	1,489	1,472	1,475	1,486	1,490	1,488	1,494
Region of Murcia	1,095	1,106	1,094	1,096	1,104	1,108	1,106	1,110
Valencian Community	1,126	1,138	1,125	1,127	1,135	1,139	1,137	1,141

$$OWHARAP \prec OWARAP \prec OWQARAP \prec POWARAP \prec PIOWARAP$$
$$\prec IOWHARAP \prec IOWARAP \prec IOWQARAP$$

As can be seen, inflation is an important element to consider when analyzing the pension adequacy. For example, the estimated average pension of Catalonia in current prices (OWA) stands at 1,247 euro, however, if we use constant prices (OWARAP) the average pension decreases dramatically to 1,153 euro. Therefore, by using the OWARAP operator and its extensions, people and governments are able to better understand old-age pension changes and thereby improve their decision-making process.

Table 4 Expert 2 aggregated results of the 2023 nominal average pensions of Spain's regions. Values in euro

Region	OWA$_2$	IOWA$_2$	OWHA$_2$	OWQA$_2$	IOWHA$_2$	IOWQA$_2$	POWA$_2$	PIOWA$_2$
Andalusia	1,095	1,140	1,094	1,096	1,138	1,141	1,108	1,124
Aragon	1,278	1,330	1,276	1,279	1,328	1,331	1,293	1,311
Balearic Islands	1,127	1,173	1,125	1,127	1,171	1,173	1,140	1,156
Basque Country	1,491	1,551	1,489	1,492	1,549	1,552	1,508	1,529
Canary Islands	1,125	1,171	1,124	1,126	1,170	1,172	1,138	1,154
Cantabria	1,299	1,352	1,298	1,300	1,351	1,353	1,314	1,333
Castile and Leon	1,209	1,258	1,207	1,210	1,257	1,259	1,223	1,240
Castile La Mancha	1,130	1,176	1,128	1,131	1,175	1,177	1,143	1,159
Catalonia	1,239	1,290	1,237	1,240	1,288	1,290	1,253	1,271
Ceuta	1,313	1,367	1,312	1,314	1,365	1,368	1,329	1,348
Community of Navarre	1,379	1,435	1,377	1,380	1,433	1,436	1,395	1,415
Community of Madrid	1,416	1,474	1,415	1,417	1,472	1,475	1,433	1,453
Extremadura	1,007	1,048	1,006	1,008	1,047	1,049	1,019	1,033
Galicia	1,029	1,071	1,027	1,030	1,069	1,072	1,041	1,055
La Rioja	1,168	1,216	1,167	1,169	1,214	1,217	1,182	1,199
Melilla	1,275	1,327	1,273	1,275	1,325	1,327	1,289	1,308
Principality of Asturias	1,459	1,519	1,457	1,460	1,516	1,519	1,476	1,497
Region of Murcia	1,084	1,129	1,083	1,085	1,127	1,129	1,097	1,112
Valencian Community	1,115	1,161	1,113	1,116	1,159	1,161	1,128	1,144

Finally, other scenarios with different assumptions could be considered. For example, one can develop a set of scenarios based on different pension revaluation rates that the government may impose in the future.

5.3 Comparison of Forecasting Methods

In the following, a comparison between the OWARAP operator and some traditional forecasting methods is conducted. To do so, the real average pension of Spain at a

Table 5 Expert 3 aggregated results of the 2023 nominal average pensions of Spain's regions. Values in euro

Region	OWA_3	$IOWA_3$	$OWHA_3$	$OWQA_3$	$IOWHA_3$	$IOWQA_3$	$POWA_3$	$PIOWA_3$
Andalusia	1,106	1,126	1,105	1,107	1,124	1,127	1,114	1,121
Aragon	1,291	1,314	1,289	1,292	1,311	1,315	1,299	1,307
Balearic Islands	1,138	1,158	1,136	1,139	1,156	1,159	1,146	1,153
Basque Country	1,506	1,532	1,503	1,507	1,530	1,534	1,516	1,525
Canary Islands	1,137	1,157	1,135	1,138	1,155	1,158	1,144	1,151
Cantabria	1,313	1,336	1,310	1,314	1,333	1,337	1,321	1,329
Castile and Leon	1,221	1,243	1,219	1,222	1,241	1,244	1,229	1,237
Castile La Mancha	1,142	1,162	1,140	1,143	1,160	1,163	1,149	1,156
Catalonia	1,252	1,274	1,250	1,253	1,271	1,275	1,260	1,268
Ceuta	1,327	1,350	1,325	1,328	1,348	1,352	1,336	1,344
Community of Navarre	1,393	1,418	1,391	1,394	1,415	1,419	1,402	1,411
Community of Madrid	1,431	1,456	1,429	1,432	1,454	1,458	1,440	1,449
Extremadura	1,017	1,035	1,016	1,018	1,033	1,036	1,024	1,030
Galicia	1,039	1,058	1,038	1,040	1,056	1,059	1,046	1,053
La Rioja	1,180	1,201	1,178	1,181	1,199	1,202	1,188	1,195
Melilla	1,288	1,310	1,285	1,289	1,308	1,311	1,296	1,304
Principality of Asturias	1,474	1,500	1,471	1,475	1,497	1,501	1,484	1,493
Region of Murcia	1,095	1,115	1,094	1,096	1,113	1,116	1,103	1,109
Valencian Community	1,126	1,146	1,125	1,127	1,144	1,147	1,134	1,141

national level is estimated for December 2020 using the OWARAP operator, Linear Trend (LT), Double Moving Average (DMA), and Holt's Exponential Smoothing (HES) [16] forecasting methods. Also, data until December 2017 is included in the training set. Table 9 shows the outcomes and the corresponding absolute values of the forecast errors. Results show that the OWARAP operator offers potentially better performance in comparison with the traditional methods.

Table 6 Absolute aggregated results of the 2023 nominal average pensions of Spain's regions. Values in euro

Region	OWA	IOWA	OWHA	OWQA	IOWHA	IOWQA	POWA	PIOWA
Andalusia	1,103	1,128	1,101	1,104	1,126	1,129	1,113	1,122
Aragon	1,287	1,316	1,285	1,287	1,314	1,317	1,299	1,309
Balearic Islands	1,134	1,160	1,133	1,135	1,158	1,161	1,145	1,154
Basque Country	1,501	1,535	1,498	1,502	1,532	1,536	1,515	1,527
Canary Islands	1,133	1,159	1,131	1,134	1,157	1,160	1,143	1,152
Cantabria	1,308	1,338	1,306	1,309	1,336	1,339	1,320	1,331
Castile and Leon	1,217	1,245	1,215	1,218	1,243	1,246	1,229	1,238
Castile La Mancha	1,138	1,164	1,136	1,139	1,162	1,165	1,148	1,157
Catalonia	1,247	1,276	1,246	1,248	1,274	1,277	1,259	1,269
Ceuta	1,322	1,353	1,320	1,323	1,350	1,354	1,335	1,345
Community of Navarre	1,388	1,420	1,386	1,390	1,418	1,421	1,401	1,413
Community of Madrid	1,426	1,459	1,424	1,427	1,456	1,460	1,439	1,451
Extremadura	1,014	1,037	1,012	1,015	1,035	1,038	1,023	1,031
Galicia	1,036	1,059	1,034	1,037	1,058	1,060	1,046	1,054
La Rioja	1,176	1,203	1,175	1,177	1,201	1,204	1,187	1,197
Melilla	1,283	1,312	1,281	1,284	1,310	1,314	1,295	1,305
Principality of Asturias	1,469	1,502	1,467	1,470	1,500	1,504	1,483	1,494
Region of Murcia	1,092	1,117	1,090	1,093	1,115	1,118	1,102	1,111
Valencian Community	1,123	1,148	1,121	1,123	1,146	1,149	1,133	1,142

6 The COVID-19 Crisis on Pensions: Applicability of the OWA Operators

The outbreak of the coronavirus (COVID-19) pandemic has caused an economic downturn across the world. Some international organizations, such as the Organization for Economic Cooperation and Development (OECD) [26, 27] and the International Monetary Fund (IMF) [10], warn about the impact of the pandemic on the future of pensions. Prior to the COVID-19, pension systems were already facing financial sustainability problems, mainly driven by the ageing of the population.

Table 7 Estimations of the 2023 CPIs of Spain's regions

Region	CPI
Andalusia	108
Aragon	107
Balearic Islands	108
Basque Country	108
Canary Islands	108
Cantabria	108
Castile and Leon	108
Castile La Mancha	108
Catalonia	108
Ceuta	106
Community of Navarre	108
Community of Madrid	108
Extremadura	107
Galicia	108
La Rioja	108
Melilla	107
Principality of Asturias	107
Region of Murcia	107
Valencian Community	107

This crisis has entailed high levels of unemployment, which leads to a reduction of government revenues from contributions and consequently makes public pension systems more unsustainable. Moreover, the declining employment among elderly people, makes it more difficult for them to find a new job because of age discrimination. Consequently, it is more likely that this population group retire early, leading to lower pension benefits. Therefore, policy makers should adopt new strategies. However, the uncertainty associated to the length of the current economic crisis and the real impact on pensions in the long-term, makes it more difficult for policy makers to choose and implement the best policy response.

In this complex environment, OWA operators can be very useful when generating forecasts of pension indicators, because they use the attitudinal character of the decision-maker. Since the COVID-19 outbreak, the pessimistic attitude towards future public pensions has further increased, and by using the OWA weights the decision-maker is able to reflect it. This method can be also helpful for assessing pension policy responses alternatives. Likewise, to address the problem POWA operators can be very useful as well, given that they allow to consider probabilistic information with the OWA operator.

Table 8 Absolute aggregated results of the 2023 real average pensions of Spain's regions. Values in euro

Region	OWARAP	IOWARAP	OWHARAP	OWQARAP	IOWHARAP	IOWQARAP	POWARAP	PIOWARAP
Andalusia	1,026	1,049	1,024	1,026	1,047	1,050	1,035	1,043
Aragon	1,198	1,225	1,196	1,199	1,223	1,226	1,209	1,219
Balearic Islands	1,055	1,079	1,053	1,056	1,077	1,080	1,065	1,073
Basque Country	1,387	1,419	1,385	1,389	1,417	1,420	1,400	1,412
Canary Islands	1,049	1,073	1,048	1,050	1,071	1,074	1,059	1,067
Cantabria	1,214	1,242	1,212	1,215	1,239	1,243	1,225	1,235
Castile and Leon	1,128	1,154	1,127	1,129	1,152	1,155	1,139	1,148
Castile La Mancha	1,057	1,081	1,055	1,058	1,079	1,082	1,067	1,075
Catalonia	1,153	1,179	1,151	1,153	1,177	1,180	1,163	1,173
Ceuta	1,249	1,277	1,247	1,250	1,275	1,278	1,261	1,270
Community of Navarre	1,290	1,320	1,289	1,291	1,318	1,321	1,303	1,313
Community of Madrid	1,323	1,353	1,321	1,324	1,351	1,354	1,335	1,346
Extremadura	945	966	943	945	965	967	953	961
Galicia	962	984	960	962	982	984	971	978
La Rioja	1,094	1,119	1,093	1,095	1,117	1,120	1,104	1,113
Melilla	1,205	1,232	1,203	1,206	1,230	1,233	1,216	1,226
Principality of Asturias	1,373	1,405	1,371	1,374	1,402	1,406	1,386	1,397
Region of Murcia	1,022	1,046	1,021	1,023	1,044	1,047	1,032	1,040
Valencian Community	1,045	1,068	1,043	1,045	1,067	1,069	1,054	1,063

Table 9 Evaluation of forecasting methods

Indicator	Real value	OWARAP	LT	2-month DMA	3-month DMA	HES
Real average pension	1,118	1,123	1,107	989	901	1,064
Absolute error	0	5	11	129	217	54

7   Conclusions

The aim of this study is to provide a new tool to improve the quality of pension information. In view of the present and future situation, it is necessary and extremely important to provide recurrent, accurate, representative, and useful data on pension adequacy for governments and citizens of a country. With this information, governments are able to improve significantly its policy decisions and people can plan properly their retirement, in an attempt to prevent old-age poverty.

This paper proposes a new method to optimize forecasts of the average pension using the OWA operator, the IOWA operator, the GOWA operator, and the IGOWA operator. Moreover, an inflation adjustment is added in order to provide a more realistic value of the average pension. The main advantage of using this method is the possibility to aggregate different estimations according to the attitudinal character of the decision-maker without losing information. As a result, forecasts are more representative and accurate. The POWA operator is also used, which shows to be very useful for situations where we find probabilistic information, and at the same time, we need to consider the attitudinal character.

The study also develops an illustrative example regarding the calculations of the projected real average pension of Spain at a regional level by using the OWARAP operator, the IOWARAP operator, the GOWARAP operator, the IGOWARAP operator, the POWARAP operator, and the PIOWARAP operator. This allows to aggregate different scenarios according to the expectations and knowledge of a selected group of experts. The information given by these operators provide a more comprehensive analysis that helps the decision-maker to deal with uncertainty.

In future works we aim to analyze the use of the OWARAP operator in other countries, like United States or Canada. Moreover, we expect to apply this new method into other pension indicators, such as the aggregate replacement ratio for pensions. Lastly, we suggest the use of other extensions of the OWA operator in the field of pensions, such as the uncertain ordered weighted averaging (UOWA) operator [29].

References

1. Blanco-Mesa, F., León-Castro, E., & Merigó, J. M. (2019). A bibliometric analysis of aggregation operators. *Applied Soft Computing, 81,* 105488. https://doi.org/10.1016/j.asoc.2019. 105488
2. Cheng, C. H., Wei, L. Y., Liu, J. W., & Chen, T. L. (2013). OWA-based ANFIS model for TAIEX forecasting. *Economic Modelling, 30*(1), 442–448. https://doi.org/10.1016/j.econmod. 2012.09.047
3. Chiclana, F., Herrera, F., & Herrera-Viedma, E. (2002). The ordered weighted geometric operator: Properties and application in MCDM problems. In B. Bouchon-Meunier, J. Gutiérrez-Ríos, L. Magdalena, & R. R. Yager (Eds.), *Technologies for constructing intelligent systems 2. Studies in fuzziness and soft computing* (Vol. 90, pp. 173–183). Heidelberg: Physica. https://doi.org/10.1007/978-3-7908-1796-6_14.
4. Chiclana, F., Herrera, F., & Herrera-Viedma, E. (2000). The ordered weighted geometric operator: Properties and applications. In *Proceedings of 8th International Conference on Information Processing and Management of Uncertainty in Knowledge-Based Systems* (pp. 985–991).
5. Dyckhoff, H., & Pedrycz, W. (1984). Generalized means as model of compensative connectives. *Fuzzy Sets and Systems, 14*(2), 143–154. https://doi.org/10.1016/0165-0114(84)90097-6
6. Emrouznejad, A., & Marra, M. (2014). Ordered weighted averaging operators 1988–2014: A citation-based literature survey. *International Journal of Intelligent Systems, 29*(11), 994–1014. https://doi.org/10.1002/int.21673
7. España, C. G. (2011). Ley 27/2011, de 1 de agosto, sobre actualización, adecuación y modernización del sistema de Seguridad Social. *Boletín Oficial del Estado, 184,* 87495–87544. Retrieved from https://boe.es/boe/dias/2011/08/02/pdfs/BOE-A-2011-13242.pdf.
8. Espinoza-Audelo, L. F., León-Castro, E., Olazabal-Lugo, M., Merigó, J. M., & Gil-Lafuente, A. M. (2020). Using ordered weighted average for weighted averages inflation. *International Journal of Information Technology & Decision Making, 19*(2), 601–628.https://doi.org/10. 1142/S0219622020500066
9. European Commission. (2017). *Communication from the Commission to the European Parliament, the Council, the European Economic and Social Committee and the Committee of the Regions: Establishing a European Pillar of Social Rights,* SWD(2017) 201 final. Retrieved April 26, 2017, from https://eur-lex.europa.eu/legal-content/EN/TXT/PDF/?uri=CELEX:520 17SC0201&from=EN.
10. Feher, C., & de Bidegain, I. (2020). *Pension schemes in the COVID-19 crisis: Impacts and policy considerations.* International Monetary Fund. Retrieved from https://www.imf. org/-/media/Files/Publications/covid19-special-notes/enspecial-series-on-covid19pension-schemes-in-the-covid19-crisis-impacts-and-policy-considerations.ashx.
11. Figuerola-Wischke, A., Gil-Lafuente, A. M., & Merigó, J. M. (2021). Herramientas para la toma de decisiones en la planificación financiera de la jubilación [Decision-making methods for retirement financial planning]. *Cuadernos del CIMBAGE, 2*(23), 33–47. Retrieved from http://ojs.econ.uba.ar/index.php/CIMBAGE/article/view/2172.
12. Flores-Sosa, M., Avilés-Ochoa, E., & Merigó, J. M. (2020). Induced OWA operators in linear regression. *Journal of Intelligent & Fuzzy Systems, 38*(5), 5509–5520.https://doi.org/10.3233/ JIFS-179642
13. Fodor, J., Marichal, J.-L., & Roubens, M. (1995). Characterization of the ordered weighted averaging operators. *IEEE Transactions on Fuzzy Systems, 3*(2), 236–240https://doi.org/10. 1109/91.388176
14. Grabisch, M., Marichal, J.-L., Mesiar, R., & Pap, E. (2009). Aggregation Functions. *Cambridge University Press.* https://doi.org/10.1017/cbo9781139644150
15. He, X., Wu, Y., Yu, D., & Merigó, J. M. (2017). Exploring the ordered weighted averaging operator knowledge domain: A bibliometric analysis. *International Journal of Intelligent Systems, 32*(11), 1151–1166. https://doi.org/10.1002/int.21894

16. Holt, C. C. (2004). Forecasting seasonals and trends by exponentially weighted moving averages. *International Journal of Forecasting, 20*(1), 5–10. https://doi.org/10.1016/j.ijforecast.2003.09.015
17. Huang, S. -F., & C. -H. Cheng. (2008). Forecasting the air quality using OWA based time series model. *2008 International Conference on Machine Learning and Cybernetics, 6*, 3254–3259. https://doi.org/10.1109/ICMLC.2008.4620967.
18. Kacprzyk, J., Yager, R. R., & Merigó, J. M. (2019). Towards human-centric aggregation via ordered weighted aggregation operators and linguistic data summaries: A new perspective on Zadeh's inspirations. *IEEE Computational Intelligence Magazine, 14*(1), 16–30. https://doi.org/10.1109/MCI.2018.2881641
19. Kacprzyk, J., & Zadrożny, S. (2009). Towards a general and unified characterization of individual and collective choice functions under fuzzy and nonfuzzy preferences and majority via the ordered weighted average operators. *International Journal of Intelligent Systems, 24*(1), 4–26. https://doi.org/10.1002/int.20325
20. Laengle, S., Loyola, G., & Merigó, J. M. (2015). OWA operators in portfolio selection. *Advances in Intelligent Systems and Computing, 377*, 53–64. https://doi.org/10.1007/978-3-319-19704-3_5
21. Laengle, S., Loyola, G., & Merigó, J. M. (2017). Mean-variance portfolio selection with the ordered weighted average. *IEEE Transactions on Fuzzy Systems, 25*(2), 350–362. https://doi.org/10.1109/TFUZZ.2016.2578345
22. León-Castro, E., Espinoza-Audelo, L. F., Merigó, J. M., Gil-Lafuente, A. M., & Yager, R. R. (2020). The ordered weighted average inflation. *Journal of Intelligent & Fuzzy Systems, 38*(2), 1901–1913. https://doi.org/10.3233/JIFS-190442.
23. Merigó, J. M. (2012). Probabilities in the OWA operator. *Expert Systems with Applications, 39*(13), 11456–11467. https://doi.org/10.1016/j.eswa.2012.04.010
24. Merigó, J. M., & Gil-Lafuente, A. M. (2009). The induced generalized OWA operator. *Information Sciences, 179*(6), 729–741.https://doi.org/10.1016/j.ins.2008.11.013
25. Merigó, J. M., Palacios-Marqués, D., & Ribeiro-Navarrete, B. (2015). Aggregation systems for sales forecasting. *Journal of Business Research, 68*(11), 2299–2304.https://doi.org/10.1016/j.jbusres.2015.06.015
26. Organization for Economic Cooperation and Development. (2019). Pensions at a Glance 2019: OECD and G20 Indicators. *OECD Publishing.* https://doi.org/10.1787/b6d3dcfc-en
27. Organization for Economic Cooperation and Development. (2020). OECD Pensions Outlook 2020. *OECD Publishing.* https://doi.org/10.1787/67ede41b-en
28. World Bank. (2019). *Age dependency ratio, old (% of working-age population)* [Data file]. Retrieved November 1, 2020, from https://data.worldbank.org/indicator/SP.POP.DPND.OL
29. Xu, Z. S., & Da, Q. L. (2002). The uncertain OWA operator. *International Journal of Intelligent Systems, 17*(6), 569–575. https://doi.org/10.1002/int.10038
30. Yager, R. R. (1988). On ordered weighted averaging aggregation operators in multicriteria decision making. *IEEE Transactions on Systems, Man, and Cybernetics, 18*(1), 183–190. https://doi.org/10.1109/21.87068
31. Yager, R. R. (1992). On generalized measures of realization in uncertain environments. *Theory and Decision, 33*(1), 41–69. https://doi.org/10.1007/BF00133982
32. Yager, R. R. (1996). Constrained OWA aggregation. *Fuzzy Sets and Systems, 81*(1), 89–101. https://doi.org/10.1016/0165-0114(95)00242-1
33. Yager, R. R. (2002). Heavy OWA operators. *Fuzzy Optimization and Decision Making, 1*(4), 379–397. https://doi.org/10.1023/A:1020959313432.
34. Yager, R. R. (2004). Generalized OWA aggregation operators. *Fuzzy Optimization and Decision Making, 3*(1), 93–107. https://doi.org/10.1023/B:FODM.0000013074.68765.97
35. Yager, R. R., & Alajlan, N. (2014). On characterizing features of OWA aggregation operators. *Fuzzy Optimization and Decision Making, 13*(1), 1–32. https://doi.org/10.1007/s10700-013-9167-8
36. Yager, R. R., & Filev, D. P. (1999). Induced ordered weighted averaging operators. *IEEE Transactions on Systems, Man, and Cybernetics, Part B: Cybernetics, 29*(2), 141–150. https://doi.org/10.1109/3477.752789

Evaluation of the Perception of Public Safety Through Fuzzy and Multi-criteria Approach

Martin Leon-Santiesteban, Alicia Delgadillo-Aguirre,
Martin I. Huesca-Gastelum, and Ernesto Leon-Castro

Abstract This work addresses the multi-criteria ordering in relation to the perception of public safety in the city of Culiacan, Sinaloa, Mexico. The objective of this study was to obtain a ranking of alternatives of the four quadrants into which the city was divided and thus knowing the social dynamics, perception of insecurity and evaluation of the authorities responsible for providing public security. To collect the data, a questionnaire was used through an exploratory study carried out on citizens in 46 neighborhoods of the city of Culiacan. The data process used was the Elimination Et Choice Translating Reality (ELECTRE) III method and the Ordered Weighted Average (OWA) operator with which an outranking relationship was constructed, and the quadrant relationship was generated. As a result of the process, quadrant 1 was the one that obtained a high evaluation of its citizen perception, followed by quadrants 3. This categorization will help in the construction of security strategies aimed at recovering public spaces and improving citizen perception.

Keywords Perception of public safety · ELECTRE III · Operator OWA

JEL Classification F52 · C39 · O29

M. Leon-Santiesteban · A. Delgadillo-Aguirre (✉) · M. I. Huesca-Gastelum
Universidad Autónoma de Occidente, Blvd. Lola Beltrán S/N Esq. Circuito Vial, Culiacán 80200,
México
e-mail: alicia.delgadillo@uadeo.mx

M. Leon-Santiesteban
e-mail: martin.leon@uadeo.mx

M. I. Huesca-Gastelum
e-mail: martin.huesca@uadeo.mx

E. Leon-Castro
Business School, Campus Sinaloa, Tecnológico de Monterrey, Culiacan, México
e-mail: ernesto.leon@tec.mx

1 Introduction

In Mexico, the issue of public security has maintained a relevance that encompasses the institutions in charge of imparting and safeguarding the public security of citizens. In this sense, studies on the perception that citizens living in the country have increased [8, 12, 22], as well as the resources allocated to combat this problem In these, it is sought to determine the feeling of insecurity that they transmit when moving freely where they live, thus recognizing the low legitimacy they have in relation to the institutions that provide public security.

However, these studies delve into the public perception in the city environment and the results are not observed in specific areas of the entity where the research is carried out, which implies that the implementation of proposals guarantees the enjoyment and exercise of rights of each individual and each group, as a joint task that can lead to satisfactory results if it is attended from a more local perspective [10].

On the other hand, at the international level, according to the Institute for Economics & Peace (IEP) [11], the concentration of violence can be seen in Central America and the Caribbean, where the three least peaceful countries are Mexico, Honduras and El Salvador, who are located to the north and the most peaceful, Costa Rica, Panama and Nicaragua, in the south.

For its part in Mexico, the National Institute of Statistics and Geography (INEGI), has published periodically and quarterly, since 2013, the Social Perception on Public Insecurity at the national level, its last publication was made in December 2018 This study includes the percentage of the population aged 18 and over residing in the cities of interest that considers that currently living in their city is unsafe [13].

In their results, 73.7% of the population aged 18 and over consider that living in their city is unsafe and the cities with the greatest sense of insecurity were Reynosa, Chilpancingo de los Bravo, Puebla de Zaragoza, Coatzacoalcos, Ecatepec de Morelos and Villahermosa [13].

Therefore, the objective of this study is to present a ranking of the alternatives, using on the one hand the ELimination and Choice Expressing Reality method (ELECTRE III) proposed by [20] for the Quadrants proposed for the analysis of the perception of public safety in the municipality of Culiacán, Sinaloa, Mexico. This method has been used in estimating safety in tourist destinations [9], evaluation of competitiveness in tourist destinations [17], competitiveness and local development, knowledge integration [14].

Additionally, the Ordered Weighted Average (OWA) operator developed by Yager [26] is used, which has proven to be a useful information aggregation operator in situations of uncertainty within the decision-making process [1, 6]. Furthermore, the OWA operator has been used in different areas, such as exchange rate forecasting [18], government transparency [2], provider selection [24], among others.

However, this operator only orders the information through maximums and minimums and underestimates the behavior of the decision maker. Hence, a proposal is made for a new operator based on the combination of the classic operators and on the same formulation. In the words of [7], the main advantage of this new technique is that it allows analyzing multiple comparisons, interrelationships and rearrangements of the data, but it also allows considering the characteristic attitudinal character of the decision-maker decisions.

In this way, to carry out the analysis of both methodologies, experts who have knowledge about public security participated, as well as those who are involved in the decisions of this social problem. This document is organized as follows: Sect. 2 reviews the ELECTRE III methodology. In Sect. 3 the calculation process is carried out with the proposed method and in Sect. 4 the conclusions are incorporated into the investigation.

2 Methodology

In this section, we briefly review Bonferroni means, Ordered Weighted Average OWA IOWA and IOWAWA operators and BON-OWA to develop new tools based on Bonferroni means in combination with IOWAWA operator.

This study was developed in two stages, in the first a questionnaire was raised and in the second, the ordering was carried out with the ELECTRE III methodology and the OWA operator, processes that are exposed below.

Regarding the first stage, a quantitative descriptive study was carried out, which involved the form of data collection, by raising a questionnaire, which included questions about the perception of public safety in the capital of the State of Sinaloa, Mexico. At the time of application, a sample of 364 surveys was determined, the selection was simple random, considering for this a margin of error of 5% and a level of reliability of 95%.

Additionally, it was necessary to consider as control variables of the interviewees, sex, age, marital status and occupation. After the pilot test, the result was a 24-item instrument and an open question in order to expose what could or should be improved in the municipal police of your locality.

On the other hand, in the second stage, the multicriteria methodology is a fundamental tool for research, because it helps to implement multidimensional analysis methods, based on the computer, such as the problem to be treated. For the particular case, the use of the ELECTRE III method proposed by [20] is chosen, explaining through the process that must be followed to execute this multi-attribute method as follows:

- Choice of alternatives.
- Identification of criteria.
- Normalization of variables through the maximum and minimum rescaling process.

$$Vn(a_i) = \frac{x(a_i) - Min}{Max - Min} \tag{1}$$

- Determination of weights through the personal construction theory of [19].
- Calculation of the decision matrix.
- Final calculations and rankings.

It is possible to measure the strength of the claim aSb (it is at least as good as b). This step consists of developing a measure of agreement, according to those contained in the index $C(a, b)$, for each of the alternatives $a, b \in A$. Where k_j is the coefficient of importance or weight of the criterion j. In this sense, the concordance index is defined by the overclassification ratio valued as follows:

$$C(a, b) = \frac{1}{k} \sum_{j=1}^{r} k_j c_j(a, b), \quad where \quad k = \sum_{j=1}^{r} k_j$$

For this study, the software called SADGAGE (Group Decision Support System with Genetic Algorithms and Electre III) [15] is used, which allows solving problems of multi-criteria ranking, in turn evaluate small to medium-sized samples of the multi-criteria ranking problem.

Therefore, this methodology is designed to select and consider the criteria of the quadrants of the city of Culiacán. Likewise, the data are normalized to make the indicators comparable and subsequently, the weights of the criteria are calculated. At that time, the ELECTRE III method, based on WEB, is used to classify these quadrants.

In order to contrast the results obtained through the ELECTRE III ranking, the Ordered Weighted Average (OWA) operator developed by Yager [26] is also used with the data obtained, the main attribute of this operator is that it allows us to obtain through of a reordering of the values of the attributes with the maximum and minimum results weights by using the following formula.

An OWA operator of dimension n is an application $F : R^n \rightarrow R$ with an associated weight vector $w = [w_1, w_2, \ldots, w_n]^T$, with $w_j \in [0, 1]$, $1 \leq i \leq n$, and

$$\sum_{j=1}^{n} w_j = w_1 + w_2 + \cdots + w_n = 1, \tag{7}$$

where

$$F(a_1, a_2, \ldots, a_n) = \sum_{j=1}^{n} w_j b_j, \tag{8}$$

with b_j denoting the jth largest element of the collection $a_1 a_2, \ldots, a_n$.

3   Results

In the evaluation of public policies, the alternatives correspond to the policy options or alternative courses of action to the different intervention strategies to solve or mitigate a problem [3]. Therefore, a decision must choose between different alternatives. In this sense, Bartolini et al. [4] consider that, in a decision, the set of alternatives can be more or less defined.

This is where the basic characteristics of multicriteria analysis represent the fact of comparing alternatives based on a series of criteria. According to [20], the alternatives have to be: (1) mutually exclusive, (2) consistent in time and space and (3) comparable for any different characteristic based on what is expressed by the evaluation criteria.

In this way, a matrix of alternatives | A | incorporating its decision label (Table 1). In this case, each alternative corresponds to a selected block of the city of Culiacán, where $A = \{a_1, a_2,....,a_m\}$ the finite set of alternatives, $|A|=m$ [21].

In the analysis process, the recommendation can take the form of the selection of a subset of alternatives in different categories or of the overall ranking. In these cases, it is necessary to identify those attributes that define each one and can be comparable.

3.1   Identification of the Criteria

For this study, it was necessary to identify the list of attributes to assess the perception of insecurity in the city of Culiacán and to assess the four quadrants. Table 2 summarizes the criteria, which were considered to generate an evaluation, which describes the performance of each quadrant.

3.2   Normalization of Variables

Before carrying out an aggregation process of the variables that have been selected for the construction of a composite indicator, to carry out the analysis, it is necessary to carry out the normalization process, the purpose of which is to avoid the congregation of variables of different measurement units and the appearance of scale-dependent phenomena [23].

Table 1 Alternatives and their location in the city

Alternative	Location
A_1	Block 1
A_2	Block 2
A_3	Block 3
A_4	Block 4

M. Leon-Santiesteban et al.

Table 2 Criteria of the blocks to be evaluated

Label	Criterion
C_1	Alcoholism problem
C_2	Drug addiction problem
C_3	Degree of violence in the area
C_4	Willingness to organize
C5	Have been a victim of crime in the last twelve months
C_6	Consider the municipality safe
C_7	Satisfied with the security provided by the municipality
C_8	Trust in the municipal police
C_9	Alcoholism problem

In order to specify this process in the research, the rescaling technique (called maximum and minimum) is detailed to normalize the simple indicators, in such a way as to facilitate the best possible comparison between units of analysis. This procedure tends to seek that the normalized scale covers the interval between 0 and 1 of the values to be rescaled [5]. In this sense, the values of the simple indicators that correspond to each alternative are converted.

3.3 Determination of Weights Through Personal Construction Theory

The weights of the criteria, unlike other methods, can be considered coefficients of importance or values of relative importance between criteria, thus avoiding problems of a compensatory type. In this study, the decision maker was assisted to define the weights of each of the criteria according to the Personal Construction Theory (PCT), proposed by [19]. Where, wj is the coefficient of relative importance attached to the criterion C_j for j = 1, 2, ...n.

Meanwhile, the data was collected from the opinion of five experts with the questionnaire that was provided to them. The weights that were obtained from the consensus of decision makers are shown in Table 3. To construct the table each expert compare the criterions between them by selecting H if the importance of the criterion is higher than the one is being compared with, S if the importance is the same or L if the importance is lower than the one being compared with.

After that, the column name Sum is the sum of the times each criterion has H as a result. Also, the column Sum + 1 is the results of the Sum column plus 1, this is because the comparison between each criterion with itself doesn't have a valid result and for that is included as an H. Finally, the weights column is Sum + 1 value for each criterion divided by the total value of the Sum + 1 multiplied by 100.

Table 3 Weights of the criteria obtained through the Personal Construction Theory

	C_1	C_2	C_3	C_4	C_5	C_6	C_7	C_8	C_9	Sum	Sum + 1	Weights (%)
C_1	–	H	S	L	L	L	L	L	L	1	2	5
C_2	H	–	L	L	L	L	L	L	L	1	2	5
C_3	L	H	–	S	H	S	S	H	L	3	4	10
C_4	H	H	H	–	H	H	L	L	H	6	7	18
C_5	H	H	S	L	–	L	L	L	L	2	3	8
C_6	H	H	H	L	S	–	S	H	H	5	6	15
C_7	H	H	H	H	H	S	–	S	S	5	6	15
C_8	S	H	H	S	H	L	S	–	H	4	5	13
C_9	H	H	H	H	S	L	L	S	–	4	5	13
									Total	31	40	100

3.4 Decision Matrix

The blocks of the city of Culiacán that will be evaluated with the criteria are all quantitative in nature. In this way, a decision matrix is generated, which is constructed with Table 4 of m alternative, by the nine decision criteria and after preparing the data collected from the experts, based on the result of calculating the weights (Table 3) of the criteria, the following steps will be followed that will show the procedure of the ELECTRE III method [20].

Table 4 Decision matrix of the alternatives

	A_1	A_2	A_3	A_4
C_1	0.145	0.050	0.094	0.043
C_2	0.131	0.030	0.150	0.066
C_3	0.340	0.113	0.289	0.211
C_4	0.490	0.169	0.477	0.445
C_5	0.127	0.092	0.075	0.031
C_6	0.104	0.185	0.162	0.000
C_7	0.488	0.139	0.429	0.225
C_8	0.500	0.185	0.281	0.125
C_9	0.486	0.111	0.346	0.129

3.5   Final Calculations and Orderings

The weights, direction, preferences and indifferences are shown in Table 5. The result allows us to exploit the relation of the organization of fuzzy approximation to obtain a ranking of alternatives of decreasing preferences.

Finally, with this information obtained, the ranking of the alternatives (calculation and final ordering) is generated, which is presented below and, for which, the SADGAGE software was used [16]. In this way, the Quadrants of the City of Culiacán are classified in descending order, according to the degree of perception of insecurity that citizens perceive, according to the weighting ranking.

Block 1 >Block 3 >Block 2 >Block 4

León and Leyva [17] point out that, in most cases, the decision support process does not end with the acceptance by whoever decides the recommendation made by the analyst; and it is generally suggested that a sensitivity analysis be necessary. This analysis allows us to interpret the effects of modifying the values of the weights, or thresholds of preference and indifference.

On the other hand, with the use of the OWA operator, the results are shown below.

Block 1 >Block 3 >Block 4 >Block 2

As can be seen with the ordering of both techniques, there is a difference at the end of the ranking, although the first places correspond, in both methods, to block 1 and 3, the decision on position 3 and 4 is not so clear, since that on the one hand with the ELECTRE III method the ordering indicates block 2, then 4, with the use of the OWA operator it is obtained that the ordering must end with block 4 and then block 3.

Table 5 Criteria parameters

	Weight (%)	Direction	Perference	Indifference
Gang problem	5	Minimize	0.004	0.069
Alcoholism problem	5	Minimize	0.005	0.060
Drug addiction problem	10	Minimize	0.019	0.170
Degree of violence in the area	18	Minimize	0.039	0.249
Willingness to organize	8	Maximize	0.003	0.055
Have been a victim of crime in the last twelve months	15	Minimize	0.009	0.046
Consider the municipality safe	15	Maximize	0.042	0.226
Satisfied with the security provided by the municipality	13	Maximize	0.047	0.219
Trust in the municipal police	13	Maximize	0.046	0.205

The variation of the ordering allows us to observe how the subjective importance given to each of the criteria based on the weight that each one has within the result allows to generate or visualize different results. The above, although it presents a limitation of the present work, since the results can be highly modified depending on the values that the experts present (See Table 3), so the result is not unique, but in turn this is the advantage of the methods proposed within the article, the possibility of generating scenarios based on different preferences and perspectives through weights assigned to the criteria will give a decision-making more attached to each of the decision makers and their expectations, therefore that the flexibility of these methods under scenarios of uncertainty is of great value.

4 Conclusions

In this paper, a new extension of the OWA operator is presented by the use of weighting average and Bonferroni means. This new formulation was called the Bonferroni induced ordered weighted average weighted average (BON-IOWAWA) operator. The characteristics of this operator is that it includes an induced ordering step of the weighting vector and an additional weighting vector to be considered in the final formulation.

An application in decision-making process in an enterprise to invest has been done. The importance of the use of different aggregation can be seen in the numerical example where it is possible to observe that depending on the operator that is used the ranking can change and in this sense the decision. Considering this, the main idea is to generate as many possible scenarios as possible and include as much information as possible and by doing this a better understanding of the environment and the problem can be achieved.

These ideas are important in different areas where uncertainty is a common element in current decisions. In addition, the application of these operators has the advantage of making complex comparisons, interrelationships and rearrangements of information, but it also makes it possible to consider the attitudinal character that is characteristic of the decision maker [1]. The foregoing is beneficial for decision makers because it allows them to visualize different orders taking into account not only the subjectivity of the subjects, but also the environment where these decisions are made.

In future research, we expect to develop further the operator by using new aggregations functions, such as heavy weights [27], heavy moving averages [9, 18] and group decision-making problems, [6, 25].

References

1. Alfaro-Garcia, V. (2019). Toma de decisiones en la incertidumbre: técnicas y herramientas ante escenarios altamente desafiantes. *Inquietud Empresarial, 19*(2), I–III.
2. Avilés-Ochoa, E., León-Castro, E., Perez-Arellano, L. A., & Merigó, J. M. (2018). Government transparency measurement through prioritized distance operators. *Journal of Intelligent & Fuzzy Systems, 34*(4), 2783–2794.
3. Bardach, E. (2001). Los ocho pasos para el análisis de políticas públicas. *Un manual para la práctica.* CIDE, Miguel Ángel Porrúa.
4. Bartolini, F., Gallerani, V., Samoggia., & Viaggi, D. (2005). Methodology for multicriteria analysis of agri-environmental schemes. *Technical report*, Sixth framework programme. University of Bologna.
5. Bas, M. (2014). *Estrategias metodológicas para la construcción de indicadores compuestos en la gestión universitaria.* Colección de tesis. Universitat poliècnica de València.
6. Blanco-Mesa, F., León-Castro, E., & Merigó, J. M. (2019). A bibliometric analysis of aggregation operators. *Applied Soft Computing, 81*(4), 105488.
7. Blanco-Mesa, F., León-Castro, E., Merigó, J., & Xu, Z. (2018). Bonferroni means with induced ordered weighted average operator. *International Journal of Intelligent Systems, 34*(1), 3–23.
8. ENVIPE. (2018). Encuesta Nacional de Victimización y Percepción sobre Seguridad Pública (ENVIPE) 2018. Consultado en: http://www.beta.inegi.org.mx/programas/envipe/2018/
9. Gamboa, S. F., López, J. C. L., & Santiesteban, M. L. (2016). Valoración de la seguridad en Municipios Turísticos de Sol y Playa con mayor actividad hotelera en México a partir de un análisis multicriterio.: Assessment of security in Tourist Municipalities of Sun and Beach with greater hotel activity in Mexico from a multicriterial analysis. *Kalpana-Revista de Investigación, 14*, 17–31.
10. Gonzáles, S., López, E., & Yañez, J. (1994). Seguridad Publica, *Problemas, perspectivas y propuestas.* Universidad Nacional Autónoma de México.
11. Institute for Economics & Peace. (2018). *Global Peace Index 2018: Measuring Peace in a Complex World*, Sydney, June 2018. Retrieved from http://visionofhumanity.org/reports.
12. Instituto Mexicano para la Competitividad A. C. (2017). *Índice de Paz México 2016* vía el Instituto para la Economía y la Paz, México. Consultado en https://imco.org.mx/seguridad/ind ice-de-paz-mexico-2016-via-el-instituto-para-la-economia-y-la-paz-2/.
13. Instituto Nacional de Estadística, Geografía e Informática [INEGI]. (2019). *Encuesta Nacional de Seguridad Pública Urbana, Diciembre 2018.* México. Disponible en https://www.inegi.org. mx/contenidos/saladeprensa/boletines/2019/ensu/ensu2019_01.pdf
14. Leon, M., & Larrañaga, A. M. (2019). Integración de conocimiento en restaurantes mediante el análisis multicriterio para la toma de decisiones. *Inquietud Empresarial, 19*(2), 25–38.
15. Leyva, J. C., Álvarez, P. A., Gastélum, D. A., & y Solano, J.J. (2017). A web-based group decision support system for multicriteria ranking problems. *Operational Research: An International Journal, 17*(2), 499–534.
16. Leyva, J. C., & Álvarez, P. A. (2013). *SADGAGE: A MCGDSS to solve the multicriteria ranking problem in a distributed and asynchronous environment.* In Leyva, J.C. Espín, R. Bello, R. & Álvarez, P. A. (Eds.), *Studies on knowledge discovery, knowledge management and decision making. Fourth International Workshop Proceedings, Eureka-2013* (pp. 247–254). Atlantis Press.
17. León, M., & Leyva, J.C. (2017). A multicriteria decision and for evaluating the competitiveness of tourist destinations in the Norwest of Mexico. *Turismo y Sociedad, XXI*, 51–67.
18. León-Castro, E., Avilés-Ochoa, E., Merigó, J. M., & Gil-Lafuente, A. M. (2018). Heavy moving averages and their application in econometric forecasting. *Cybernetics and Systems, 49*(1), 26–43.
19. Roger, M., Bruen, M., & Maystre, L. (2000). *ELECTRE and decision support. Methods and applications in engineering and infrastructure investment.* Springer US.
20. Roy, B. (1991). The outranking approach and the foundations of ELECTRE methods. *Theory and Decision, 31*(1), 49–73.

21. Roy, B. (2010). Two conceptions of decision-aid science? *International Journal of Multicriteria Decision Making, 1*(1), 184–203.
22. Sani y Nunes (2016). Diagnóstico de seguridad/inseguridad. Un estudio exploratorio en una comunidad urbana. *Anuario de Psicología Jurídica. 26*(1), 102–106.
23. Schuschny, A. y Soto, H. (2009). Comisión económica para América Latina y el Caribe (CEPAL). *Diseño de indicadores compuestos de desarrollo sostenible.* Colección documentos de proyectos, guía metodológica.
24. Xiao, P., Liu, M., Li, H., Wu, Q., Zhou, L., Chen, H., & Li, L. (2019). Approach for multi-attribute decision making based on gini aggregation operator and its application to carbon supplier selection. *IEEE Access, 7*, 164152–164163.
25. Xu, Z. (2006). Induced uncertain linguistic OWA operators applied to group decision making. *Information fusion, 7*(2), 231–238.
26. Yager, R. R. (1988). On ordered weighted averaging aggregation operators in multicriteria decisionmaking. *IEEE Transactions on systems, Man, and Cybernetics, 18*(1), 183–190.
27. Yager, R. R. (2002). Heavy OWA operators. *Fuzzy optimization and decision making, 1,* 379–397.

A Multicriteria Hierarchical Approach to Investment Location Choice

Laura Arenas⊙, Manuel Muñoz Palma⊙,
Pavel Anselmo Alvarez Carrillo⊙, Ernesto León Castro⊙,
and Anna M. Gil-Lafuente⊙

Abstract This chapter presents an analysis of investment location choice based on decision-maker preferences in consideration of the specific financial and macroeconomic criteria of different Latin American countries. The analysis considers different subgroups of criteria and uses a hierarchical version of the ELECTRE-III method to apply different aggregated rankings using the dimensions of Financial Market, Economic Situation and Growth, Labor Market and Purchasing Power Indicators, Foreign Commercial Operations, and Fiscal Indicators. The results show that Peru and Chile are the best countries in which to invest, and Argentina is the least attractive based on different ranking schemes and investor preferences.

Keywords Multiple criteria hierarchy process · ELECTRE-III · Investment targeting

1 Introduction

Latin America has undergone significant transformation in recent decades in terms of trade, labor market, tourism, technology, and financial markets [1, 2] driven by major changes that have happened worldwide [3], making the region more attractive as an investment location. As of 2020, Latin America accounts for about 8.4% of the global population, with an increasing middle class and an average share of the world's GDP of 7.18% over the period from 2010 to 2020.

L. Arenas (✉) · A. M. Gil-Lafuente
Universidad de Barcelona, 08034 Barcelona, CAT, Spain
e-mail: laura.arenas@ub.edu

M. M. Palma
Universidad de Sonora, 83000 Hermosillo, SON, Mexico

P. A. A. Carrillo
Autómoma de Occidente, 80020 Culiacán, SIN, Mexico

E. L. Castro
Universidad Autónoma de Baja California, 21100 Mexicali, BC, Mexico

© The Author(s), under exclusive license to Springer Nature Switzerland AG 2023
Y. P. Kondratenko et al. (eds.), *Artificial Intelligence in Control and Decision-making Systems*, Studies in Computational Intelligence 1087,
https://doi.org/10.1007/978-3-031-25759-9_15

According to the International Finance Corporation (IFC), portfolio flows to emerging markets have continued to rise since the early 1980s, even after several financial crises [4]. Most Latin American countries are emerging markets. Perhaps their most outstanding characteristic is their high volatility compared to more developed markets, see [5] and [6], and to compensate for the additional risk, investors expect higher returns. Caballero [7] reports that despite the impressive success in stabilizing inflation and implementing economic reforms in Latin American countries, high volatility of real macroeconomic aggregates has persisted over the years.

For investors who are looking to allocate their economic resources geographically, the decision-making process is complex, especially under the lens of an increasingly interconnected economic environment and a growing number of conflicting criteria that need to be considered. Beside the traditional objectives of maximizing shareholder value and minimizing business risks, many other criteria should be considered. According to [8], "A firm cannot maximize value if it ignores the interests of its stakeholders", that is, the value maximization objective cannot be achieved unless it is complemented by other objectives that unite participants in the organization. Therefore, most models do not incorporate the multidimensional nature of the problem of investment decisions. Hence, investors are required to adopt more sophisticated methods in order to include more criteria in their decisions.

In this context, Multi-criteria decision-making (MCDM) is an alternative that provides a range of techniques and methods to sort and choose the best or a set of good alternatives from a set that considers multiple criteria as targets or goals. MCDM helps decision-makers to find the most satisfactory alternative as a solution to their decision-making [9], considering the requirements, limitations, and intuition, imposed by the process, in this case, by the investor.

This chapter addresses investment location choice in the form of a multicriteria ranking of problems by adapting the multiple criteria hierarchical process to include considerations of country-specific financial and economic criteria.

In this context, an analysis of financial and macroeconomic indicators ranked in terms of various subgroups of criteria is performed by applying the comprehensive ranking process.

The results of this research can help investors to consider and/or include individual preferences in a preselection of potential locations for consideration in an investment decision-making process.

The remainder of the article is structured as follows. Section 2 provides relevant background by reviewing the related studies in the literature. Section 3 describes the methodology of the multiple criteria hierarchy process. Section 4 presents the data, results, and analysis by country. Section 5 summarizes the conclusions and provides certain directions for future research.

2 Theoretical Framework

MCDM methods are Operations Research tools that can be used to solve complex problems with high uncertainty, conflicting objectives, different forms of data and information, and multiple interests and perspectives, and can account for complex and evolving biophysical and social economic systems [10]. For investors who are looking to allocate their economic resources geographically, the decision-making process is complex.

Within this framework, traditional assessment methods seem to be limited, since the parameters of their decisions are expected return and risk only, omitting relevant information. Therefore, most models do not incorporate the multidimensional nature of the problem of investment decisions, such as investor intuition, which researchers have found to play a critical role in expert decision making [11]. Dane and Pratt [12] define core intuition as affectively charged judgments that arise through rapid, nonconscious, and holistic associations between different elements, including experience-based complex patterns, which financial specialists seek and use to reach investment decisions.

Under these circumstances, investors are required to adopt more sophisticated methods so as to include more criteria.

An MCDM framework is suitable for multidimensional economic activities because it considers all factors involved in reducing risk while, for example, evaluating investment in projects [13]. The abovementioned and other advantages explain the growing interest in multidimensional methods for the analysis and evaluation of economic decisions [14]. Guerrero-Baena [15] perform a bibliometric analysis of the international research on the application of MCDM techniques to corporate finance issues during the 1980–2012 period. Recent papers have empirically targeted certain aspects related to the central theme of this chapter, but there has been surprisingly little study of the constellation of decisions involved in the choice of investment location based on both economic and financial variables.

Stock market, financial variables and the sensitivity of macroeconomic news are highly discussed by academics, since investors closely monitor economic data, announcements, political events, and regulatory mandates. In fact, there are several theoretical justifications for a relationship between macroeconomic variables and stock returns [16–18]. The perception of an economic slowdown is enough to generate large changes in stock market prices, as argued by [19].

Of the different applications, MCDM approaches have been widely used in the location selection process, presenting interesting results in a variety of problem-solving domains [20]. However, there is still a gap in the literature on business location decision-making tools [21].

Location selection is also one of the most crucial and influential choices that managers and business leaders must tackle when a company decides to relocate its supply chain in foreign countries [22]. All decision-making processes related to investment activities involve a broad complexity of factors, including geographical choices. Several criteria are used, mainly of a financial and economic nature. Essentially, scientific, and analytical methods based on multiple decision-making criteria are used to support final investment decisions.

3   Data Description

Financial and economic indicators are key factors for investment decision-making, including the problem of investment location choice. Most indicators are used to facilitate comparisons made over time and between countries at the macroeconomic scale. These indicators also help to judge the overall health of an economy.

3.1   Financial Market Indicators

Stock market benchmark indices: The stock market cycle is not only correlated to the economic business cycle but is also a leading indicator of it. The benchmark indices are created to include multiple securities representing the performance of the total market of, in this context, a country.

Interest rates: Interest rates are known to be lagging indicators of economic growth. They reflect the investment opportunities of various countries. Higher interest rates mean a higher rate of return on investments. Interest rates are based on a federal reference or funds rate, which is determined by the Central Bank. When the reference or funds rate increases, the interest rates increase too. The federal reference or funds rate increases or decreases as a result of economic and market events.

Country risk: Country risk refers to the uncertainty associated with investing in a particular country and may be affected by several sources, including political, economic, exchange-rate, and technological influences. In a broader sense, country risk is the degree to which political and economic unrest affect the securities of issuers doing business in a particular country.

3.2   Economic Situation and Growth Indicators

Gross Domestic Product (GDP): This is known as a lagging indicator and is one of the first indicators used to gauge the health of an economy. It represents economic production, growth, and size.

Public Debt: Public debt is the total amount, including total liabilities, borrowed by the government to meet its development budget. Growing debt has a direct effect on the economic opportunities available to society. The government debt-to-GDP ratio compares a country's public debt to its GDP and is used as a key indicator of the sustainability of government finance. The debt-to-GDP ratio reliably indicates a country's ability to pay back its debts.

3.3 Labor Market and Purchasing Power Indicators

Unemployment rate: Unemployment is a lagging indicator that is expected to rise or fall under changing economic conditions. Monthly estimates are released in most countries as the cumulative number of jobs created or lost in the previous month. meaning that it generally rises or falls in the wake of changing economic conditions, rather than anticipating them.

Inflation rate: The Consumer Price Index (CPI) is a lagging indicator and relies on the best indicators of inflation. Variation in inflation can spur the Federal Reserve to make changes to its monetary policy. CPI measures changes in prices paid for goods and services by urban consumers for a specified month and is essentially a measure of the cost-of-living.

3.4 Foreign Commercial Operations Indicator

Balance of trade: Balance of trade is a lagging indicator. It is the net difference between a country's value of imports and exports and shows whether there is a trade surplus or deficit. A trade surplus is generally desirable and shows that there is more money coming into the country than leaving.

Current account: The current account is an important indicator of an economy's health. It is defined as the sum of the balance of trade (goods and service exports minus imports), net income from abroad, and net current transfers. A positive current account balance indicates that the nation is a net lender to the rest of the world, while a negative current account balance indicates that it is a net borrower.

3.5 Fiscal Indicators

Fiscal result in terms of GDP: The tax-to-GDP ratio is a gauge of a nation's tax revenue relative to the size of its economy as measured by gross domestic product (GDP). This ratio provides a useful overview of a country's tax revenue because it reveals potential taxation relative to the economy. It also offers a view of the overall direction of a nation's tax policy, as well as international comparisons between the tax revenues of different countries.

Corporation tax and Income tax: Tax revenue is defined as the revenues collected from taxes on income and profits, social security contributions, taxes levied on goods and services, payroll taxes, taxes on the ownership and transfer of property, and other taxes. Total tax revenue as a percentage of GDP indicates the share of a country's output that is collected by the government through taxes. It can be regarded as a measure of the degree to which the government controls the economy's resources. The tax burden is measured by taking the total tax revenues received as a percentage of GDP. This indicator relates to all government levels and is measured in millions of USD and percentage of GDP.

For the indicators mentioned above, the information was retrieved from Cesla.com [23] Circulo de Estudios Latinoamericanos for data as of December 31, 2020.

4   The Multiple Criteria Hierarchy Process

The MCHP is based on Multiple criteria decision-making methodology. It is used to deal with a problem characterized by a set of attributes/criteria but presenting a structure with hierarchy. It is frequently used to find practical applications with a hierarchical structure of criteria [24]. The addressed problem presents a similar structure, whereby subsets of decision criteria belong to dimensions related to the investment location choice. Hence, its structure is observed to be hierarchical in nature. It therefore seems appropriate to use the methodology based on the MCHP.

Some real applications with MCHP have been carried out to analyze the competitiveness of regions [25], to evaluate sustainable rural development [26], to study healthcare in the Italian co-payment system [27], and to evaluate stock market assets [28].

The MCDA process involves defining a set of alternatives $A = \{a_1, a_2, \ldots, a_m\}$ and a coherent family of criteria $G = \{g_1, g_2, \ldots, g_m\}$. Preference is comprehensively modeled using any MCDM method as an aggregation procedure. The method generates a recommendation in the form of a ranking of alternatives in descending order from the best to the worst.

The MCDM methodology includes four main approaches, namely first, the full aggregation, second the outranking, third the goal, aspiration, or reference level, and fourth the non-classical MCDM approach [29], each of which can be materialized by different methods. Since each approach and method uses a particular technique to evaluate or compare the alternatives in a ranking problem, the resulting ranking is directly associated to the selected technique.

The ELECTRE multicriteria aggregation procedure is adequate for use in the presence of heterogeneous scales and is designed not to allow for compensation of performances among criteria [30].

In this context, the outranking approach with the ELECTRE III method is suitable for application to investment location choice, especially since, in the ELECTRE III method, the expert explicitly considers situations where country deterioration or downgrading is driven by poor performance in specific criteria.

Corrente et al. [31] proposed a hierarchical version of the ELECTRE III method. The notation used in this process is explained hereinafter.

G is a comprehensive set of all criteria at all considered levels in the hierarchy.

G_0 is the root of the criterion.

I_G is the set of indices of the criteria in G.

$E_G \subseteq I_G$ is the set of indices of elementary criteria.

g_r is the generic non-root criterion (where r is a vector of equal length to the level of the criterion).

$g_{(r,1)}, \ldots, g_{(r,n(r))}$ are the immediate subcriteria of criterion g_r (located at the level below g_r).

$E_{(g_r)}$ is the set of indices of all the elementary criteria descending from g_r.

$E(F)$ is the set of indices of the elementary criteria descending from at least one criterion in the subfamily $F \subseteq G$ (that is, $E(F) = U_{g_r \in F} E(g_r)$).

G_r^I is the set of subcriteria of g_r located at level 1 in the hierarchy (belonging to g_r).

L is the number of levels of the hierarchy $I = 1, \ldots, L$.

The ELECTRE method is carried out in two steps. The first step involves developing the aggregation of preference information, constructing a decision model in the valued outranking relation. In the second step, the valued outranked relation is exploited by the distillation process, generating a complete or partial ranking of alternatives.

For each elementary criterion $g_t \in E(g_r)$.

The elementary concordance index, for each elementary criterion g_t.

$$\varphi_t(a, b) = \begin{cases} 1 & if \ g_t(b) - g_t(a) \le q_t, (aS_tb) \\ \frac{p_t - [g_t(b) - g_t(a)]}{p_t - q_t} & if \ q_t < g_t(b) - g_t(a) < p_t, (bQ_ta) \\ 0 & if \ g_t(b) - g_t(a) \ge p_t, (bP_ta) \end{cases} \tag{1}$$

The discordant elementary index, for each elementary criterion g_t.

$$d_t(a, b) = \begin{cases} 1 & if \ g_t(b) - g_t(a) \ge v_t \\ \frac{[g_t(b) - g_t(a)] - p_t}{v_t - p_t} & if \ p_t < g_t(b) - g_t(a) < v_t \\ 0 & if \ g_t(b) - g_t(a) \le p_t \end{cases} \tag{2}$$

The partial concordance index for each non-elementary criterion g_r,

$$C_r(a, b) = \frac{\sum\limits_{t \in E(g_r)} w_t \varphi_t(a, b)}{\sum\limits_{t \in E(g_r)} w_t} \tag{3}$$

Partial credibility index,

$$\sigma_r(a, b) = \begin{cases} C_r(a, b) \times \prod\limits_{g_t \in E(g_r)} \frac{1 - d_t(a, b)}{1 - C_r(a, b)} & if \ d_t(a, b) > C_r(a, b) \\ C_r(a, b) \ otherwise \end{cases} \tag{4}$$

The distillation method is used to exploit the fuzzy outranking relation. For the pairs, $a, b \in A$ in the hierarchical process, the alternatives are ranked in a partial or complete pre-order on the non-elementary criterion g_r. For a further description of the distillation process, see [32].

5 Investment Location in Latin American Countries with an MCHP

5.1 Definition of Investment Location Choice as a Multiple Criteria Decision-Making Problem

The MCHP is applied to some countries of Latin America to deal with the investment location problem, which is evaluated by five main domains, called macrocriteria, in a hierarchy of criteria. The Financial Market macrocriteria (g_1) integrate four elementary criteria; Economic Situation and Growth Indicators (g_2) integrate three elementary criteria; Labor Market and Purchasing Power Indicators (g_3) integrate two elementary criteria; Foreign Commercial Operations Indicators (g_4) integrate three elementary criteria; Fiscal Indicators (g_5) integrate three elementary criteria.

The investment location problem is analyzed by 15 elementary criteria and is structured in a hierarchy of two levels. Figure 1 shows the hierarchy of the criteria. On Level 1, 5 macrocriteria (non-elementary criteria) are defined. On Level 2, 15 elementary criteria represent the comprehensive problem. Table 1 lists the macrocriteria and elementary criteria used to analyze the investment location problem.

This chapter applies the decision on investment location to some Latin American countries. Table 2 show the list of the countries that have been analyzed and compared.

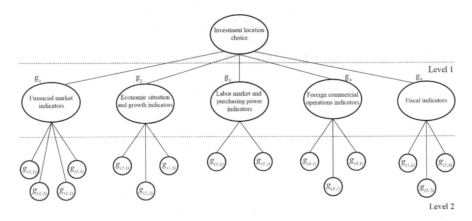

Fig. 1 Hierarchical structure of the investment location problem

Table 1 Decision criteria for investment location choice

Macrocriteria	Elementary criteria
Financial Market Indicators (g_1)	$g(1,1)$ Stock market benchmark indices
	$g(1,2)$ Interest rate of the 10-year sovereign bonds
	$g(1,3)$ Interest rates
	$g(1,4)$ Country risk
Economic Situation and Growth Indicators (g_2)	$g(2,1)$ GDP quarterly growth rate
	$g(2,2)$ GDP annual growth rate
	$g(2,3)$ Public debt in terms of GDP
Labor Market and Purchasing Power Indicators (g_3)	$g(3,1)$ Unemployment rate
	$g(3,2)$ Inflation rate
Foreign Commercial Operations Indicators (g_4)	$g(4,1)$ Balance of trade
	$g(4,2)$ Current account
	$g(4,3)$ Current account in GDP terms
Fiscal Indicators (g_5)	$g(5,1)$ Fiscal result in terms of GDP
	$g(5,2)$ Corporation tax
	$g(5,3)$ Income tax

Table 3 shows the performances of countries when evaluated by each elementary criterion.

The MCHP described in Sect. 3 can be used to evaluate the performance of each macrocriterion in terms of the elementary criteria immediately descending from it.

Table 2 Latin American countries as investment options

Code	Country
A_1	Argentina
A_2	Brasil
A_3	Bolivia
A_4	Colombia
A_5	Costa Rica
A_6	Chile
A_7	Ecuador
A_8	Mexico
A_9	Peru
A_{10}	Uruguay

Table 3 Performances of Latin American countries

		A_1	A_2	A_3	A_4	A_5	A_6	A_7	A_8	A_9
g_1	$g(1,1)$	−7.76	9.38	0	0.79	−14.83	−0.16	−1.27	11.78	13.38
	$g(1,2)$	1	0.65	1	0.37	0	0.1	1	0.41	1
	$g(1,3)$	38	2	3.82	1.75	2.5	0.5	9.02	4	0.25
	$g(1,4)$	1499	267	491	223	507	133	1209	355	147
g_2	$g(2,1)$	12.8	3.2	−6.2	6	5.6	5.2	4.51	3.3	8
	$g(2,2)$	−10.2	−1.1	−21.68	−3.6	−4.5	−9.1	−8.8	−4.3	−1.7
	$g(2,3)$	89.4	75.79	57.7	48.4	65.8	27.9	49.4	45.5	27.5
g_3	$g(3,1)$	11.7	13.9	3.6	17.3	0.1	10.2	6.6	4.7	13
	$g(3,2)$	38.5	4.56	1.2	1.6	0.9	3.1	3.59	3.54	2.4
g_4	$g(4,1)$	1068	1152.3	7.62	−1.1	13,540	1739.8	12,174	−1236	1079.6
	$g(4,2)$	1163	−7250	−311.4	−1774	−1554	−25	1213.6	17,409	738.68
	$g(4,3)$	−0.8	−0.9	−3.4	−4.3	2.3	−3.9	−0.1	2.4	−1.5
g_5	$g(5,1)$	−3.8	−5.9	−7.2	−7.8	0.9	−1.6	−3	−1.6	−2.7
	$g(5,2)$	30	34	25	33	10	27	25	30	29.5
	$g(5,3)$	35	27.5	25	39	25	35	35	35	30

This is done by generating a preferential model and a ranking for each macrocriterion to understand how the countries perform against other countries and, at the same time, how this impacts the investment location choice problem in terms of each macrocriterion.

5.2   Application of the Hierarchical ELECTRE III

The Hierarchical ELECTRE III was applied to solve and compare the countries in each macrocriterion g_i and the comprehensive level. In Table 3, the parameters used are represented. Column 2 describes the type of criteria, Column 3 lists the relative weights of the elementary criteria, Column 4 shows the indifference threshold (q) and Column 5 shows the preference threshold (p). It is essential to describe the relative importance of the most important macrocriteria. This is $g_1 > g_2 > g_3 > \{g_4 = g_5\}$, with the weights 0.2934, 0.2410, 0.1912, 0.1737, 0.1737 respectively. These are computed with the sum of the values from the elementary criteria belonging to each macrocriterion.

Table 4 contains the rankings of each macrocriteria (g_i) and the comprehensive problem (g_0). Each macrocriterion is evaluated by a subset of subcriteria (elementary criteria). The generated ranking is the result of the evaluation of the elementary criteria corresponding to each macrocriteria. For the location of investment countries, the influence of the interaction of the subset of elementary criteria on the countries in Latin America is analyzed based on the macrocriteria (Level 2 of the hierarchy) and

Table 4 Parameters of the Hierarchical ELECTRE III method

Decision criteria	Type	Weight	q	p
Financial Market Indicators (g_1)		**0.2934**		
$g(1,1)$ Stock market benchmark indices	Cost	0.0978	4	10
$g(1,2)$ Interest rate of the 10-year sovereign bonds	Benefit	0.0816	0.1	0.3
$g(1,3)$ Interest rates	Benefit	0.0490	1	1.5
$g(1,4)$ Country risk	Benefit	0.0650	100	200
Economic Situation and Growth Indicators (g_2)		**0.2410**		
$g(2,1)$ GDP quarterly growth rate	Cost	0.0803	0.5	1
$g(2,2)$ GDP annual growth rate	Cost	0.0803	0.5	1
$g(2,3)$ Public debt in terms of GDP	Benefit	0.0803	5	10
Labor Market and Purchasing Power Indicators (g_3)		**0.1902**		
$g(3,1)$ Unemployment rate	Benefit	0.0634	2	4
$g(3,2)$ Inflation rate	Benefit	0.1268	1	2
Foreign Commercial Operations Indicators (g_4)		**0.1377**		
$g(4,1)$ Balance of trade	Cost	0.0529	100	500
$g(4,2)$ Current account	Cost	0.0424	100	500
$g(4,3)$ Current account in GDP terms	Cost	0.0424	0.5	1
Fiscal Indicators (g_5)		**0.1377**		
$g(5,1)$ Fiscal result in terms of GDP	Benefit	0.0529	0.5	1
$g(5,2)$ Corporation tax	Benefit	0.0424	2	4
$g(5,3)$ Income tax	Benefit	0.0424	1	2

then the impact of the interaction of macrocriteria on the comprehensive investment location choice problem (Level 1). Table 5 shows the ranking for each macrocriterion and the comprehensive problem.

In what follows, the macrocriteria Financial Market Indicators (g_1) and Economic Situation and Growth Indicators (g_2) are analyzed, considering the most important dimensions.

The ranking generated in Financial Market Indicators (g_1) is represented in Fig. 2 as an investment map with four categories. The best countries to invest in are Chile, Peru and Brazil. Chile (A_2) shows better performance in sovereign bonds $g(1,2) = 0.65$, interest rates $g(1,3) = 2$ and country risk $g(1,4) = 267$. Peru (A_9) shows good performance in the country risk criterion $g(1,4) = 147$ and stock market $g(1,1) = 13.38$ (index of the Peruvian stock market in 2020), despite the COVID 19 pandemic. The countries with most potential for investment (medium category) are Mexico and Colombia. The worst country in which to invest is Argentina. The ranking of g_1 supports the analysis for someone interested in Financial Market Indicators (g_1).

The investment map of Economic Situation and Growth Indicators (g_2) is depicted in Fig. 3. The best countries in which to invest are Peru (A_9), Colombia (A_4) and Chile (A_6). The annual GDP growth criterion $g(2,2)$ is negative in all countries. However,

Table 5 Comprehensive and individual rankings of the location of investment countries

Position	g_0	g_1*	g_2*	g_3	g_4	g_5
1	A_9	A_6	A_9	A_5	A_7	A_3
2	A_5	A_9	A_4	A_3	A_{10}	A_{10}, A_9, A_7
3	A_6, A_8	A_2	A_6	A_8, A_7	A_1	A_1
4	A_4	A_4	A_{10}	A_9	A_5	A_2
5	A_{10}, A_2	A_8	A_8	A_6	A_6	A_5
6	A_7	A_{10}	A_5	A_2	A_9	A_4
7	A_3	A_3	A_7	A_{10}	A_3	A_6
8	A_1	A_7	A_2	A_4	A_8	A_8
9		A_5	A_1	A_1	A_2	
10		A_1	A_3		A_4	

Note *The most important macrocriteria defined by the expert

Fig. 2 Investment map of the macrocriterion Financial Market Indicators (g_1)

Peru has a better performance (-1.7). Colombia (A_4) shows good performance in the quarterly GDP growth rate $g(2,1) = 6\%$ while for Chile it is $g(2,1) = 5.2\%$. Peru has less public debt in terms of GDP $g(2,3) = 27.5$, followed by Chile with 27.9 and Colombia with 48.9. Regarding the macrocriterion g_2, the worst countries in which to invest are Bolivia (A_3) and Argentina (A_1), which present annual GDP growth rates $g(2,2)$ of -21.68% and -10.2%, respectively. Argentina is one of the worst performers in the management of public debt in terms of GDP $g(2,3) = 89.4$.

The Financial Market Indicators (g_1) show the first positions for $A_6 > A_9 > A_2 > A_4$. The macrocriterion Economic Situation and Growth Indicators (g_2) show $A_9 > A_4 > A_6 > A_{10}$. Labor Market and Purchasing Power Indicators (g_3) show $A_5 > A_3 > \{A_8, A_7\} > A_9$. Foreign Commercial Operations Indicators (g_4) show $A_7 > A_{10}, > A_1 > A_5$. Fiscal Indicators (g_5) show $A_3 > \{A_{10}, A_9, A_7\} > A_1 > A_2$.

Peru (A_9) is in top place for the comprehensive problem (g_0). The country is shown in the top positions in the rankings for the most important dimensions (g_1) and (g_2). Costa Rica (A5) is second in the comprehensive ranking (g_0) and is allocated

Fig. 3 Investment map of the macrocriterion Economic Situation and Growth Indicators (g_2)

a significant position in Labor Market and Purchasing Power Indicators (g_3) and Foreign Commercial Operations Indicators (g_4).

Regarding the comprehensive problem, where all elementary criteria are considered, the ranking from g_0 shows that Peru (A_9), Costa Rica (A_5), Mexico (A_8) and Chile (A_6) are the best countries to invest in. Argentina (A_1) is in the last place in the g_0 ranking and in the macrocriteria g_1 and g_3; and is in second-to-last for g_2. Argentina is in third place for macrocriteria g_4 and g_5, which are considered the least important. The investment map regarding the whole set of elementary criteria for investment location choice in Latin America is depicted in Fig. 4.

An interesting observation arising from analysis of the interaction with elementary criteria is that some less important countries in the comprehensive problem (Level 1) are the best countries for a specific macrocriterion (Level 2). For example, Ecuador (A_7) and Bolivia (A_3) are in the top places for the macrocriteria Foreign Commercial Operations Indicators (g_4) and Fiscal Indicators (g_5), respectively. But those countries are not very well positioned in the comprehensive problem because they perform poorly in other important macrocriteria.

Fig. 4 Investment location map in Latin America (g_0)

6 Discussion

The result obtained with the proposed methodology was compared against financial experts' recommendations and 'intuition' with regard to investment in Latin American countries. Recall that the proposed methodology is highly intuitive since the investor bases decisions on hierarchical criteria, as well as their intuition regarding the model.

Financial experts refer to Rating Agencies, the Milken Institute Global Opportunity Index, the World Bank Doing Business Index and other types of reports in this field.

The Standard & Poor, Moody, and Fitch Ratings are the three most influential agencies wherefrom many countries seek ratings to encourage investor confidence. A sovereign credit rating is an independent assessment of the creditworthiness of a country. Data on the countries' degrees of investment were retrieved from tradingeconomics.com [33].

The Global Opportunity Index published by the Milken Institute is an over-all reference for the attractiveness of countries to international investors. It uses a combination of economic, financial, institutional, and regulatory factors that are divided into five categories, namely business perception, economic fundamental, financial services, institutional framework, and international standards and polity measure. The lower the index, the greater the investors' perception of attractiveness. The indices are publicly available at milkeninstitute.org [34].

The World Bank Doing Business 2020 report is a ranking of 190 countries according to best regulatory practices for opening and running a business that is also used to contrast our results and report the best and worst countries in which to do business. The Doing Business reports are available at Worldbank.org [35]. The financial expert recommendations are reported on Table 6.

Focusing on the details, the S&P Global Ratings, Fitch Ratings Global, and Moody's Investors Service are aligned with the overall results that our methodology indicates for recommended Latin American countries for investment location in the 2020 period.

The best countries in which to invest as per the Rating Agencies in 2020 were Chile and Peru followed by Mexico and Uruguay. Chile was rated by S&P Global Ratings with an A, Global Fitch Rating with an A- and by Moody's Investor Services with an A1, with a descriptive assignment of 'upper medium grade'.

Also, the Global Opportunity Index and the Ease for Doing Business Index agree with the aforesaid outlook for 2020, whereby Chile is the best performer in the region with a strong result across the five categories that compose the Global Opportunity index. Chile is the highest ranked Latin American country in 59th position, followed by Mexico in 60th and Colombia in 67th. Chile's best category was the Institutional Framework and Financial Services.

The World Bank [36] defined Chile as one of the two Latin American high-income countries with a favorable regime for investment stability compared to their Latin American neighbors.

Table 6 Financial expert recommendations, 2020 for the Latin American counties

Code	Country	SP	FITCH	Moody's	Global Opportunity Index	Doing Business Index
A	Argentina	CCC+[5]	CCC[5]	Ca[5]	89	126
A	Brazil	BB-[3]	BB-[3]	Ba2[3]	69	124
A	Bolivia	B+[4]	B[4]	B2[4]	107	150
A	Colombia	BB-[3]	BB-[3]	Ba2[3]	67	67
A	Costa Rica	B[4]	B[4]	B2[4]	55	74
A	Chile	A+[1]	A-[1]	A1[1]	39	59
A	Ecuador	B-[4]	B-[4]	Caa3[5]	111	129
A	Mexico	BBB[2]	BBB-[2]	Baa1[2]	54	60
A	Peru	ND	BBB+[2]	ND	62	76
A	Uruguay	ND	BBB-[2]	ND	48	101

Source 33, 34, 35
Note 1 [1]upper medium grade; [2]lower medium grade; [3]not speculative investment grade; [4]highly speculative grade; [5]extremely speculative grade
Note 2 ◼Best, ◼Good, ◼Medium, ◼Worst

IMF analysts [37], based on their projections published in 2020, estimated Peru to be the Latin American country with the greatest real GDP growth rate. On the other hand, Peru was rated by the Global Opportunity Index and the Doing Business Index in a medium-good position, ranked overall 62nd with its best category being Economic Fundamentals.

The comprehensive result (g_0) of our proposed methodology indicates that Peru is the best country in which to invest followed by Costa Rica and Chile. In this context, the results agree with the financial experts' recommendations and 'intuition'.

On the other hand, the expert criteria in our model define g_1 Financial Market Indicators and g_2 Economic Situation and Growth Indicators, as the most important macroctieria.

For g_1 the best country in which to invest is Chile, followed by Peru, which is in full accordance with the Rating Agencies' recommendations and partially aligned with the Global Opportunity Index and the Doing Business Index. The Seeking Alpha portal also recommended Chile and Peru, but also included Colombia among its top recommendations [38].

Colombia was a special case in 2020, since it reported negative performances due to the rise in oil prices during the COVID 19 pandemic. However, in the first and second quarters of 2020 it had GDP growth rates of −2.1% and −15.9%, followed in the third and fourth quarters by a significantly improved 10% and 6.6%, respectively, which explains the expectation of a quick recovery and room for investment opportunities. Colombia was rated the third best of the countries studied in this chapter with which to do business with an overall position of 67th. Our model assigned Colombia as the second-best country in which to invest based on the Economic Situation and Growth Indicators (g_2) macrocriterion, S&P Global Rating and Global Fitch Rating of BBB- and Moody's Investor Services rating of Baa2 quoted in the 'lower medium' bracket. In this context, the applied methodology identified a sudden improvement in Colombia that reflects these circumstances.

The worst country in which to invest according to the Rating Agencies in 2020 was Argentina, which the S&P Global Rating CCC+ and Global Fitch Rating gave CCC and Moody's Investor Services gave Ca and quotes in the 'substantial risk' bracket. Argentina was also reported as the worst county in which to invest by the comprehensive vision g_0, financial market indicators g_1 and Labor Market and Purchasing Power Indicators g_3 as per the applied methodology. Argentina was described as Latin America's second worst country in terms of Economic Situation and Growth Indicators g_2 after Bolivia. The Seeking Alpha portal highlighted the monetary crisis that Argentina is experiencing [38]. Argentina was also the worst Latin American country in which to do business in 2020 as per the Doing Business Index, in 126th position. Brazil was quoted as the second worst country by the Doing Business Index in 124th position, being assigned the medium–low rating for the Opportunity Index, and the lowest category for Business Perception.

Bolivia is also not a recommended country in which to invest, being rated by S&P Global Ratings with B+, by the Global Fitch Rating with B2 and by Moody's Investor Services with a B, where it is classed as 'highly speculative'. Additionally, Bolivia was rated the worst county in which to do business for the main factors of Business Perception and Institutional Framework. The general need for legal security, allegations of corruption, and vague investment incentives are some of the barriers to investment in Bolivia.

Our proposed methodology shows for the comprehensive results g_0 and for Economic Situation and Growth Indicators g_2, that the above appreciations are in line with the g_2 macrocriterion results for Argentina, Brazil, and Bolivia that suggest that they are the worst countries in which to invest.

7 Conclusion

From the perspective of an increasingly interconnected economic environment and growing number of conflicting criteria, beside the traditional objectives of maximizing shareholder value and minimizing business risks, a number of other criteria should be considered in the choice of investment location. For example, in recent decades Latin America has undergone significant transformation in terms of trade, labor market, tourism, technology, and financial markets, making it more attractive as an investment target. However, the investment location choice is a complex matter, especially since Latin American countries have experienced hyperinflation and political instability over the years, among other multidimensional aspects that need to be considered when viewing the problem through the lens of decision-maker preferences.

In the present chapter, the investment location choice was analyzed for ten Latin American countries as a multiple criteria problem (MCHP) approached using the hierarchical ELECTRE III method, as an alternative to the decision-making problem considering unique and case-specific requirements, in this case, by the investor.

The hierarchical structure of the problem involves analysis of the following dimensions: Financial Market Indicators (g_1), Economic Situation and Growth Indicators (g_2), Labor market and Purchasing Power Indicators (g_3), Foreign Commercial Operations Indicators (g_4), and Fiscal Indicators (g_5).

The MCHP generated a ranking for each of these dimensions and provides an individual analysis of the most important of these in order to understand the performances of the best and the worst countries in terms of the investment location choice problem.

As can be seen in Table 5 and Figs. 2, 3 and 4, different rankings can be generated based on the relative importance of each criterion in terms of investor preferences. There are some countries, such as Chile and Peru, that are still well placed even when the criteria are changed.

The results obtained were compared against financial experts' recommendations and 'intuitions' with regard to investment in Latin American countries in 2020 in order to provide evidence for the reliability of the recommendation obtained by the model. The overall direction of the financial experts' recommendations is aligned with the results obtained by the model for selecting investment locations in Latin America in 2020.

This information is important to the decision-maker because even when there are changes in preferences, location selection remains the same.

The results of this analysis may help an investor to include individual preferences when making a preselection of countries to consider in the investment decision-making problem. The chapter presents several considerations that have not been used in traditional methods and uses a hierarchical approach to analyze the performance of the countries' financial and economic parameters.

Future research could use a more complex method to aggregate the information, such as using the ordered weighted average (OWA) operator and some of its extensions, such as the heavy OWA operator induced OWA operator or prioritized OWA operator as well as applications to different fields such as innovation, sustainability, and agriculture, among others.

References

1. Oyewole, P. (2009). Prospects for Latin America and Caribbean region in the global market for international tourism: A projection to the year 2020. *Journal of Travel & Tourism Marketing, 26*(1), 42–59.
2. Vendrell-Herrero, F., Gomes, E., Mellahi, K., & Child, J. (2017). Building international business bridges in geographically isolated areas: The role of foreign market focus and outward looking competences in Latin American SMEs. *Journal of World Business, 52*(4), 489–502.
3. Lowenthal, A., & Baron, H. (2014). A transformed Latin America in a rapidly changing world. In J. I. Domínguez & A. Covarrubias (Eds.). *Routledge handbook of Latin America in the world.*
4. IFC. (2000). *Emerging stock markets factbook.* International Finance Corporation.
5. Aggarwal, R., Inclan, C., & Leal, R. (1999). Volatility in emerging stock markets. *Journal of Financial and Quantitative Analysis, 34*(1), 33–55.
6. Bekaert, G., & Harvey, C. (1997). Emerging equity market volatility. *Journal of Financial economics, 43*(1), 29–77.
7. Caballero, R. J. (2000). Macroeconomic volatility in Latin America: A conceptual framework and three case studies. *Estudios de Economia, 28*, 5–52.
8. Jensen, M. (2010). Value maximization, stakeholder theory, and the corporate objective function. *Journal of Applied Corporate Finance, 22*(1), 32–42.
9. Stewart, T. (1992). A critical survey on the status of multiple criteria decision-making theory and practice. *Omega, 20*, 569–586.
10. Wang, J. J., Jing, Y. Y., Zhang, C. F., & Zhao, J. H. (2009). Review on multi-criteria decision analysis aid in sustainable energy decision-making. *Renewable and Sustainable Energy Reviews, 13*(9), 2263–2278. https://doi.org/10.1016/j.rser.2009.06.021.
11. Salas, E., Rosen, M. A., & Diaz Granados, D. (2010). Expertise-based intuition and decision making in organizations. *Journal of Management, 36*(4), 941–973.
12. Dane, E., & Pratt, M. G. (2007). Exploring intuition and its role in managerial decision making. *Academy of Management Review, 32*(1), 33–54.
13. Shvetsova, O., Rodionova, E., & Epstein, M. (2018). Evaluation of investment projects under uncertainty: Multi-criteria approach using interval data. *Entrepreneurship and Sustainability Issues, 5*(4), 914–928.
14. Brigham, E. F., & Ehrhardt, M. C. (2015). *Financial management: Theory and practice.* (South-Western, Ed.) College.
15. Guerrero-Baena, M.D.-L. (2014). Are multi-criteria decision-making techniques useful for solving corporate finance problems? A bibliometric analysis. *Revista de Metodos Cuantitativos para la Economia y la Empresa, 17*, 60–79.
16. Boudoukh, J., & Richardson, M. (1993). Stock returns and inflation: A long-horizon perspective. *The American economic review, 83*(5), 1346–1355.
17. Chen, N., Roll, R., & Ross, S. (1986). Economic forces and the stock market. *Journal of business*, 383–403.
18. Mandelker, G. (1985). Common stock returns, real activity, money, and inflation: Some international evidence. *Journal of International Money and Finance, 4*, 267–286.
19. Peralta-Alva, A. (2012). New technology may cause stock volatility. *The Regional Economist,* April.
20. Abdel-Basset, M., Gamal, A., & Chakrabortty, R. (2021). A new hybrid multi-criteria decision-making approach for location selection of sustainable offshore wind energy stations: A case study. *Journal of Cleaner Production, 280*(124462).
21. Sabat, W., & Pilewicz, T. (2019). Business location decision: the behavioural aspect in empirical research. *Journal of Management and Financial Sciences, 37*, 99–109. https://doi.org/10.33119/JMFS.2019.37.6
22. Berger, S. (2005). *How we compete: What companies around the world are doing to make it in today's global economy.* Doubleday.

23. Circulo de estudios latinoamercianos CESLA. (2022). Estadísiticas. Retrieved November 12, 2021, from https://www.cesla.com/.
24. Corrente, S., Greco, S., & Słowiński, R. (2012). Multiple criteria hierarchy process in robust ordinal regression. *Decision Support Systems, 53*(3), 660–674. https://doi.org/10.1016/j.dss.2012.03.004.
25. Alvarez, P., Muñoz-Palma, M., & Miranda-Espinoza, L. (2020). Enfoque multicriterio jerárquico para el análisis de la competitividad de las regiones en México. *Inquietud empresarial, 20*(2), 29–51. https://doi.org/10.19053/01211048.11408.
26. Angilella, S., Catalfo, P., Corrente, S., Giarlotta, A., Greco, S., & Rizzo, M. (2018). Robust sustainable development assessment with composite indices aggregating interacting dimensions: The hierarchical-smaa-choquet inte- gral approach. *Knowledge-Based Systems*, 136–153. https://doi.org/10.1016/j.knosys.2018.05.041.
27. De Matteis, D., Ishizaka, A., & Resce, G. (2019). The 'postcode lottery' of the italian public health bill analysed with the hierarchy stochastic multiobjective acceptability analysis. *Socio-Economic Planning Sciences, 41*. https://doi.org/10.1016/j.seps.2017.12.001.
28. Bernal , M., Alvarez, P., Leon-Castro, E., & Gastélum-Chavira, D. (2021). A Multicriteria hierarchical approach for portfolio selection in stock exchange. *Journal of Intelligent & Fuzzy Systems*, 1945–1955. https://doi.org/10.3233/JIFS-189198.
29. Alvarez, P. A., Ishizaka, A., & Martínez, L. (2021) Multiple-criteria decision-making sorting methods: A Survey, *Expert Systems with Applications, 183*. https://doi.org/10.1016/j.eswa.2021.115368.
30. Figueira, J.R., Greco, S., Roy, B., & Slowinski, R. (2013). An overview of electre methods and their recent extensions. *Journal of Multi-Criteria Decision Analysis, 20*, 61–85. https://onlinelibrary.wiley.com/doi/pdf/10.1002/mcda.1482.
31. Corrente, S., Figueira, J., Greco, S., & Slowinski, R. (2017). A robust ranking method extending ELECTRE III to hierarchy of interacting criteria, imprecise weights and stochastic analysis. *Omega, 73*, 1–17. https://doi.org/10.1016/j.omega.2016.11.008.
32. Marzouk, M. M. (2011). ELECTRE III model for value engineering applications. *Automation in Construction, 20*(5), 596–600. https://doi.org/10.1016/j.autcon.2010.11.026.
33. Trading Economics. (2022). *Credit rating*. Retrieved March 05, 2022, from https://tradingeconomics.com/.
34. Milken Institute. (2020). *Global opportunity index 2021: Attracting international investors.* Retrieved January 22, 2022, from https://milkeninstitute.org/reports/latin-america-global-opportunity-index/white-paper.
35. World Bank. (2022). *Doing business 2020*. Retrieved March 05, 2022, from https://documents1.worldbank.org/curated/en/688761571934946384/pdf/Doing-Business-2020-Comparing-Business-Regulation-in-190-Economies.pdf.
36. World Bank. (2022). *Data Country profile*. Retrieved January 14, 2022, from https://data.worldbank.org/country/CL.
37. IMF Blog. (2022). *Latin America's strong recovery is losing momentum, underscoring reform needs*. Retrieved January 22, 2022, from https://blogs.imf.org/2022/01/31/latin-americas-strong-recovery-is-losing-momentum-underscoring-reform-needs/.
38. Henao, R. (2018) *The 3 best Latin American countries to invest in*. Seeking Alpha. Retrieved January 22, 2022, from https://seekingalpha.com/article/4212379-3-best-latin-american-countries-to-invest-in.

Uncertainty in Computer and Decision-Making Sciences: A Bibliometric Overview

Carlos J. Torres-Vergara, Víctor G. Alfaro-García,
and Anna M. Gil-Lafuente

Abstract Publications on uncertainty and decision-making systems have grown exponentially. This chapter aims to apply bibliometric tools and techniques to analyze the evolution in the last decades of research documents focused on uncertainty and decision-making systems in computer sciences. The analysis is divided into key sections including journals, institutions, key scholars, countries, institutions, collaborations, and word frequency. This chapter utilizes the SCOPUS scientific database, identifying 11,122 papers with an average of 19.94 citations per document and an average of 7.2 years from publication between 1964 and 2021. Results show the fundamental metrics for the evolution and growth of academic research in this field.

Keywords Bibliometrics · Uncertainty · Decision-making · Citation analysis

1 Introduction

Decision-making is an inherent and determinant process in every aspect of life. Taking decisions requires the assessment of a wide range of factors, as variables and alternatives are singular and unique to the areas where they arise and develop [33]. Decision-making requires a deep understanding of the reality and the context where problems are framed to interpret and model them [8].

Sciences like mathematics seek to model reality by making abstractions and representations of the environment [9]. In this process, precise and well-defined data is required [22]. It is imperative to establish a series of conditions on the specific occurrences of the initial scenarios, as many engineering and scientific problems consisted of partly deterministic and partly random situations [20]. When the behavior of the

C. J. Torres-Vergara · V. G. Alfaro-García (✉)
Facultad de Contaduría y Ciencias Administrativas, Universidad Michoacana de San Nicolás de Hidalgo, 58030 Morelia, Mexico
e-mail: victor.alfaro@umich.mx

A. M. Gil-Lafuente
Facultat d'Economia i Empresa, Universitat de Barcelona, 08034 Barcelona, Spain

© The Author(s), under exclusive license to Springer Nature Switzerland AG 2023
Y. P. Kondratenko et al. (eds.), *Artificial Intelligence in Control and Decision-making Systems*, Studies in Computational Intelligence 1087,
https://doi.org/10.1007/978-3-031-25759-9_16

phenomena becomes uncertain, the establishment of probability law becomes one of the most frequent tools, mainly in the economic field [27].

These mathematical tools enabled the emergence of a proto theory of decision-making back in the XVII and XVIII centuries by Blaise Pascal and Daniel Bernoulli [14]. But it was in 1939 with the publication of Contributions to the Theory of Statistical Estimation and Testing Hypotheses by Abraham Wald [31] where the field reignited, as Wald synthesized many concepts of statistical theory that are still relevant nowadays, including loss functions, risk functions, admissible decision rules, antecedent distributions, Bayesian procedures, and minimax procedures [11].

However, getting accurate data to model reality for the use of statistics and probability is difficult to accomplish due to uncertainty [15]. A wide range of possibilities, disturbances, changes, complexities, non-linearities, chaos, and evolution must be considered in the analysis of the particular phenomenon [8].

Funtowicz and Ravetz [12] describe uncertainty as a situation of inadequate information, which can be of three sorts: inexactness, unreliability, and the border with ignorance. Uncertainty can prevail in situations where such information is available, new information can either increase or decrease the amount of uncertainty [32]. It is important to distinguish between randomness and uncertainty. Randomness is the measurable uncertainty with the help of the concept of probability [21].

Research in uncertainty is difficult to catalog as uncertainty is a topic that does not fall neatly within a single discipline [15]. Despite academicians' attempts to study this topic, it sprawls across a considerable variety of disciplines, professions, and problem domains [28].

Uncertainty has been present in all history [5]. There have been pandemics, wars, and natural disasters since we have records. In the twentieth and twenty-first centuries, we've already seen two world wars [4], stock market crashes like the Wall Street Crash of 1929 [23] or Black Monday in 1987 [29], several natural disasters like Hurricane Katrina [3], the 2004 Indian Ocean Tsunami [26], and even several pandemics like the COVID-19 [18] that caused several periods of uncertainty.

This chapter aims to analyze a comprehensive amount of research documents focused on uncertainty and decision-making systems, visualizing trends and presenting a general overview of key scholars, journals, words, countries, and the general evolution of the field in the last decades using bibliometric tools and techniques.

The rest of the chapter is structured as follows. Section 2 presents the methodology followed for gathering the information and the criteria of the search for the treatment of data. Section 3 presents the retrieved structured information offering 9 tables with the aggregated results and describing them. Finally, Sect. 4 present the analysis and the concluding comments.

2 Methodology

Bibliometric studies have grown exponentially in the last decades since Pritchard [25] defined the term. Bibliometrics can be defined as the statistical or quantitative description of literature taken here to mean, simply, a group of related documents [13].

The main advantage of employing a bibliometric approach relies on the methodical quantification of the information concentrated in scientific databases [7]. This representation allows a quick understanding of the connections, relations, and impact of the studied phenomena, thus proposing both a clear picture of all the relevant elements that shape the field and the areas of opportunity that allow the possibility of generating novel research and synergies [1]. To the best of our knowledge, there are no bibliometric studies that use the keyword "uncertainty" in the field of decision and computer sciences.

When writing a bibliometric analysis, it is vital to properly describe the methods and tools employed for the retrieval of information [30]. This chapter retrieves data from the SCOPUS scientific database. SCOPUS is currently owned by Elsevier and covers nearly 25,100 titles. 23,452 represent peer-reviewed journals, more than 5,500 international publishers in top-level subject fields. SCOPUS includes around 9.8 million conference papers from over 120,000 worldwide events and more than 210,000 books. In total, the selected database includes more than 77.8 million records, over 71.2 million records post-1969 with references and over 6.6 million records pre-1970, being the oldest record from 1788. Note that many other databases could be considered, including Google Scholar and Web of Science. However, in this chapter, the focus will be given to the SCOPUS scientific database.

The methodological procedure for the retrieval of information is the following. We have used the keyword "uncertainty" for every title, abstract, and keyword of matching documents and limited the search for the subareas computer science and decision sciences, the advanced search for the replicability of this procedure is SUBJAREA (COMP) TITLE-ABS-KEY (Uncertainty) AND SUBJAREA (DECI). The timespan limit of the search includes all the papers between 1964 and December 2021.

3 Structured Results and Observations

The methodological procedure for the retrieval of information yields 11,122 documents. This number includes all the publications covered by the SCOPUS database and includes 6,005 articles, 575 book chapters, 4,270 conference papers, 96 reviews, and 115 conference reviews.

As shown in Fig. 1, 8,196 of these papers come from the last decade between 2012 and 2021, representing 73.6% of the total and showing that the topic has captured the attention of researchers through time. The first two decades show almost no research

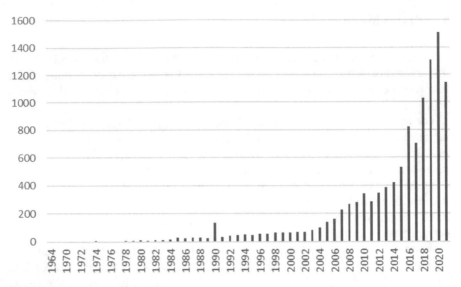

Fig. 1 Number of annual publications in uncertainty in decision-making systems research (articles + reviews) since 1964

activity, with just 185 papers between 1964 and 1987, the majority of them from the last years as we can see in Fig. 1. 1990 showed a peak in the topic with 137 papers. After 1990 the paper production returns to previous levels but continues to grow until 2002, where the production started to grow exponentially. 2020 was, by far, the year with the most uncertainty and decision-making systems productions with a record of 1,509 papers. This is probably related to the emergence of the COVID-19 pandemic, which has since then caused a period of great instability and uncertainty.

Table 1 shows the general citation structure of uncertainty in decision-making systems. Regarding citation count, this area has a normal citation rate according to computer science and engineering standards [24]. There are only three papers with more than 2,000 cites, which is a remarkable achievement, and 29 papers with more than 500 citations, which represents only 0.261% of total papers. 1,043 papers

Table 1 General citation structure of uncertainty in decision-making systems in computer sciences

Number of citations	Number of papers	% Papers (%)
≥2,000 citations	3	0.027
≥1,000 citations	9	0.081
≥500 citations	29	0.261
≥250 citations	96	0.863
≥100 citations	420	3.776
≥50 citations	1,043	9.378
≤50 citations	10,079	90.622
Total	11,122	

have more than 50 citations, which represents 9.378% of the total, leaving the total of papers with less than 50 citations to 10,079, representing the vast majority of 90,622% of the total papers. The average citation per document is 19.94, raising the average citation per document per year to 2.223.

Research in the uncertainty in decision-making systems is published in a wide range of journals and books, with a total of 1,126 unique sources. Although there is no strict journal devoted to uncertainty in decision-making systems, some of them are strongly influenced by it, with a high number of publications. To classify the journals with uncertainty in decision-making systems publications, the 30 most influential journals in this field are presented in Table 2, ordered by the total citations in the field. It is important when analyzing the published research in any field to classify the papers according to the total number of citations received [10].

According to Table 2, the most influential journal in the field is the European Journal of Operational Research with a total of 1,330 papers that represent 11.96% of the total 11,122. It also has the biggest number of citations in the field with 59,779, which represents 43% more citations than the second most influential journal, Information Sciences, with 792 papers and 34,050 citations, and 65% more than Operations Research, the third most influential journal with 310 papers and 21,129 citations.

Other key outlets are Knowledge-Based Systems and Information Systems Research, completing the top 5 with 275 and 65 papers in uncertainty in decision making research and 9,207 and 9,073 citations respectively. It is worth noting that the distance between the top 3 influential papers and the rest is remarkable in both papers and citations.

European Journal of Operational Research has the highest H-index for uncertainty in decision-making systems research with 114, compared to the 88 of Information Systems and the 73 of Operations research. The H-index is a measure that aims to represent the importance of a set of papers defined the largest number H for which an author has H papers with at least H citations each [19]. For example, if a journal has an H-index of 50, that means that 50 of the papers published by the journal in the set have received at least 50 citations each. The H-Index for uncertainty in decision-making systems is lower than the general H-Index for every publication, revealing that uncertainty in decision-making systems has less influence than other fields in the journals citations. Note that since its introduction, the H-index has been extended and generalized by many authors [2].

Comparing the total papers of journals and uncertainty in decision-making systems papers, only the Journal of Environmental Informatics has more than 20% articles related to the field, with other 4 journals having more than 10%. It can be concluded that, even with the rise of publications in the last years, it's not one of the main fields of the journals.

Table 3 presents the most cited papers in uncertainty in decision-making research in computer sciences of all time. Continuing with the parameters established for Table 2, the papers are ordered from most citations to the least. Although many aspects may influence the value of a paper, the number of citations reflects the popularity and influence that each one has in the scientific community [6].

Table 2 Most influential journals in uncertainty in decision-making research in computer sciences

R	Name	TP-UDMS	H-UDMS	TC-UDMS	PY_start	TC	TP	IF	H	%TP-UDMS/TP (%)
1	EJOR	1,330	114	59,779	1977	121.004	12,869	5.334	260	10.33
2	IS	792	88	34,050	1969	123.410	9,355	6.795	184	8.47
3	OR	310	73	21,129	1970	16.502	2,131	4.137	140	14.55
4	KBS	275	53	9,207	1991	56.184	4,342	8.038	121	6.33
5	ISR	65	34	9,073	1992	14.491	904	5.207	159	7.19
6	COR	258	49	8,562	1976	43.405	4,351	4.008	152	5.93
7	MQMIS	39	30	8,465	1982	31.838	948	12.234	230	4.11
8	DSS	162	42	6,722	1986	32.647	2,679	5.795	151	6.05
9	EO	215	28	3,473	1985	7.728	1,658	3.230	64	12.97
10	MOR	59	24	3,294	1981	6.721	1,175	3.832	83	5.02
11	JMIS	49	27	2,932	1990	12.540	983	8.319	144	4.98
12	ORL	67	19	2,897	1985	6.657	2,290	1.154	77	2.93
13	TD	174	30	2,860	1971	2.097	1,043	1.206	37	16.68
14	PPC	117	27	2,473	1992	10.394	1,902	6.791	76	6.15
15	IM	57	28	2,466	1982	21.192	1,450	7.555	162	3.93
16	SC	57	19	2,255	1992	8.484	1,191	3.571	77	4.79
17	JSIS	14	8	2,065	1993	5.983	494	11.022	88	2.83
18	JGO	61	20	2,017	1999	11.044	2,444	2.540	86	2.50
19	JEI	66	26	1,775	2007	2.572	319	6.789	31	20.69
20	JF	80	19	1,533	1985	2.950	1,013	2.395	59	7.90
21	ITOR	89	20	1,450	1994	4.509	1,330	3.753	52	6.69

(continued)

Table 2 (continued)

R	Name	TP-UDMS	H-UDMS	TC-UDMS	PY_start	TC	TP	IF	H	%TP-UDMS/TP (%)
22	IJC	56	18	1,311	1997	5.529	952	2.833	80	5.88
23	JS	41	21	1,220	2000	3.983	852	2.352	61	4.81
24	JSS	21	10	1,193	2005	21.198	1,110	6.488	145	1.89
25	SD	38	17	1,044	2014	16.924	1,525	8.987	64	2.49
26	ISORMS	105	17	1,035	2005	3.050	2,345	1.149	43	4.48
27	IFIP	195	13	984	2002	17.451	10,881	0.665	53	1.79
28	IEEE	22	11	934	2008	10.853	925	6.791	70	2.38
29	IJIM	20	13	916	1988	22.728	1,616	14.098	114	1.24
30	IPM	44	16	861	1975	15.696	1,794	6.222	101	2.45

Acronyms TP-UDMS, Total papers only with uncertainty in decision-making systems. H-UDMS, H-rating only with uncertainty in decision-making systems. TC-UDMS, Total citations about uncertainty in decision-making systems. PY-Start, Publishing starting year. TC, Total citations after 1999. TP, total papers. TC/PY, total citations per year after 1999. H, H-rating. IF, impact factor 2020; EJOR, European Journal of Operational Research. IS, Information Sciences. OR, Operations Research. KBS, Knowledge-Based Systems. ISR, Information Systems Research. COR, Computers and Operations Research. MQMIS, Mis Quarterly: Management Information Systems. DSS, Decision Support Systems. EO, Engineering Optimization. MOR, Mathematics of Operations Research. JMIS, Journal of Management Information Systems. ORL, Operations Research Letters. TD, Theory and Decision. PPC, Production Planning and Control. IM, Information and Management. SC, Statistics and Computing. JSIS, Journal of Strategic Information Systems. JGO, Journal of Global Optimization. JEI, Journal of Environmental Informatics. IF, Journal of Forecasting. ITOR, International Transactions in Operational Research. IJC, Informs Journal on Computing. JS, Journal of Scheduling. JSS, Journal of Statistical Software. SD, Scientific Data. SAC, Statistics and Computing. IFIP, IFIP Advances in Information and Communication Technology. IEEE, IEEE Transactions on Services Computing. IJIM, International Journal of Information Management IPM, Information Processing and Management.

Table 3 30 Most cited papers in uncertainty in decision-making research in computer sciences of all time

R	Journal	Title	Authors	TC	Year	TC/PY
1	ISR	Developing and Validating Trust Measures for E-Commerce: An Integrative Typology	McKnight, D; Choudhury, V; Kacmar, C	2,955	2002	147.75
2	OR	The Price of Robustness	Bertsimas, D; Sim, M	2,461	2004	136.722
3	MQMIS	Review: Information Technology and Organizational Performance: An Integrative Model of It Business Value	Melville, N; Kraemer, K; Gurbaxani, V	2,152	2004	119.556
4	MOR	Robust Convex Optimization	Ben-Tal A;Nemirovski A	1,589	1998	66.208
5	ISR	Building Effective Online Marketplaces with Institution-Based Trust	Pavlou, P; Gefen, D	1,485	2004	82.5
6	Oper Res	Regret in Decision Making Under Uncertainty	Bell Davide	1,471	1982	36.77
7	MQMIS	Understanding and Mitigating Uncertainty in Online Exchange Relationships: A Principal-Agent Perspective	Pavlou P; Huigang, L; Yajiong, X	1,466	2007	97.733
8	ORL	Robust Solutions of Uncertain Linear Programs	Ben-Tal, A; Nemirovski, A	1,232	1999	53.565
9	JSIS	The Impact of Initial Consumer Trust on Intentions to Transact with A Web Site: A Trust Building Model	McKnight, D; Choudhury, V; Kacmar, C	1,131	2002	56.55
10	IS	Toward A Generalized Theory of Uncertainty (Gtu)- An Outline	Zadeh, LA	953	2005	56.058
11	IS	Is There A Need for Fuzzy Logic?	Zadeh, LA	940	2008	67.142
12	MQMIS	A Cross-Cultural Study on Escalation of Commitment Behavior in Software Projects	Keil, M; Tan Bcy; Wei, K;Saarinen T;Tuunainen V;Wassenaar A	910	2000	41.363
13	EJOR	Strategic Facility Location: A Review	Owen SH; Daskin MS	863	1998	35.958

(continued)

Table 3 (continued)

R	Journal	Title	Authors	TC	Year	TC/PY
14	DSS	Combining Belief Functions When Evidence Conflicts	Murphy CK	812	2000	36.909
15	IS	Soft Sets and Soft Groups	Akta, H; Aman, N	760	2007	50.667
16	EJOR	A Stochastic Programming Approach for Supply Chain Network Design Under Uncertainty	Santoso, T; Ahmed, S; Goetschalckx, M; Shapiro, A	732	2005	43.058
17	OR	Distributionally Robust Optimization Under Moment Uncertainty with Application to Data-Driven Problems	Delage, E; Ye, Y	701	2010	58.4167
18	EJOR	Project Scheduling Under Uncertainty: Survey and Research Potentials	Herroelen, W; Leus, R	691	2005	40.6471
19	EJOR	Operating Room Planning and Scheduling: A Literature Review	Cardoen, B; Demeulemeester, E; Belin, J	686	2010	57.167
20	MQMIS	The Role of Espoused National Cultural Values In Technology Acceptance	Srite, M; Karahanna, E	681	2006	42.562
21	OR	Disappointment in Decision Making Under Uncertainty	Bell, DE	633	1985	17.108
22	OR	Reducing the Cost of Demand Uncertainty Through Accurate Response to Early Sales	Fisher, M; Raman, A	614	1996	23.6154
23	EJOR	Rule and Utility Based Evidential Reasoning Approach for Multiattribute Decision Analysis Under Uncertainties	Yang, JB	566	2001	26.952
24	EJOR	Rough Set Approach to Knowledge-Based Decision Support	Pawlak, Z	562	1997	22.48
25	EJOR	Multi-Criteria Decision-Making Methods Based on Intuitionistic Fuzzy Sets	Liu, HW; Wang, GJ	550	2007	36.667

(continued)

Table 3 (continued)

R	Journal	Title	Authors	TC	Year	TC/PY
26	EJOR	Managing Non-Homogeneous Information in Group Decision Making	Herrera, F; Martínez, L; Sánchez, PJ	527	2005	31
27	EJOR	The Design of Robust Value-Creating Supply Chain Networks: A Critical Review	Klibi, W; Martel, A; Guitouni, A	521	2010	4.,417
28	DSS	Accessing Information Sharing and Information Quality in Supply Chain Management	Li, S; Lin, B	517	2006	32.312
29	SAC	A Markov Chain Monte Carlo Version of The Genetic Algorithm Differential Evolution: Easy Bayesian Computing for Real Parameter Spaces	Ter Braak, CJF	512	2006	32
30	EJOR	Executing Production Schedules in The Face of Uncertainties: A Review and Some Future Directions	Aytug, H; Lawley, M Mckay, K; Mohan, S	499	2005	29.352

Acronyms TC/PY, total citations per year.

The most cited paper is "Developing and Validating Trust Measures for E-Commerce: An Integrative Typology" by D. Harrison McKnight, Vivek Choudhury, and Charles Kacmar with 2,955 citations. It was by published Information System Research in 2002, leaving a citation per year ratio of 147,75. The authors also wrote the 9th most cited paper, "The Impact of Initial Consumer Trust on Intentions to Transact with A Web Site: A Trust Building Model" with 1,131 cites. It is noticeable that only other 2 authors, Paul Pavlou and Lofti Zadeh appear twice in the table as well.

The second most cited paper is "The Price of Robustness" by Dimitris Bertsimas and Melvyn Sim, published by Operations Research in 2004, with 2,461 citations and a citation per year ratio of 136,722. The third most cited paper is "Review: Information Technology and Organizational Performance: An Integrative Model of It Business Value" by Nigel Melville, Kenneth Kraemer, and Vijay Gurbaxani, published by MIS Quarterly: Management Information Systems in 2004, with a total of 2,152 total citations, with a citation per year ratio of 119,556.

Despite these 3 papers having specific thematic, they have become reference papers in their specific field since their publication, and are the only 3 papers in the uncertainty in decision-making systems field with more than 2,000 citations, as can be seen in Table 1.

33% of the most cited papers in Table 3 (10 papers out of 30) are published in the European Journal of Operational Research, which is the most influential journal as seen in Table 2. The second most represented journal is Operations Research with 4 articles, followed by Information Systems with 3.

The most recent papers in Table 3 were published in 2010, having 11 citable years and a high ratio of citations per year of 58.41, 57.16, and 40.64, respectively. As seen in Fig. 1. The majority of articles are published after the year 2000. Thus, there are only 7 articles written before the year 2000, representing 23.3% of the 30 most cited ones, and having a ratio of citations per year between 66.2 and 17.1.

Table 4 presents a list of the 25 most prolific authors in the field of uncertainty and decision-making systems. It has to be noted that the number of papers is only

Table 4 The most productive and influential authors in uncertainty in decision-making research

R	Name	Country	TP-UDMS	TC-UDMS	H-UDMS	H	TP	TC
1	Huang, G.H	Canada	70	3,474	32	78	1,289	32,850
2	Shi, P	Australia	38	700	12	136	1,417	64,153
3	Bertsimas, D	USA	29	4,376	13	60	255	17,055
4	Li, Y.P	PR China	29	873	18	44	428	8,368
5	Escudero, L.F	Spain	24	796	13	27	161	2,471
6	Xu, Z	PR China	22	1,999	20	112	813	49,250
7	Dezert, J	France	21	447	7	31	195	3,260
8	Yager, R.R	Saudi Arabia	20	430	10	93	894	45,237
9	Yang, J.B	PR China	19	2,552	16	59	203	12,709
10	Goerigk, M	Germany	18	157	5	17	69	815
11	Kreinovich, V	USA	16	75	2	32	1,019	5,021
12	Dolgui, A	France	15	275	7	51	448	10,504
13	Schöbel, A	Germany	15	404	19	30	134	2,814
14	Li, T	PR China	14	392	9	59	203	12,356
15	Mendel, J.M	USA	14	1,539	9	74	457	33,380
16	Pedrycz, W	Canada	14	941	6	88	1,457	40,456
17	Sim, M	Singapore	14	4,310	13	33	69	13,513
18	Tavakkoli-Moghaddam, R	Iran	14	297	9	55	622	11,961
19	Bruni, M.E	Italy	13	344	5	18	62	1,108
20	Demeulemeester, E	Belgium	13	1,545	11	40	103	6,056
21	Leus, R	Belgium	13	1,109	9	25	87	2,987
22	Martínez, L	Spain	13	1,821	12	52	332	16,223
23	Moshkov, M	Saudi Arabia	13	35	3	15	252	1,054
24	Ahmed, S	USA	12	1,409	10	32	116	4,915
25	Beraldi, P	Italy	12	405	8	22	85	1,525

Acronyms USA, United States of America. PR China, People's Republic of China.

indicative because numerous limitations must be considered, such as the size of each paper, the quality of the journal, and the number of authors per paper [17]. Therefore, other indicators like H-index and total citations are shown to obtain a general overview.

Gordon Huang is the most prolific author with 70 papers, 84% more than the second most prolific author, Peng Shi with 38 total papers. Although Huang is not the most cited with 3,474 total citations, he has the higher H-Index in the uncertainty in the decision-making systems field with 32. No other author has 30 papers published, with Y.P. Li being the third most prolific author with 29 papers. It is worth noting that Y.P. Li and Gordon Huang are coauthors in many papers, 21 just in the last year. Therefore, must be assumed that they appear on the database as coauthors of some papers.

The most cited author is Dimitris Bertsimas with a total of 4,376 total citations and a specific H-Index of 13. As seen in Table 3, 2,461 of these citations comes from the same article "The Price of Robustness" published in 2004, coauthored with Melvyn Sim that also appears in the table as the second most cited author.

USA and China are the countries with more authors on the list with 4 each representing 13.33% of the total. Canada, Spain, France, Italy, Saudi Arabia, and Belgium, have 2 authors, representing 6.67% each. In terms of quantity, PR China is the strongest research country with 4,335 papers, which accounts for 29.88% of the top 10 in the field. The USA is the second one with 3,841 representing 26.47%. Table 5 shows the total number of papers published. United Kingdom, France, and Germany complete the top 5 with 1,228; 976 and 920, far beyond the first 2.

In terms of citations, the USA is the most influential country with 64,866 citations, 44.45% of the total citations of the top 10. It is also 100.29% more than PR China, the second one with 32,385 citations. USA, PR China, and the United Kingdom are the only countries with more than 10,000 total citations, being the total of the United Kingdom 11,691. On average article citations, Israel has the higher ratio with 83.17, followed by the USA with 50.56, Belgium with 44.07, Singapore with 42.89, and Canada with 36.32.

Institutions from all over the world have performed research in uncertainty in many fields, some of them being very well-known universities that appear on top of university rankings [16]. Table 6 presents a list of the 30 most influential institutions in uncertainty in decision-making studies ordered by the total papers published in the field.

The University of California obtains the most prominent results in all criteria with 108 total papers published in the field with 12,970 and 47 H-Index specific ratio. It must be noted that the University of California is not a single campus, but a system of universities with 10 campuses affiliated. The University of Singapore comes second with 99 total papers and 5,035 citations, 62% less than the University of California. Completing the top 3 Tsinghua University has published 99 papers with 1,482 total citations and 20 H-Index.

Table 5 The most influential countries in uncertainty in decision-making research.

R	Name	TP-UDMS	TC-UDMS	AAC
1	PR China	4,335	32,385	22.182
2	USA	3,841	64,866	50.558
3	United Kingdom	1,228	11,691	28.103
4	France	976	7,268	23.674
5	Germany	920	4,851	15.699
6	Canada	808	9,987	36.316
7	India	780	2,580	12.113
8	Italy	601	3,339	15.317
9	Spain	517	6,079	28.948
10	Australia	501	2,881	18.954
11	Iran	467	3,170	14.813
12	Netherlands	438	5,544	36
13	Brazil	339	1,563	16.453
14	Turkey	319	4,795	36.053
15	Japan	305	1,736	20.667
16	Portugal	281	1,644	22.216
17	Singapore	235	2,445	42.895
18	South Korea	221	2,879	22.492
19	Belgium	202	3,790	44.07
20	Poland	181	1,904	25.387
21	Greece	166	1,205	16.067
22	Ukraine	159	68	3.579
23	Norway	156	1,175	21.759
24	Switzerland	156	1,142	20.035
25	Sweden	148	632	15.048
26	Finland	146	1,086	20.885
27	Austria	140	949	17.906
28	Malaysia	117	881	29.367
29	Indonesia	112	8	0.533
30	Israel	110	3,493	83.167

Acronyms AAC, Average Article Citations.

PR China has 11 institutions in the table, representing 36.67%, followed by the USA, which has 9 institutions representing 30% of the total. Behind these two research potencies, Singapore, Netherlands, and Iran have 2 institutions representing 6.67% each.

Table 7 shows the collaboration between countries in uncertainty in decision-making research. As deducible from Table 5 and Table 6, the collaboration between

Table 6 The most influential institutions in uncertainty in decision-making research

R	Name	Country	TP-UDMS	TC-UDMS	H-UDMS
1	U of California	USA	108	12,970	47
2	National U of Singapore	Singapore	99	5,035	30
3	Tsinghua U	PR China	95	1,482	20
4	U of Regina	Italy	90	3,083	32
5	Northeastern U	USA	89	1,089	19
6	North China Electric Power U	PR China	84	1,819	24
7	City U of Hong Kong	PR China	72	3,044	29
8	Northwestern Polytechnical U	USA	67	1,159	17
9	U of Michigan	USA	63	2,306	25
10	Xi'an Jiaotong U	PR China	63	845	13
11	Islamic Azad U	Iran	61	646	11
12	Beihang U	PR China	60	653	12
13	Delft U of Technology	Netherlands	60	838	17
14	Massachusetts Institute of Technology	USA	58	1,627	22
15	Tongji U	PR China	55	743	11
16	U of Electronic Science and Technology of China	PR China	55	1,628	18
17	Southeast U	USA	54	1,454	14
18	Carnegie Mellon U	USA	53	982	19
19	Southwest Jiaotong U	PR China	53	603	8
20	Iran U of Science and Technology	Iran	52	832	13
21	Stanford U	USA	51	1,933	19
22	Tilburg U	Netherlands	51	2,222	20
23	Huazhong U of Science and Technology	PR China	49	1,489	19
24	Shanghai Jiao Tong U	PR China	49	1,173	14
25	U of Southampton	United Kingdom	49	871	13
26	Sichuan U	PR China	48	1,521	19
27	Nanyang Technological U	Singapore	47	1,075	15
28	Norwegian U of Science and Technology	Norway	47	1,712	18
29	U of Florida	USA	46	2,158	20
30	U of Toronto	Canada	46	1,333	18

Acronyms U, University.

Table 7 Collaborations between countries in uncertainty in decision-making research

From	To	Frequency
PR China	USA	223
PR China	Canada	130
PR China	United Kingdom	119
USA	United Kingdom	108
PR China	Hong Kong	99
PR China	Australia	94
USA	Canada	93
USA	Germany	57
United Kingdom	Germany	54
USA	France	52
PR China	Singapore	51
PR China	France	49
USA	Hong Kong	48
USA	Singapore	46
Germany	Netherlands	43
USA	Italy	43
USA	Netherlands	42
USA	Turkey	42
France	Italy	40
United Kingdom	Italy	38
United Kingdom	Australia	37
China	Japan	36
France	Tunisia	35
USA	Spain	33
USA	Australia	31

PR China and USA is the most common, as both are the most influential countries in the field. The United Kingdom, Germany, and France have as well a strong net of collaborations with other countries, but the frequency is lower than the one of the USA and PR China.

Table 8 shows the word frequency in the field. As expected, the most frequent words are "Uncertainty Analysis" with a frequency of 2,283 words, "Decision Making" with 1,380 words, "Optimization" with 1,009 words, "Stochastic Systems" with 856 words, and "Uncertainty" with 686 words.

As seen in the introduction, uncertainty sprawls across a considerable variety of disciplines and fields. In Table 8, it can be observed the frequency of words of different fields and disciplines related with uncertainty like "Fuzzy Sets", "Risk Assessment, "Mathematical Models", "Forecasting", "Integer Programming", "Artificial Intelligence", "Investments", "Sales" or "Supply Chains".

Table 8 Word frequency in uncertainty in decision-making research

Terms	Frequency
Uncertainty analysis	2,283
Decision making	1,380
Optimization	1,009
Stochastic systems	856
Uncertainty	686
Costs	506
Stochastic models	465
Fuzzy sets	445
Risk assessment	432
scheduling	428
Decision theory	425
Mathematical models	417
Forecasting	398
Decision support systems	383
Integer programming	377
Robust optimization	376
Artificial intelligence	372
Operations research	367
Commerce	357
Stochastic programming	346
Information systems	336
Problem solving	335
Investments	334
Sales	332
Supply chains	322
Probability distributions	281
Algorithms	270
Information management	270
Supply chain management	256
Probability	250
Random processes	248
Controllers	241
Linear programming	239
Monte carlo methods	238
Fuzzy logic	235
Learning systems	234
Risk management	231

(continued)

Table 8 (continued)

Terms	Frequency
Intelligent systems	230
Economics	225
Bayesian networks	205
Inventory control	205
Information use	203
Dynamic programming	199
Sensitivity analysis	199
Competition	195
Numerical methods	186
Computer simulation	183
Manufacture	183
Economic and social effects	181
Classification (Of Information)	176

To obtain a better picture of when the field started to become relevant, Table 9 shows the frequency of the keywords "Uncertainty Analysis" and Decision Making" between 1969 and 2018. It can be observed that before 1990 the frequency of "Uncertainty Analysis" was almost 0, but in 1990 the frequency increased by 3,733.33% and then start to grow slowly until 2006 when the growth accelerates until it reaches the maximum in 2018 with a frequency of 1,405. The keyword "Decision Making" follows a similar pattern but with a more stable growth until its maximum also in 2018 with a frequency of 721.

4 Conclusions

An analysis of research documents focused on uncertainty and decision-making systems using bibliometric tools and techniques has been presented. The results obtained are by our common knowledge and show that the last decade has been the most prolific both in articles and citations, representing 73.6% of total papers and having the most citations per year ratio.

The European Journal of Operational Research is the most influential journal in total papers, citations, and H-Index and has 10 of the 30 most influential articles in the field. The USA and PR China are the most influential countries, having together 40.44% of total papers and 50.43% of total citations between all countries. They also have 66.67% of total institutions, 2 out of 3. It should be noted that speaking about authors they only represent 26% of the total, and the most influential and prolific author, Gordon Huang, is from Canada, although he collaborates often with Chinese authors.

Table 9 Word frequency of keywords "uncertainty" and "decision making" per year

Year	Uncertainty analysis	Decision making
1969	0	0
1970	0	0
1971	0	0
1972	0	0
1973	0	0
1974	0	0
1975	0	0
1976	0	0
1977	0	0
1978	0	0
1979	0	0
1980	0	0
1981	0	0
1982	0	0
1983	0	0
1984	0	0
1985	0	1
1986	0	1
1987	0	2
1988	2	6
1989	3	9
1990	115	18
1991	115	19
1992	116	19
1993	117	21
1994	120	22
1995	121	25
1996	122	35
1997	122	45
1998	123	52
1999	127	66
2000	128	77
2001	129	90
2002	130	115
2003	131	133
2004	133	158
2005	140	194

(continued)

Table 9 (continued)

Year	Uncertainty analysis	Decision making
2006	150	229
2007	198	266
2008	288	332
2009	346	355
2010	420	394
2011	471	433
2012	559	473
2013	630	502
2014	709	548
2015	833	636
2016	1006	721
2017	1180	819
2018	1405	943

It should be underlined that one of the limitations of this bibliometric analysis is that many papers that address uncertainty in decision-making systems do not use the keyword "uncertainty", as it is considered too generalist and non-specialized. As uncertainty is becoming a more relevant topic, this can limit the number of papers that appear in the first years of the search of this study, omitting relevant papers. We must also note that some papers that use the keyword "uncertainty" are not very related to uncertainty research, instead, they focus on other topics that have some level of uncertainty.

From a general perspective, it has been assumed that the general numbers representing the total production are valid due to the high number of papers analyzed. We must consider as well that the two limitations should be compensated between them, as one adds papers and the other subtracts them.

It is not the purpose of this article to make a political analysis of the environment and how it affects uncertainty, but if we saw the peak years in total papers, and the turning point for the growing ratio, we saw that 1990 and 2002 are both very influential years. If we match these years with historical and political events, we can relate that 1990 was probably influenced by the fall of the Berlin Wall and the revolutions of 1989, which eventually caused the dissolution of the Soviet Union and a period of great uncertainty. The same can be inducted with 2002, probably influenced by the September 11 attacks, which caused several changes in the way of how reality is perceived.

Acknowledgements Research supported by Red Sistemas Inteligentes y Expertos Modelos Computacionales Iberoamericanos (SIEMCI), project number 522RT0130 in Programa Iberoamericano de Ciencia y Tecnología para el Desarrollo (CYTED).

References

1. Alfaro-García, V. G., Merigó, J. M., Pedrycz, W., & Gómez Monge, R. (2020). Citation analysis of fuzzy set theory journals: Bibliometric insights about authors and research areas. *International Journal of Fuzzy Systems, 22*, 2414–2448. https://doi.org/10.1007/s40815-020-00924-8

2. Alonso, S., Cabrerizo, F. J., Herrera-Viedma, E., & Herrera, F. (2009). H-Index: A review focused on its variants, computation, and standardization for different scientific fields. *Journal of Informetrics, 3*, 273–289. https://doi.org/10.1016/j.joi.2009.04.001

3. Ascough, J. C., Maier, H. R., Ravalico, J. K., & Strudley, M. W. (2008). Future research challenges for incorporation of uncertainty in environmental and ecological decision-making. *Ecological Modelling, 219*, 383–399. https://doi.org/10.1016/j.ecolmodel.2008.07.015

4. Blossfeld, H.-P., Golsch, K., & Rohwer, G. (2007). *Event history analysis with stata.* Psychology Press.

5. Booker, J. M., & Ross, T. J. (2011). An evolution of uncertainty assessment and quantification. *Scientia Iranica, 18*, 669–676. https://doi.org/10.1016/j.scient.2011.04.017

6. Bornmann, L., & Williams, R. (2013). How to calculate the practical significance of citation impact differences? An empirical example from evaluative institutional bibliometrics using adjusted predictions and marginal effects. *Journal of Informetrics, 7*, 562–574. https://doi.org/10.1016/j.joi.2013.02.005

7. Broadus, R. N. (1987). Toward a definition of "bibliometrics." *Scientometrics, 12*, 373–379. https://doi.org/10.1007/BF02016680

8. Castiblanco Ruiz, F.A. (2013). La incertidumbre y la subjetividad en la toma de decisiones: una revisión desde la lógica difusa. Lúmina 116–141. https://doi.org/10.30554/lumina.14.1086.2013

9. Doerr, H. M., Ärlebäck, J. B., Misfeldt M. (2017). Representations of modelling in mathematics education. Mathematical Modelling and Applications. In International Perspectives on the Teaching and Learning of Mathematical Modelling (pp. 71–81). Springer

10. Donner, P. (2018). Effect of publication month on citation impact. *Journal of Informetrics, 12*, 330–343. https://doi.org/10.1016/j.joi.2018.01.012

11. French, S., Maule, J., & Papamichail, N. (2009). *Decision behaviour, analysis, and support.* Cambridge University Press.

12. Funtowicz, S. O., & Ravetz, J. R. (1990). *Uncertainty and quality in science for policy.* Kluwer Academic Publishers.

13. Garfield, E., Malin, M. V., Small, H. (1978). Citation data as science indicators. In Y. Elkana, J. Lederberg, R. K. Merton, A. Thackray, Harriet Zuckerman J.W.S. (Eds.), *Essays of an information scientist* (pp. 580–608). New York, NY, USA

14. Gilboa, I. (2009). *Theory of decision under uncertainty.* Cambridge University Press.

15. Hansen, L. P. (2014). Nobel lecture: Uncertainty outside and inside economic models. *Journal of Political Economy, 122*, 945–987. https://doi.org/10.1086/678456

16. Hazelkorn, E. (2011). *Rankings and the reshaping of higher education.* Palgrave Macmillan UK.

17. He, M., Zhang, Y., Gong, L., et al. (2019). Bibliometrical analysis of hydrogen storage. *International Journal of Hydrogen Energy, 44*, 28206–28226. https://doi.org/10.1016/j.ijhydene.2019.07.014

18. He, P., Sun, Y., Zhang, Y., & Li, T. (2020). COVID–19's Impact on stock prices across different sectors—an event study based on the Chinese stock market. *Emerging Markets Finance & Trade, 56*, 2198–2212. https://doi.org/10.1080/1540496X.2020.1785865

19. Hirsch, J. E. (2005). An index to quantify an individual's scientific research output. *Proceedings of the National Academy of Sciences, 102*, 16569–16572. https://doi.org/10.1073/pnas.0507655102

20. Kabir, H. M. D., Khosravi, A., Hosen, M. A., & Nahavandi, S. (2018). Neural network-based uncertainty quantification: A survey of methodologies and applications. *IEEE Access, 6*, 36218–36234. https://doi.org/10.1109/ACCESS.2018.2836917

21. Kauffman, A., & Gil-Aluja, J. (1987). *Técnicas operativas de gestión para el tratamiento de la incertidumbre*. Hispano Europea.
22. Kouser, H. N., Barnard-Mayers, R., & Murray, E. (2021). Complex systems models for causal inference in social epidemiology. *Journal of Epidemiology and Community Health, 75*, 702–708. https://doi.org/10.1136/jech-2019-213052
23. Lavoie M (2014) Post-Keynesian Economics. Edward Elgar Publishing
24. Merigó, J. M., Gil-Lafuente, A. M., & Yager, R. R. (2015). An overview of fuzzy research with bibliometric indicators. *Applied Soft Computing, 27*, 420–433. https://doi.org/10.1016/j.asoc.2014.10.035
25. Pritchard, A. (1969). Statistical bibliography or bibliometrics? *J Doc, 25*, 348–349.
26. Richard Eiser, J., Bostrom, A., Burton, I., et al. (2012). Risk interpretation and action: A conceptual framework for responses to natural hazards. *International Journal of Disaster Risk Reduction, 1*, 5–16. https://doi.org/10.1016/j.ijdrr.2012.05.002
27. Sigel, K., Klauer, B., & Pahl-Wostl, C. (2010). Conceptualising uncertainty in environmental decision-making: The example of the EU water framework directive. *Ecological Economics, 69*, 502–510. https://doi.org/10.1016/j.ecolecon.2009.11.012
28. Smithson, M. (2008). The many faces and masks of uncertainty. *Uncertainty and risk: Multidisciplinary perspectives* (pp. 13–25). Routledge.
29. Sornette, D. (2009). *Why stock markets crash: Critical events in complex financial systems*. Princeton University Press.
30. Tvaronavičienė, M., Razminienė, K., Piccinetti, L. (2015) Approaches towards cluster analysis. Econ Sociol *8*, 19–27. https://doi.org/10.14254/2071-789X.2015/8-1/2
31. Wald, A. (1939). Contributions to the theory of statistical estimation and testing hypotheses. *The Annals of Mathematical Statistics, 10*, 299–326. https://doi.org/10.1214/aoms/1177732144
32. Walker, W. E., Harremoës, P., Rotmans, J., et al. (2003). Defining uncertainty: A conceptual basis for uncertainty management in model-based decision support. *Integrated Assessment, 4*, 5–17. https://doi.org/10.1076/iaij.4.1.5.16466
33. Wang, J., Ma, X., Xu, Z., & Zhan, J. (2021). Three-way multi-attribute decision making under hesitant fuzzy environments. *Information Sciences (Ny), 552*, 328–351. https://doi.org/10.1016/j.ins.2020.12.005

Intelligent Traffic Signal Control Using Rule Based Fuzzy System

Tamrat D. Chala and László T. Kóczy

Abstract Over the past decades, there has been an ever-increasing saturation of traffic networks due to the growing number of road vehicles, and due to the available limited. To solve these problems, adaptive, (semi-) intelligent traffic control has been used widely for the last decades. These systems nevertheless, have some shortages, the most obvious one being that these systems use the presence of vehicles at the lanes immediately before reaching the intersections. The real queue size cannot be taken into consideration. In the present approach, the input values are supposed to come from cameras connected with image processing systems and directed microphones. We propose a new traffic signal control system with a hierarchical structure based on similarly Mamdani control, however, containing essentially novel elements and having more intelligent features. This new model and the connected algorithmic approach allow rather complex control strategies, but only a simple case study has been implemented. Compared with existing fuzzy traffic controls, the novel approach has more adaptability and flexibility, by having the potential to differentiate an arbitrary number of traffic directions and by increasing general safety by the additional emergency vehicle handling feature. In addition, the calculation with queues, and individual vehicles weighted with the waiting time makes the system more flexible than any existing intelligent model.

Keywords Fuzzy traffic light control · Fuzzy inference · Heavy traffic · Traffic weighted with waiting time

T. D. Chala (✉) · L. T. Kóczy
Department of IT, Széchenyi István University, Győr, Hungary
e-mail: chala.tamrat.delessa@hallgato.sze.hu

L. T. Kóczy
e-mail: koczy@tmit.bme.hu

L. T. Kóczy
Budapest University of Technology and Economics, Budapest, Hungary

Y. P. Kondratenko et al. (eds.), *Artificial Intelligence in Control and Decision-making Systems*, Studies in Computational Intelligence 1087,
https://doi.org/10.1007/978-3-031-25759-9_17

1   Introduction

Traffic jams are abundant in the modern world and are getting attention by many researchers, due to the rapidly increasing traveling population and vehicle usage. These problems, mainly occur in urban regions, suburban and main highway junction areas, where traffic signals are the typical control mechanism. To resolve this problem the present infrastructure is often limited in its flexibility and its expansion is infeasible due to spatial, environmental, and economic constraints [1]. Therefore, the design of intelligent traffic light control systems is very important to enhance the performance of the current fixed program systems and to improve the shortage of common fuzzy control via conducting some targeted research. Conventionally, the traffic signal controller is deterministic in nature, and there are two main types widely used, such as pre-timed and actuated traffic control systems. A pre-timed traffic control (PTC) system consists of a series of intervals that are fixed in duration for green, yellow (amber) and red light signals. However, this offline approach is not responsive to the dynamicity of the traffic conditions. In contrast to PTC, the actuated traffic control has the capability to respond to the presence of vehicles or pedestrians at the intersection [2, 3]. This paper lays emphasis on the problem of the limitations of the PTC system, deficiency of existing intelligent traffic signal control, and the shortages of existing fuzzy traffic controllers, putting forward the aspect of improving fuzzy traffic control efficiency, plus safety issues based on prioritizing emergency even in heavy traffic situations.

Sometimes, while there is no traffic at all at certain junctions, the pedestrians must wait for getting the green signal to cross the road. Another problem with the widespread traditional traffic signal systems is that the traffic light remains red for a long time while there is no intersecting traffic at all and the drivers in the crossing direction must wait quite long until the green light comes. This way, both travelers in the vehicles, and pedestrians may face an annoying waste of time. Due to the fact that the PTC approach is not responsive to the dynamic nature of the traffic conditions, it is the root of various problems, including heavy traffic jams, and as a consequence, considerable loss of time, too high fuel consumption, and heavy environmental pollution, particularly at main junctions and during rush hours [1].

Often, due to traffic congestion, emergency cars (such as ambulance and police cars, and fire trucks) get stuck or considerably delayed in the traffic bottlenecks caused by the inflexible traffic light control, which is reason for decreased general safety security in high traffic areas. It is also annoying when convoys (e.g. military, or VIP) are separated because of the rigidity of traffic light control (which may also occur with up to date commercial intelligent systems, as the whole convoy may not be sensed by the built in inductive sensors, while in a different direction conflicting with the green light necessary for the passing of the whole convoy may indicate a longer queue already waiting, this way switching the lights before the convoy's last vehicle is crossing the intersection. In order to overcome this problem, a hierarchical intelligent adaptive control system is proposed that has a specific module and structure for taking care of this problem, where the input information is based on traffic cameras (and possibly, directed microphones, complete with the necessary intelligent recognition software).

To this end, this study attempts to explore and answer the following research questions:

- How to reduce traffic congestion efficiently by using a Mamdani-type rule-based fuzzy system?
- To what extent the proposed fuzzy control is able to reduce the traffic congestion problem?
- How can general safety and security be increased by prioritizing emergency vehicles, convoys, etc., including the problem of conflicting priority issues.

Fuzzy Control (FC) may be efficiently deployed in solving complex problems that cannot be analytically optimized, because of the uncertain and non-deterministic elements in them, but can be controlled quite well by skilled human experts. The basic idea behind FC is to incorporate human expert knowledge in the design of the controller, whose input–output mapping is generated by a collection of fuzzy control rules (so-called production rules) involving linguistic variables rather than a complicated analytical and dynamic model, where a proper fuzzy inference engine evaluates the inputs and compares them with the information in the knowledge base, and thus produces a proper quasi-optimal (crisp or defuzzified) output [4, 5]. Thus, fuzzy controllers have four main components, namely the degree of the matching unit, the fuzzy rule (knowledge) base, the inference engine, and the defuzzification module. Here, the fuzzy rule base embraces the available expert knowledge (or statistical information mined by some machine learning technique), containing a fuzzy set based quantification of the experts' linguistic description of how to achieve "as good as possible" control. The inference engine emulates the experts' decision-making in interpreting and applying the knowledge on the control of the system in the best way. The degree of matching unit evaluates the inputs and transforms them into a set of fuzzy degrees of matching telling which of the individual rules fire, and to what degree; so that these inputs can be interpreted in the context of the rules in the rule-base. The defuzzification module interface converts the fuzzy conclusions calculated by the inference engine into actual inputs in the controlled process [6].

2 Overview of the Related Work

In the past five decades, a number of fuzzy control based models have been proposed to solve the dynamic traffic problems. Some research work in the field of traffic light systems using fuzzy systems is reviewed in this section.

Pappis and Mamdani, presented fuzzy traffic control on an isolated traffic intersection with simple one-way east to west or north to south traffic control with random vehicle arrivals and no turning movements [7]. Similarly, Niittymaki et al. introduced the new concept of fuzzy control to signalized pedestrian crossings. The discovered result demonstrated that fuzzy control provided pedestrian-friendly control, keeping vehicle delay smaller than with conventional control [8].

Trabia et al. discussed a fuzzy traffic signal controller for an isolated four-way intersection with through and left-turning movements. In that study, two parameters were used as input, namely, vehicle flows on the green light phase and vehicle queues on the red side, they were described using a trapezoidal fuzzy membership functions. The traffic intensity of the green light phase was decided as output value. Finally, they granulated traffic flow rats into three fuzzy levels (labels) and compared this with a traffic actuated controller. The results showed that the fuzzy controller had the ability to adjust its signal timings in response to changing traffic conditions on a real-time basis [9]. All these approaches applied Mamdani-systems.

In another study, Zaied et al. developed a system that was applied and tested using real data collected from signalized intersections. The proposed system received the data and mapped it to the appropriate membership function and combined the results of the values by the fuzzy inference mechanism. Here again, the Mamdani model was used for building the control system. The results from those studies showed that the proposed fuzzy traffic system provided better performance in terms of total moving time and vehicle queue than traditional traffic signals [10].

Alam et al. and Askerzade et al. also proposed intelligent fuzzy traffic control systems. Both developed one module of fuzzy inference, which was responsible for extending the current green light phase or switching it to red, depending on the sensed traffic. In these researches, again Mamdani rules and Sugeno type fuzzy systems were used, respectively. In both cases, based on the observed results, the proposed system had better performance in terms of total waiting time and total moving time both with than actuated "semi-intelligent" and fixed time traffic control systems [11, 12].

Salehi et al. conducted research on multi-agent-based autonomous fuzzy traffic light control systems to overcome problems like congestion, accidents, further speed and traffic irregularity. In that rather advanced study, real traffic parameters like traffic density and queue length were obtained by using image-processing techniques on data obtained from wireless sensors. This research is one of those which gave some initial ideas for our own investigation. As inputs, fuzzy variables with Gaussian shaped labels were used, whereas triangular fuzzy membership functions were used for the output variables. In this study, based on the available input data, the prioritization of emergency vehicles was included. The method used was though quite different from our proposed technique. The results showed that the proposed agent-based approach provided a very efficient solution by minimizing the vehicles' waiting time for the emergency vehicles mentioned above [13].

Yan developed fuzzy traffic control taking in account the degree of traffic urgency to improve the performance at isolated four-way intersections. The author designed and developed a fuzzy inference module that calculated the urgency degree of traffic for the red phase and decided whether the red-light should be switched to green. Queue lengths and average waiting times were used as input fuzzy variables. The obtained result was compared with pre-timed traffic control (PTC) by classifying the vehicle arrival rates into three classes, namely, low, middle and high traffic flow, and also here, the fuzzy control method performed better than PTC [14].

Zacharia et al. conducted research on the same topic and they implemented a decision support system using real-time road traffic data and a dynamic round-robin allocation of right-of-way to road users. As with many other authors, queue lengths and vehicle waiting times were used as input, and extension time of the green phase was the output. The proposed system determined the duration of the continuous traffic flow under given conditions, based on the lengths of the respective queues and the waiting times. As expected, the solution compared with a non-optimal traffic control system showed in superior performance in terms of the average waiting time [15].

Chabchoub et al. designed an intelligent traffic light controller using a combined fuzzy system and image processing approach, which is again one of the inspiring researches for our investigation. The designed fuzzy controller's inputs were the images captured from cameras fixed at each road direction. The input and output membership functions were similar to the ones used in [14]. The results of the simulation concerning the waiting times were similarly good, compared to non-adaptive solutions [16].

The above reviewed literature equivocally reported that fuzzy traffic control had better performance compared with PTC and also with actuated semi-intelligent traffic control. However, in most of the reviewed approaches, the traffic control phases changed in sequential (circular) order, and did not consider any unusual situation, such as the approach of one or several emergency cars, or the sudden appearance of a heavy traffic load in one of the directions ("traffic bursts").

In the present research, we tried to lay a stress on the latter problem, while keeping all the advantages of the earlier published fuzzy system based intelligent traffic signal control systems. In addition, we used a refined granulation of the traffic situation, everywhere applying five linguistic labels for differentiating the levels of traffic intensity, waiting time, etc. The structure of the proposed fuzzy control system is novel in the sense that it is built up as a hierarchical (two-level) decision system, and it combines Mamdani-style control with one or several classic production rule(s) based decision conditionally overruling the Mamdani-controller, if necessary. This architecture shows rather advantageous behaviour in every respect, both in reducing the waiting times, and in increasing general safety and security in the neighbouring urban area.

We also realized a simple case study, where only the main directions were considered. It is easy to realize that a real life system with more directions (3×4, for example), may be developed based on the same principles, although it requires much more resources. The implementation of the model was done in the Python programming language, and a structured, modular software was developed that can be extended to the real life control system structure. Thus, three modules were created, the Heavy Traffic Evaluation Module (HTEM), the Prioritize Emergency Car Module (PECM), and the Extension Time Decision Module (ETDM) (with the extension of Calculating Waiting Time Sub-Module which switching the green light for the long time weighted vehicle(s) if the heavy traffic forces the non-stop maintaining only one phase for a long time). The developed system was compared with PTC under the assumption of Poisson distribution for the traffic in every direction.

3   The Basic Model, the Methodology and the Formal Tools in the Proposed Approach

Traffic congestion in urban or metropolitan areas may occur for various reasons, such as excess demand, signal interference, incidents, work zones, weather, or special events [17]. However, this study considers only the traffic congestion that occurs due to the limitations of the deployment of traditional traffic signals, considering also special events (emergency cars, military convoys, etc.). Traffic signals offer control at intersections to ensure the safe and orderly flow of vehicles and passing of pedestrians, and to reduce the number of accidents. In this phase of the study, we have not yet assumed the processing of pedestrian related information yet. With traditionally controlled traffic signals, the duration of green light in every direction, and all the cycle times are predetermined [2]. The main disadvantage of traditionally controlled traffic signals relates the time loss that vehicles experience when crossing the intersection, maybe, by turning left or right. We claim that the average vehicle delay, which was selected to quantify the time loss of the vehicles crossing the intersection is considerably decreased when applying the proposed fuzzy system based method, using camera inputs and image processing.

Vehicle delay may be expressed in absolute and relative terms, which latter degree is mostly associated with the driving style and behavior of the driver. The latter is the relative measure of the additional (unnecessary) time consumption measured when passing the traffic intersection. This delay is a parameter that is not easily determined due to the non-deterministic nature of arrival and departure times at the intersection. Here, two types of uncertainty occur: the one caused by the vagueness of the notions used for description and measured for featuring the new system, and the one caused by the probabilistic-statistical behavior of the vehicles and the whole traffic flow under investigation. To model the latter, the statistical type non-deterministic nature of the arrival and departure processes, a Poisson distribution based model was used in this investigation. For simplicity, four isolated intersection traffic flows coming from North, South, West, and East were considered in this simple case study implemented system.

3.1   The Distribution of Vehicle Arrivals

Vehicle arrivals can be modeled in two interrelated ways; namely modeling how many vehicles arrive in each interval of time, or modeling what is the time interval between the successive arrivals of the vehicles. The distribution of the arrivals of the vehicles is a discrete random distribution. Two statistical models are most frequently applied for calculating the distribution of vehicle arrivals/departures, namely, the Poisson and the binomial distributions [18]. The arrival of the vehicles in a Poisson process follows the next equation:

$$P(x) = \frac{(\lambda t)^x e^{-\lambda t}}{x!} x = 0, 1, 2, \ldots \tag{1}$$

where $P(x)$ is the probability of the arrival of the vehicles x during the observation (counting) interval t, $\lambda = rt$, where r is the average rate at which arrivals occur; λ being the expected value of the events in t. The assumed range of the vehicle arrival traffic flow rate is $0 \sim 1$ per second.

3.2 Computation of the Average Delay Time of the Vehicles

As mentioned above, the most essential parameter used in evaluating the performance of a signalized intersection is the vehicle delay [3, 14]. So, in this paper, the average delay time of vehicles is proposed for the performance evaluation of the fuzzy traffic control. The average delay time of the vehicle in cycle x is calculated as shown in Eq. (3). A signal cycle is called one complete rotation of all the light phases.

The total delay time of the vehicle in cycle x can be calculated as follows:

$$D_x = D_r + D_g \tag{2}$$

where D_r is the delay time of the red-light phase and D_g is the delay time of the green light phase. From the total delay times of the vehicles, the average vehicle delay time is calculated as shown below:

$$D_{avg} = \sum_{x=1}^{n} D_x / A \tag{3}$$

where n is the number of cycles and A is the number of directions vehicles arrive at an intersection [14].

3.3 The Software Tools

All the studies reviewed above were developed and simulated using MATLAB programming language. In this investigation, however, the case study system was developed in a Windows environment and for the simulation, the Python programming language was used for both implementing and testing a simple hierarchical fuzzy traffic control system, and the simulation of the PTC, which latter served as a base for comparison. Python was chosen as it is a very powerful general-purpose programming language that has many toolkits which allow simulation using numerical methods, and the development of graphical illustration [19], and further work will be much more supported in this language.

3.4 The Case Study and the Fuzzy Model for the Proposed Novel Fuzzy Traffic Control System

Nowadays, in many countries (especially in developed countries) traffic intersections are controlled by an intelligent traffic signal control with some limitations. The majority of existing traffic monitoring technologies at junctions rely on high-cost sensors buried in the pavement that are difficult to install and maintain [20]. See e.g. the review by Yau et al. on reinforcement learning models and algorithms for traffic signal control systems. The author states that existing intelligent traffic signal controllers typically use inductive loop detectors and are installed in the cover of the lanes in order to detect the presence of vehicles in practice. Note that the road sections with built in loops are always of bounded length. However, during congestion, the traffic flowing in is higher than the traffic flowing out in an intersection, the queue size of the vehicles may not be reduced despite the green signal being activated, which causes vehicles move slowly or stop. A more problematic situation may be cross-blocking, when no vehicles can cross at all, although the green signal is being activated, when the respective lane of the downstream intersection has become fully occupied [2]. Similarly, Stephen et al. also assume the presence of buried detector loops. One of the limitations of intelligent traffic signal control using only loops as input is that they may fail to detect at all motorcycles or bicycles, and thus cause them wait indefinitely [21]. Tomar et al. conducted a very recent (2022) review on the state of the art of the research on intelligent traffic light synchronization, including the current status, the challenges, and the emerging trends. The author listed here most of the existing intelligent commercial traffic systems, including SCOOT, SCATS, and PRODYN, mentioning also their drawbacks. They are examples of adaptive traffic control systems that have been already implemented in several cities. SCOOT and SCATS use loop/magnetic detectors for gathering traffic data. The limitation of all of these methods is that they do not provide any information on the vehicles' speed, intended heading and exact position. While SCATS is a fully computer-based system (using regional and central computers), it is very expensive to use.

Unlike SCOOT and SCATS, PRODYN uses both cameras and loop/magnetic detectors to track the vehicles, but this systems also has significant drawbacks in terms of cost and functionality. Likewise, these systems need a large communication infrastructure that is capable of supporting a centralized control with a high data rate [20]. The above-reviewed literature reported that most existing intelligent traffic control systems use loop/magnetic detectors buried in the pavement. These systems usually have some inconvenient limitations, such as these systems activate the green signal based on the presence of vehicles at the lanes immediately before reaching the intersections, while the real vehicle queues cannot be taken into consideration as heavy traffic may generate (much) longer queues than lengths of the sections with sensors.

In the new proposed model, the input values are supposed to come from cameras connected with image processing systems, and this way, the amount and the type of traffic can be taken in account from an if not unlimited, but very long distance. The cameras can detect unusual vehicles approaching and recognize blinking blue and yellow lights on the cars, and a directed microphone can detect the sound of a siren, this way the system may act upon much more information than the existing ones. As the system implemented is modular, new modules can be added at any time without affecting the main structure. We claim that the new approach is robust and scalable, much more than the existing intelligent traffic signal controls.

The general structure of an isolated traffic intersection is depicted in Fig. 1 and is characterized by four incoming directions, namely North, South, East and West assuming 12 lanes for all outgoing combinations; each incoming road having three lanes (or lane groups), and from each direction going straight through, turning left, and turning right is possible. There may be no conflicting directions getting green at the same time. For example, if the incoming North direction is enabled with the lanes going through, turning left, and turning right, additionally vehicles driving East to North, West to South, and South to East are allowed to go, and all the other six lanes will have red. The proposed system takes into account the degree of heavy traffic for all directions getting red, considers traffic conditions under the current green light, and checks if any emergency cars (unusual urgency traffic) are detected, when extending or switching the green light. The other novel element of the proposed system is the calculation of the waiting times for the stacked vehicle(s) and the action upon. It also resolves the situation when heavy traffic forces the continuous maintaining only one non-conflicting directions for a long time, and thus one or several vehicle(s) have to wait for a too long time, the drivers have less and less patience for waiting so long, and anyway, such a situation is not fair, and must be avoided.

The development of the model is the first step for the construction of a real life really intelligent traffic control system. Fuzzy modeling offers the possibility to deal with highly complex, vague, and uncertain systems, in which conventional mathematical models may fail in giving satisfactory results [5]. As the most-wide spread

Fig. 1 General architecture of the proposed traffic intersection

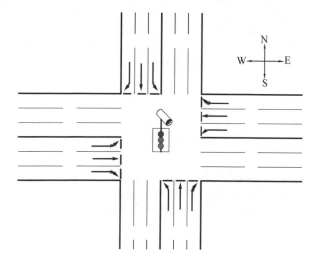

Fig. 2 The architecture of
the hierarchical three
modules system of the
proposed fuzzy traffic control

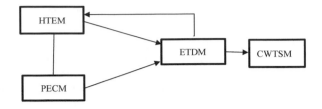

and successfully deployed fuzzy rule based model, the Mamdani model was chosen for developing the intended traffic control system. Mamdani style fuzzy inference is based on a set of linguistic production rules (IF–THEN rules), where the knowledge represented by the rule base is usually obtained from experienced human operators [6] (but may also be obtained by machine learning methods applied on observations or measurements of input–output data) .

In the proposed hierarchical system, a two level architecture was constructed, based on three fuzzy inference modules, as mentioned above, which are connected in a cascade style (Fig. 2). In theory, each module may have an arbitrary number of input and output variables, but in the simple implemented case study, each module has a simple structure with two input variables and one output variable. For the linguistic values, and the corresponding fuzzy partition (more correctly, fuzzy cover) a finer granulation was applied, compared with the majority of the above overviewed approaches. There are five linguistic values used in every dimension, and so, five membership functions are defined for each of the input variables, as well as the output variable in each of the heavy traffic evaluation module, and extension time decision module (see Figs. 3, 4, 5, 8 and 9). This finer granulation allows a more sensitive reaction of the control system to changes in the traffic flow, and also allows a more informative comparison of two different direction traffic flows competing for getting (longer) green light. The emergency car prioritizing module has only three values for both input variables and the output variable (including the no–no combination), as there is just the question of any emergency car arriving or not arriving from any direction. Of course, by extending the model into a finer one, with 12 directions for any intersection, this number must be accordingly increased. For simplicity, all membership functions used for the fuzzy values were triangular.

4   A Simple Case Study of the Implementation of the Proposed Traffic Signal Control System

As mentioned above, three modules form the basic structure of the proposed system. All three were implemented as told before. In the next, some details of the software and some simulation results will be presented. The three modules are:

Fig. 3 Vehicle queue membership functions for the red light phase (Q_r)

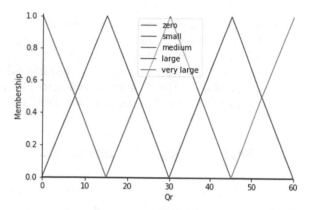

Fig. 4 Duration of red light membership functions (T_r)

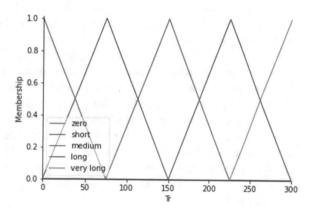

1. Heavy Traffic Evaluation Module (HTEM): this module is responsible for calculating the degree of heavy traffic for the red light phase and decide whether to switch on the next green light phase.
2. Prioritize Emergency Car Module (PECM): this module detects any emergency car(s) approaching from both the red and the green light directions. If there is an emergency car arriving, the module gives priority to the direction indicated by the lane of approaching of the detected emergency car(s). In the case of conflicting directions of several such cars, it will choose the most reasonable decision of keeping the current phase until the passing of the emergency car(s) unless there is a considerably longer convoy coming in the conflicting lane, and then switching the light so that the next prioritized car(s) may pass.
3. Extension Time Decision Module (ETDM): it is responsible for the decision on the length of extension time of the current green light phase, depending on the traffic situation, depending on the degree of heavy traffic presence in the lanes being currently in the red light phase. In addition, this module has a function (a submodule) which calculates the waiting times (CWTSM), and weighting the individual vehicle(s)' importance in influencing the time of switching for green

light. In the case when there are no emergency cars approaching and the degree of heavy traffic comes continuously only in one conflicting direction, when the weighted importance "weighted virtual queue length" is surpassing the importance value of the real heavy traffic, nevertheless the real amount of vehicles, a switch is initiated. The weighting function may be determined depending on the importance of intersecting roads and local urban priorities.

4.1   The Heavy Traffic Evaluation Module (HTEM)

This fuzzy inference module evaluates the degree of heavy traffic for the red light phase. The first input variable is the queue length of the vehicles waiting under the red light phase (Q_r) while the second one is the duration of the red light since the last end of the green light (T_r). The output variable is the degree of heavy traffic of the red-light phase (HT_r).

There are five linguistic values, and accordingly, five membership functions defined for all the input variables Q_r and the output variable HT_r, namely *zero*, *small*, *medium*, *large*, and *very large* and input variable. T_r also has five linguistic values, namely *zero*, *short*, *medium*, *long*, and *very long*. The membership functions of Q_r, T_r, and HT_r are depicted in Figs. 3, 4 and 5, respectively.

Accordingly, $5*5 = 25$ fuzzy inference rules of the heavy traffic evaluation module have been defined. For example, 'IF the current red-light phase of the vehicle queues is *small* AND the duration of red-light phase time is *short* THEN the heavy traffic of the red phase is *small*.' Rule base developed in the same manner illustrated in Figs. 10 and 11.

Fig. 5 Heavy traffic membership functions (HT_r)

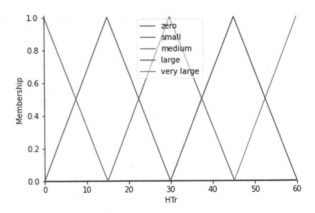

4.2   The Prioritize Emergency Car Module (PECM)

The main role of the prioritizing emergency car module is to detect the emergency cars, such as ambulance cars, fire trucks, police cars, convoys of military and VIP vehicles, etc. using siren and red light, or just yellow light, and to provide priority to them. This module has two input and one output variable, namely, the number of emergency cars detected from the direction of the actual red-light phase (E_{cr}) and of the actual green-light phase (E_{cg}). The output variable is the indicator of prioritizing emergency cars (P_{ec}). Since the emergency cars are occasional, only three linguistic values were selected, in the case study, thus three membership functions were defined for both input variables such as *zero, few,* and *medium* (see Figs. 6 and 7). There are three possibilities of the values of the output variable: *no emergency car, emergency car from the red light phase,* and *emergency car from the green phase.* In the case of conflict, it will decide depending on the number of emergency cars detected, for example, if there is an approximately equal number of emergency cars approaching on both sides, it will take the most realistic decision, and give the priority for the green phase. However, if the number of emergency cars detected from the red phase side is considerably greater than those in the conflicting direction, obviously the green light will be switched to red. Therefore, in the case study, only nine fuzzy inference rules have been set for this module, as it is shown in Table 1.

A sample of the fuzzy rules in Table 1 is illustrated by "IF emergency cars appear from the red phase direction is a *medium* AND emergency cars appear from the green phase direction is also a *medium*, THEN keep the green light till all the emergency cars pass." Another example is "IF emergency cars appear from the red phase direction is a *medium* AND emergency cars appearing from the green phase direction is *few*, THEN switch the green light to the red phase immediately." This latter rule shows clearly that the output is a crisp decision.

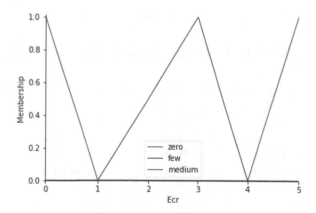

Fig. 6 Input membership functions of E_{cr}

Fig. 7 Input membership functions of E_{cg}

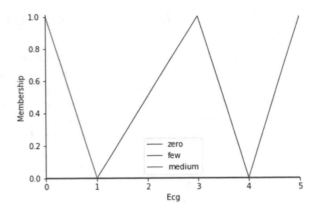

Table 1 The fuzzy rules of the prioritize emergency car module

Emergency car of red phase	Emergency car of green phase	Prioritize emergency car
Zero	Zero	No emergency
Zero	Few	Emergency for green phase
Zero	Medium	Emergency for green phase
Few	Zero	Emergency for red phase
Few	Medium	Emergency for green phase
Medium	Zero	Emergency for red phase
Medium	Few	Emergency for red phase

4.3 Extension Time Decision Module (ETDM)

The extension time decision module is the third, the core module of fuzzy traffic control that is accountable for switching the green light phase by considering the current vehicle queue length of the green light phase, the degree of heavy traffic of the red light phase (considering also potential emergency cars). This module has two main responsibilities. First, if there is no emergency car, it takes the decision to extend the green light or switch to the red phase by considering the degree of heavy traffic and the current traffic situation of the green phase. Second, it either switches the green light to red, or keeps the green light based on the PECM output criteria. The output variables of the HTEM and the PECM are used as input variables for the ETDM. In addition to these input variables, it has the input variable Q_g, which counts the number of vehicles of the current green light. It has one output variables, extending or

Fig. 8 Input membership functions of Q_g

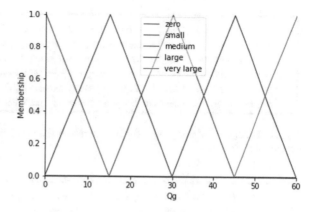

Fig. 9 Output membership functions of G_e

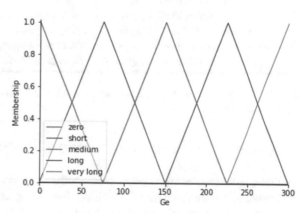

switching off the current green light (G_e). The triangular fuzzy membership functions of Q_g and G_e are depicted in Figs. 8 and 9, respectively.

There are three types of fuzzy inference rule bases that have been set for this module. For the first inference system, $5*5 = 25$ rules have been produced from the heavy traffic and the current green traffic situation input variables, which consider the degree of heavy traffic and the vehicle queue of the current green light to extend or switch off the green light time. Another $5*3 = 15$ inference rules have been set that consider the number of emergency cars and the degree of heavy traffic to prioritize the emergency car if necessary (see Fig. 10). Finally, $5*5*3 = 75$ rules have been generated which are the source of the hierarchical decision on whether to extend or to switch of the green light considering the arrival of any emergency car(s), the degree of heavy traffic, and the traffic conditions of the current green light (see a subset of the rules in Fig. 11). All fuzzy inference rule base was developed in the same fashion (as shown in Figs. 10 and 11) using Python Programming language.

Rule1 = ctrl.Rule(HTr['zero'] & Ec['No emergency'], Pe['no prioritization'])

Rule2 = ctrl.Rule(HTr['zero'] & Ec['emergency of greenphase'], Pe['keep green-light'])

Rule3= ctrl.Rule(HTr['zero'] & Ec['emergency of redphase'], Pe['Switcht to redphase'])

Rule4 = ctrl.Rule(HTr['zero'] & Ec['No emergency'], Pe['no prioritization'])

Rule5 = ctrl.Rule(HTr['small'] & Ec['emergency of greenphase'], Pe['keep green-light'])

Rule6 = ctrl.Rule(HTr['small'] & Ec['emergency of redphase'], Pe['Switcht to redphase'])

Fig. 10 Examples for fuzzy rules for the extension time decision module for emergency cars developed in Python

rule1 = ctrl.Rule(Qr['zero'] & Qg['zero'] & P['No emergency'], Ge['zero'])

rule2 = ctrl.Rule(Qr['zero'] & Qg['small'] & P['No emergency'], Ge['short'])

rule3 = ctrl.Rule(Qr['zero'] & Qg['medium'] & P['No emergency'], Ge['short'])

rule4 = ctrl.Rule(Qr['zero'] & Qg['large'] & P['No emergency'], Ge['medium'])

rule5 = ctrl.Rule(Qr['zero'] & Qg['very large'] & P['No emergency'], Ge['long'])

rule6 = ctrl.Rule(Qr['zero'] & Qg['zero'] & P['emergency of greenphase'], Ge['short'])

Fig. 11 Examples for fuzzy rules for the time extension decision module with three input variables developed in Python

4.3.1 Calculating Waiting Time Submodule (CWTSM)

In this simple case study implementation, the extension time decision module has one submodule that calculates the waiting time of one or some individual vehicle(s). Although it is occasional, sometimes the heavy traffic continues for a long time in one direction and some vehicles in conflicting direction(s) may wait for a very long time to get the green light. To avoid such a deadlock, CWTSM calculates the waiting time of individual vehicle(s) and switches to green light if the weighted "importance" of the waiting cars surpasses the normal importance of the heavy traffic direction (cars in a green light direction have basic unit weight). This submodule has two input variables and one output variable, such as the waiting time of the red light phase (weighted vehicle(s)) (W_{tr}) and the vehicle queues of waiting time of the actual green light phase (W_{tg}). The output variable is switching the green light (S_{gt}) to the red phase or keeping the current green phase. Both input variables have five linguistic values, and consequently, five membership functions are defined, including *zero, short, medium, long*, and *very long*. For simplification, only two linguistic values were defined for the output variable, such as *yes* and *no*. Accordingly, $5*5 = 25$ fuzzy inference rules of the CWTSM have been developed. Here also rule base developed and defined in the same manner as shown in (Figs. 10 and 11) and Table 1.

5 Simulation Analysis and Discussion

The developed hierarchical system of the case study, consisting of the three main modules and one submodule was implemented and then tested by simulation experiments.

In Table 2, some examples for the decision of the HTEM are shown. So, if the value of input Q_r is *small* and the value of input T_r is *small* the degree of heavy traffic is also *small*. If the values of Q_r and T_r are increased, the value of HT_r is also increased. For instance, if the input value of Q_r is 16.5, that is, it is in the range of the label *small* and the input value of Tr is 25.5 which falls in the *zero* label value, then the result obtained from the heavy traffic output value during defuzzification is 11.7, that is categorized as *zero (almost zero)* for the heavy traffic output variable (see Fig. 12). This latter is formulated as the fuzzy production rule follows:

IF the vehicle queue number of the red phase is *small* AND the duration of the red-light phase is *zero*, THEN the degree of heavy traffic is *zero*.

Correspondingly, the ETDM is evaluated in a similar way, based on the fuzzy inference rules on extending or switching the green light time period. According to the results demonstrated in Table 3, if the queue of the vehicles in the red phase is *small* and the vehicle queue of the green light phase is *large*, the extended time for the current green time is *long*. On the other hand, if the value of H_{tr} input is increased

Table 2 Examples for some results of the HTEM

Input variables		Output variable
Q_r	T_r (per second)	HT_r
5.0	90.0	16.6
16.5	25.5	11.7
30.5	80.5	17.6
55.0	200.0	44.9
61.0	305.0	55.0

Fig. 12 Result of output variable HT_r if $Q_r = 16.5$ and $T_r = 25.5$

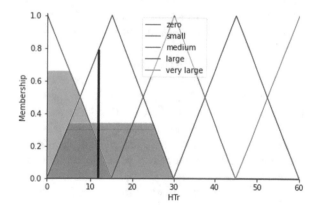

Table 3 Examples for some results of the ETDM	Input variables		Output variable
	HT_r	Q_g	G_e
	5.0	60.0	225.0
	10.5	75.5	224.9
	30.5	15.5	78.6
	55.0	25.0	71.08
	61.0	10.0	27.1

Fig. 13 Result of output variable G_e if $Q_r = 5$ and $T_r = 60$

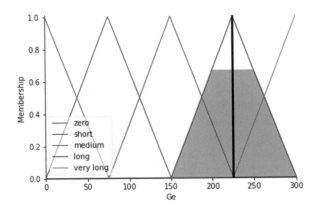

and the value of Q_g is decreased, then the value of extended time will also decrease. For instance, if the value of the input variables H_{tr} is 5 and Q_g is 60, the result of G_e will be 225 (see Fig. 13). The equivalent production rule is 'IF the vehicle queue of the red phase is *zero* (*almost zero*) AND the current green light queue is *very large* THEN the extension time is *long*.'

Similarly, the PECM was tested based on the defined fuzzy inference rules. The linguistic values of the input variables of this module were set to *zero, few,* and *medium*. For simplicity and defuzzification process purpose, the output value range [0, 1] represents *no emergency*, (1, 3] *keeping the green light* till the emergency car pass, and >3 means *switching the green light to red*. As observed from the result shown in Table 4, if the value of emergency car detected from the red phase is 2.5 which falls into the label *few* and the value of emergency car detected from the green phase is also 2.5 (also *few*), emergency cars were detected from both directions, but the output result is 2, which gives priority for the emergency cars detected from the green light direction. This is formulated as "IF the number of emergency cars detected from the red phase direction is *few* AND the number of emergency cars detected from the green phase direction is also *few* THEN *keep the green light* till the green phase emergency cars pass." In another example, if the value of the input variable of E_{cr} is 5 and E_{cg} is 1.5, the result is 3.68. The linguistic rule firing in this case, is "IF the number of emergency cars detected from the red phase direction is

Table 4 Examples for some results of the emergency car module

Input variables		Output variable
E_{cr}	E_{cg}	P_{ec}
0	0	0.33
1.5	4.5	1.99
2.5	2.5	2.0
5	1.5	3.68
3.5	3.5	1.99

medium AND the number of emergency cars detected from the green phase direction is *few* THEN *switch the green light to red* immediately."

Overall, the ETDM ranks the order of traffic light changes from one phase to another, based on the presence of emergency cars and the degree of heavy traffic. For example, if the value of HT_r is 3 (the degree of heavy traffic is almost *zero*) and if the value of Q_g is 65 (the current green light vehicle queue is *very large*) and the value of P_{ec} is 5.5 (emergency cars appear on the red side direction) G_e is calculated 25.8 as depicted in Fig. 14. The dominant firing linguistic rule is "IF the emergency car is detected from the red phase direction AND the heavy traffic of the red phase direction is almost *zero* and the current green light vehicle queue is *very large* THEN *switch the green light to red*." For the easiness of representation and uniformity in the system, these crisp decision values have been represented by pseudo-fuzzy triangular membership functions, however, without overlap in the support: *no emergency* is equivalent to [0,1], *emergency of the green phase* direction is denoted by (1,3], and *emergency in the red-phase* direction is denoted by >3.

In the same fashion, the CWTSM was analysed, based on the defined fuzzy inference rules the waiting time of the weighted vehicle(s) in the red phase was calculated, while the actual green light has been kept extended because the degree of heavy traffic is high on the green side, in the case of no emergency cars detected. According to the results demonstrated in Table 5, if the waiting time of stacked vehicle(s) in the red phase is *long* and the waiting time of the vehicle queues of the green light phase is

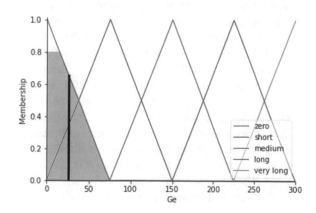

Fig. 14 Result of Ge output variable if $HT_r = 3$, $P_{ec} = 5.5$ and $Q_g = 65$

Table 5 Examples for some results of the calculated waiting time submodule

Input variables		Output variable
W_{tr}	W_{gt}	S_{gt}
25	50	0.36
150	200	0.36
250	50	1.64
305	220	1.67
310	310	0.33

short, then the green light switches to the red phase. In this simple study case, for the simplification of representation and consistency in the system; these crisp decision values have been represented by the pseudo-fuzzy triangular membership functions, but, without overlap in the support: *no* (keeping the green light) is equivalent to [0, 1], *yes* (switch the green light to the red phase) is represented by >1. For instance, if the value of W_{tr} is 305 (waiting a *very long* time) and if the value of W_{gt} 220 (the current green vehicles queues waiting time is a *medium*), then the value of S_{gt} is 1.67 (switching the green light to red phase) as depicted in Table 5.

In order to be able to evaluate the performance of the proposed system by comparison with a traditional non-intelligent system, PTC was simulated using the SimPy Python. Here, the vehicle arrival distribution also followed the Poisson distribution. The simulation counted the vehicle arrivals, and calculated the average delay time of vehicles based on Eqs. (2) and (3). In order to be able to compare the traffic flow rate was classified into five traffic flow rates such as *low, medium, high, very high*, and *extreme* traffic flow. The range of intervals of arrival rate was set as 0 ~ 0.2 for low, 0.2 ~ 0.4 for *medium*, 0.4 ~ 0.6 for *high*, 0.6 ~ 0.8 for *very high* and 0.8 ~ 1 for *extreme* traffic flow.

Improvement of Congestion (%) =

(Average of PTC - Average of FTC)/Average of PTC)*100 (4).

According to the results presented in Table 6, the average delay time with *low* traffic flow of the proposed fuzzy traffic control system decreased by 28.76% compared to the PTC. With *medium* traffic flow (see Table 7), the average delay time of the fuzzy control system is reduced by 31.50% compared with the PTC. Under *high* traffic flow conditions, the developed fuzzy approach resulted in a decrease by 34.43% compared with the PTC (see Table 8). The results under *very high* traffic flow (see Table 9), the proposed system achieved a reduction by 40.31% Table 10, and with *extreme* traffic flow rate the average delay time, the fuzzy control system showed a 42.86% compared with the PTC. From these simulation results, we may conclude that the higher is the traffic flow in one direction, the better the proposed system performs compared with the traditional PTC timing system. This may be similar to the other "semi-intelligent" fuzzy rule based approaches, but the weighted calculation of queue importance offers a more realistic and driver friendly solution, and in

Table 6 Average delay times for low traffic flow

Simulation case	PTC	Fuzzy traffic control
1	34.32	25.10
2	57.69	43.59
3	75.27	50.06
4	80.10	55.36
5	80.95	59.78
Average	65.67	46.78

Table 7 Average delay times for medium traffic flow

Simulation case	PTC	Fuzzy traffic control
1	87.1	63.49
2	91.49	64.78
3	97.56	67.26
4	103.22	69.35
5	111.07	71.08
Average	98.09	67.19

Table 8 Average delay times for high traffic flow

Simulation case	PTC	Fuzzy traffic control
1	115.36	73.57
2	121.99	74.36
3	127.80	75.00
4	133.29	93.60
5	134.75	98.65
Average	126.64	83.04

addition, the proposed new approach allows for intelligent recognition and action in case of special situations, such as the arrival of emergency vehicles or convoys. It is obvious that the same solution can be extended to any other special circumstance, such as accidents, irregular blocking the road, etc.

Table 9 Average delay times for very high traffic flow

Number of simulation	PTC	Fuzzy Traffic Control
1	152.89	103.41
2	171.15	107.99
3	193.96	110.50
4	201.01	118.55
5	224.49	122.73
Average	188.70	112.64

Table 10 Average delay times for extreme traffic flow

Number of simulation	PTC	Fuzzy traffic control
1	237.94	131.90
2	260.80	143.10
3	267.52	156.90
4	281.98	162.85
5	286.923	168.10
Average	267.03	152.57

Table 11 Improvement under the various traffic flow types compared to PTC

	Low	Medium	High	Very High	Extreme
PTC	65.67	98.09	126.64	188.70	267.03
Fuzzy traffic control	46.78	67.19	83.04	112.64	152.57
Improvement (%)	28.76	31.50	34.43	40.31	42.86

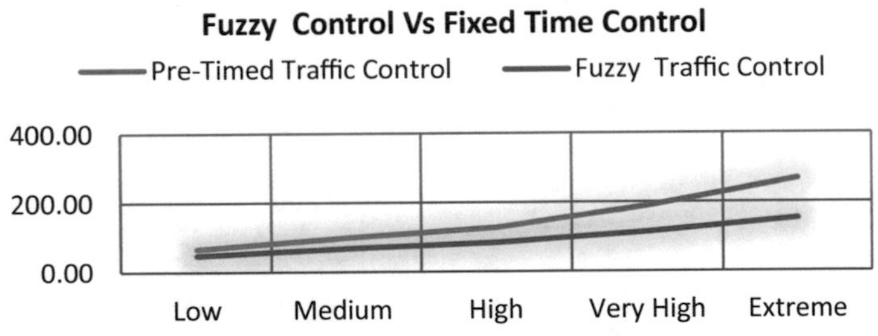

Fuzzy Control Vs Fixed Time Control

Pre-Timed Traffic Control　　Fuzzy Traffic Control

Fig. 15 Improvement of the performance of the new FTC compared with PTC in the throughput

Looking at Table 11 and Fig. 15, it is obvious that the advanced fuzzy traffic control is more efficient than the pre-timed traffic control. If existing intelligent (or, in reality, semi-intelligent) fuzzy traffic control systems are considered, the novel approach has more flexibility and driver friendly features when the weighted queues are taken as input for the decision, and it is an additional intelligent feature that the switching (or extension) of the green light phase in case of special events or unusual traffic flow elements, like the arrival of emergency cars, the heavy traffic induced decisions may be adaptively modified (overruled, if necessary). In our experiment, finer granulation was applied compared to other researches.

At this point, one more important aspect of the implementation must be considered. Because in the real system, the number of variables is much higher compared to the case study, the number of rules is growing exponentially, and may reach sizes where the real time calculations become a challenge (especially, when one controller takes care of several intersections). In the future, the Mamdani algorithm may be exchanged with a more resource efficient inference engine, like e.g. some type of fuzzy rule interpolation [22].

6 Conclusions and Future Work

The problem of ever increasing traffic jams strongly affects daily life activities, especially in cities and urban regions. This study has proposed a novel approach for controlling traffic lights, a crucial component both in the cause and in the potential reduction of those traffic jams, considering the strong limitations of the generally used traditional (PTC) traffic light control systems and shortage of existing intelligent traffic control system by having a more advanced set of intelligent features. In order to solve the problem encountered by the PTC, and also the existing commercial "semi-intelligent" systems, such as minimizing waiting times (in a flexible and intelligent manner) and travel time as well as maximizing safety and security, and increasing the efficiency and economic benefit, some developed countries use nowadays intelligent (in our interpretation, semi-intelligent) traffic control. These existing "intelligent traffic signal controls" are better than PTC, however, they still have considerable shortages. Since the most existing intelligent traffic control systems use induction loop detectors buried in the pavement, these systems activate the green signal based on the presence of vehicles at the lanes directly before reaching the intersections (at least in a bounded section of the road, with a bounded number sensors). Therefore, they cannot take into consideration the real degree of heavy traffic, and in reality, there may be much longer queues compared to the lengths of the sections with sensors. The proposed novel model supposes that cameras connected with image processing systems and potentially, also directed microphones will enable the real system to sense at a long distance for each lane. Hence, the proposed approach may consider even very long queues sensed, and may take proper decisions based on all available information coming in the fuzzy inference engine. In addition, this approach takes into consideration the psychology and behavior of drivers; and in the case heavy traffic forces continuously maintaining only one light phase for a long time and one or some vehicle(s) wait for a too long time, obviously, the drivers can get annoyed. The system calculates the waiting times of individual vehicles, and weights and compares this with the actual green light phase vehicle queues, and takes the proper decision of switching the green light to the red phase if necessary, which makes the proposed system more intelligent than any existing intelligent traffic control system. Summarizing this, by reviewing the relevant literature, we have identified some weak points of all existing fuzzy traffic controllers, and thus, we proposed in this study a more flexible and adaptive solution, with the goal of further reducing traffic congestion, reducing driver annoyances and of increasing traffic safety and general security by adding the hierarchically connected emergency car priority unit that is able to handle even multiply conflicting situations.

In the presented simple case study, a refined rule-based fuzzy control system was applied with a richer granulation of the traffic situations. It is obvious that this simple model can be formally extended to more sophisticated models based on the same principle, e.g., by increasing the number of traffic flow directions from four to twelve (N–E, N–S, N–W, E–N, etc.) and the same principle of setting up the control rule base can be applied. Also, the number and variety of "unusual" situations, where in this case study, only the emergency car prioritization was included, can be arbitrarily extended 8 such as accidents, VIP and military convoys, etc.). In the future, a more complex and more generally applicable extension of the present implementation will be elaborated. As we mentioned it, in the case of larger systems, the inference engine may be changed to a less resource demanding one, such as interpolative reasoning.

The simulation experiments have clearly demonstrated that the thus developed fuzzy traffic control is more efficient and has more intelligent features than other existing and proposed systems. For comparison purposes, a traditional PTC system was implemented and included in the simulated (for the implementation, using the SimPy library in Python programming) and so, comparison was made between the proposed refined fuzzy traffic control algorithm and the PTC in terms of the average delay times of the road vehicles. Under various traffic conditions. The simulation results showed that the novel fuzzy traffic control system outperformed PTC for all traffic flow types, with increasing advantage under heavier traffic. Thus, the new system combines improved safety with an increased efficiency of the transportation by reducing the delay time. Obviously, less traffic congestion and less delay time at traffic lights reduce not only the wasted time, but also the cost of fuel consumption, reduce the need for expensive macro level decisions such as the immediate need for constructing new infrastructure (extending the street and road network), and reduce also air pollution.

The future plans of this investigation are to integrate the fuzzy inference system with machine learning techniques like neural networks, or evolutionary meta-heuristic, which enable the fuzzy traffic control system to predict traffic conditions. Also, it will be possible to extend the system to arbitrary structure intersections and to arbitrary topologies of complex networks of intersecting roads, where either direct connection in the control or indirect connection by the traffic flow will be possible.

Acknowledgements L. T. Kóczy acknowledges the support given by the National Research, Development and Innovation Office (NKFIH), grant no. K108405.

References

1. Bull, A., Thomson, I., Pardo, V., & Thomas, A. (2003). Traffic Congestion, The Problem and how to deal with it. Santiago, Chile: United Nations Publication (p. 198). ISSN: 1727-0413, ISBN: 92-1-121432-7.
2. Alvin Yau, K.-H., Qadir, J., Khoo, H. L., Ling, M. H., Komisarezuk, P. (2017). A survey on reinforcement learning models and algorithms for traffic signal control. *Acm Computing Surveys Journal, 50*(34), 1–38. http://dx.doi.org/https://doi.org/10.1145/3068287.
3. Alam, J., Pandey, M. K. (2015). Design and analysis of a two stage traffic light system using fuzzy logic. *Information technology & Software Engineering, 5*(3), 1–10.

4. Lee, C.-C. (1990). Fuzzy logic in control systems: fuzzy logic controller-1. *IEEE transactions on systems, man, and cybernetics, 20*, 404–418.
5. Ross, T. J. (2010). Fuzzy logic with engineering applications, West Sussex, PO19 8SQ. A John Wiley and Sons, Ltd, p. 606. ISBN 978-0-470-74376-8.
6. Passin, K. M., Yurkovich, S. (1998). Fuzzy control. Menlo Park, California: Addison Wesley Longman, Inc., p. 522. ISBN: 0-201-1874.
7. Pappis, Mamdani. (1977). A fuzzy logic controller for a traffic junction. *IEEE transactions on systems, man, and cybernetics, 7*(10), 707–717. doi: https://doi.org/10.1109/TSMC.1977.430 9605.
8. Niittymäki, J., Pursula, M. (2000). Signal control using fuzzy logic. *Fuzzy Sets and Systems, international journal of soft computing, 116*(1), 11–22.
9. Trabia, M. B., Kaseko, M. S., Ande, M. (1999). A two-stage fuzzy logic controller for traffic Signal. *Transportation Research Part C: Emerging Technologies, 7*(1), 353–367. https://www.journals.elsevier.com/locate/trc.
10. Zaied, A. N. H., Othman, W. A. (2011). Development of a fuzzy logic traffic system for isolated signalized intersections. *Expert Systems With Applications, 38*(8), 9434–9441. doi:https://doi.org/10.1016/j.eswa.2011.01.130.
11. Alam, J., Pandey, M.K., Ahmed, H. (2013). Intellegent traffic light control system for isolated intersection using fuzzy logic. In *Conference on Advances in Communication and Control Systems 2013 (CAC2S 2013)*, DIT University, Dehradun, India, 2013. https://doi.org/10.13140/RG.2.1.4854.6406.
12. Askerzad, I., Mahmood, M. (2010). Control the extension time of trafic light in single junction by using fuzzy logic. *International Journal of Electrical & Computer Sciences, 10*(2), 48–55.
13. Salehi, M., Sepahvand, I., Yarahmadi, M. (2014). A traffic lights control system based on fuzzy logic. *International Journal of U-and E-Service, Science and Technology, 7*(3), 27–34. http://dx.doi.org/https://doi.org/10.14257/ijunesst.2014.7.3.03.
14. Ge, Y. (2014). A two-stage fuzzy logic control method of traffic signal based on traffic urgency degree. *Modelling and Simulation in Engineering, 2014*(694185), 1–6. http://dx.doi.org/https://doi.org/10.1155/2014/694185.
15. Zachariah, B., Ayuba, P., Damuut, L. P. (2017). Optimization of traffic light control system of an intersection using fuzzy inference system. *Science World Journal, 12*(4), 27–33. ISSN 1597-6343.
16. Chabchoub, A., Hamoud, A., Al-Ahmadi, S., Cherif, A. (2021). Intelligent traffic light controller using fuzzy logic and image processing. *International Journal of Advanced Computer Science and Applications (IJACSA), 12*(4), 396–399.
17. Afrin,T., Yodo, N. (2020). A survey of road traffic congestion measures towards a sustainable and resilent tranportation system." *MDPI, 12*(11), 1–23. https://doi.org/10.3390/su12114660.
18. Mathew, T. V. (2019). Transportation systems engineering: vehicle arrival models. Indian Institute of Technology, Bombay, India, 2019. https://www.civil.iitb.ac.in.
19. Halvorsen, H.-P. (2020). Python Programming, p. 140. ISBN:978-82-691106-4-7 2020.[Online]. Available: https://www.halvorsen.blog/documents/programming/python/ [Accessed 23 December 2021].
20. Tomar, I., Sreedevi, I., Pandey, N. (2022). State-of-art review of intelligent traffic light synchronization for vehicles: current status, Challenges and Emerging Trends. *MDPI, 11*(3), 465. doi: https://doi.org/10.3390/electronics11030465.
21. Inkporo Stephen, C., & Ume Leonard, E., Sensor-based intelligent traffic light control system: A panaceato traffic congestion in nigerian motorways. *Computer Engineering and Intelligent Systems, 8*(6), 14–20. ISSN 2222-2863 (online).
22. Johanyák, Z.C., Kovács, S. (2006). A brief survey and comparison on various interpolation based fuzzy reasoning methods. *Acta Polytechnica Hungarica, 3*(1), 91–105.

Generative Adversarial Networks in Cybersecurity: Analysis and Response

Oleksandr S. Striuk and Yuriy P. Kondratenko

Abstract Cybersecurity is one of the key problems of the twenty-first century, as the digital environment has already become equally important to the real world, and in some situations, perhaps, even surpasses it. Especially when it comes to classified data or information with limited access. The main challenge for cybersecurity is a timely and adequate response to any type of threat since every day there are more and more malicious scenarios for compromising information technology data. For cyber-attack defense algorithms to be effective, it is important to maintain a high level of awareness of the widest range of state-of-the-art threat types and malicious technologies. Machine learning and AI in general are considered as both a helpful tool and a threat. Generative adversarial networks (GANs) and their modifications can pose a serious threat to the entire field and thus should be properly and thoroughly researched, especially in light of the fact that GANs can be used to improve known attack types so that even AI-based detection system cannot identify them. The purpose of this article is a comprehensive analysis and structuring of existing GAN methodologies, with the subsequent development of approaches for an adequate response to potential threats.

Keywords Cybersecurity · Information security · Artificial intelligence, AL · Machine learning, ML · Deep learning, DL · Generative adversarial networks, GAN

1 Introduction

Generative adversarial network (GAN) is a machine learning framework developed by Goodfellow et al. that allows the model that is trained on a large dataset to generate

O. S. Striuk (✉) · Y. P. Kondratenko
Department of Intelligent Information Systems, Petro Mohyla Black Sea National University, Mykolaiv, Ukraine
e-mail: oleksandr.striuk@gmail.com

Y. P. Kondratenko
Institute of Artificial Intelligence Problems of MES and NAS of Ukraine, Kyiv, Ukraine

© The Author(s), under exclusive license to Springer Nature Switzerland AG 2023 373
Y. P. Kondratenko et al. (eds.), *Artificial Intelligence in Control and Decision-making Systems*, Studies in Computational Intelligence 1087,
https://doi.org/10.1007/978-3-031-25759-9_18

new data samples that are indistinguishable from real data. This is achieved due to the fact that the model actually consists of two separate neural networks that compete with each other in a zero-sum game (where the gain of one agent is equal to the loss of the other agent) [1].

Mathematically GAN can be represented with the following equation of its loss function [1]:

$$\min_{G}\max_{D} V(D, G) = \mathbb{E}_{x \sim p_{\text{data}}(x)}[\log D(x)] + \mathbb{E}_{z \sim p_z(z)}[\log(1 - D(G(z)))] \quad (1)$$

where G represents the generator network; D represents the discriminator network; x is a scalar—sample of real training data; z is noise or a latent space vector extracted from a standard normal distribution; $p_z(z)$ stands for a prior on input noise variables; $p_{data}(x)$ stands for a prior on input real data variables; $\mathbb{E}$ is an expectation; $D(x)$ is the discriminator network output that represents the probability that x actually came from the data rather than from the generator; $V(D, G)$ is the value function of D and G in the two-player minimax game [1].

GAN functions as follows: the first network, the generator, produces samples (candidates), and the second network, the discriminator, estimates them, trying to distinguish real data from synthesized candidates. The generator network attempts to generate a new sample by combining the primary samples using latent space variables, the discriminator network learns to distinguish between real and fake samples (Fig. 1).

Since GANs have been publicly introduced and have continued to evolve, it became obvious that the technology poses a serious threat in terms of cybersecurity, as its scope includes deepfakes, facial and fingerprint recognition, malware detection, password cracking, and more. While security systems use machine learning to recognize new threats by extrapolating their knowledge towards scenarios that have common patterns with training data, GANs are the exact opposite of such systems, since their objective function is to outsmart such systems.

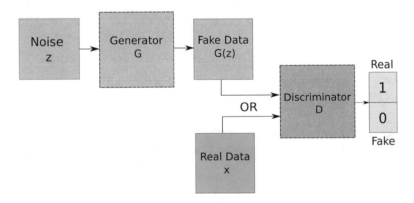

Fig. 1 Structure of interactions between Generator and Discriminator networks in GAN

Given the above, the purpose of this work is to analyze the possible risks and propose protection approaches in the context of GANs. The number of possible applications is too large to review them all comprehensively within one article, so we focused on the most crucial ones.

2 GAN Application Analysis in the Context of Cybersecurity

2.1 Malware

GAN demonstrated the ability to create malicious software (or malware) signatures that can bypass ML-based detecting systems. Weiwei Hu and Ying Tan described the model called MalGAN that is able to generate adversarial malware for black-box attacks, which means that the attacker has no information about the detection algorithms being used [2]. In addition, this work showed that the GAN outperformed even other adversarial ML-based models.

The success of this approach shows that today an attacker needs to know much less information about a security system in order to attack or break it and that even cutting-edge malware detectors with machine learning algorithms has its own vulnerabilities [3].

Shahpasand et al. presented a new approach, which showed that adversarial samples generated by GAN can avoid detection in 99% of attempts [4]. When it comes to black-box attacks, attackers aren't aware of the network architecture (specification and parameters), but they still can obtain the detection classifier's results and use this data for altering the malware codes.

Kargaard et al. proposed a method against intelligent malware for Windows and Linux that utilizes GAN. The authors of the paper turned malware binaries into grayscale images and resized them to 32 × 32 pixels. The obtained images can be used to create arrays for the deep learning models, potentially increasing the capability of detection of new malware types [5]. Taheri et al. used GAN-based algorithms to ensure more secure IoT-device communication. The paper describes a highly effective defensive approach towards Android mobile data privacy and its robustness [6].

Kim et al. described a method for automatic classification and detection of the zero-day attacks using a generative model—transferred generative adversarial network or tGAN [7]. Later for the same purposes they used adjusted model with deep convolution layers and deep autoencoder—tDCGAN that is capable of creating fake malware samples and learning to distinguish them from true malware [8]. tDCGAN demonstrated 95.74% average classification accuracy that outperforms other models' results.

2.2 Password Guessing

Password safety and its uniqueness is the key to the security of digital data and documents. But what if there is already a method that can effectively crack passwords by learning from real-life user passwords that have become publicly available due to data leaks and hacker attacks?

Hitaj et al. presented an approach for password guessing that is based on generative adversarial network architecture called PassGAN [9]. Modern password cracking software, such as CrackStation and RainbowCrack, enables hackers to guess a huge number of passwords per second using password hashes. These instruments are good enough for practical application, but when it comes to more sophisticated password modeling, they can face real issues due to the need for specific expertise and knowledge. GAN allows cracking passwords more efficiently since by consuming data from large datasets it learns to recognize hidden patterns and compound information.

PassGAN utilizes the Rockyou.txt wordlist as a dataset, which is an industry benchmark and the number one password dictionary used by cybersecurity experts and ethical hackers. The experiment showed that the model was able to not only mimic the distribution of the dataset but also predict new password examples that, by their features, may be suitable for password guessing. The researchers tested results on two different datasets—a pre-prepared subset of the Rockyou.txt wordlist, as well as the dataset based on leaked passwords published from the Linkedin data breach [9]. Results showed that PassGAN was able to guess 43.6% of unique passwords from the Rockyou.txt wordlist (1,350,178 out of 3,094,199), and 24.2% of unique passwords from the LinkedIn dataset (10,478,322 out of 43,354,871).

For evaluation purposes, researchers excluded from the testing set all passwords that were in the training set. After that, PassGAN showed 34.6% and 34.2% match for Rockyou.txt and LinkedIn samples, respectively. Furthermore, even those passwords that didn't match the samples from the testing set still contained all features of real passwords created by human intelligence; hence may still be applicable for accessing real user accounts despite not being present in the reviewed datasets.

It's also worth noting that PassGAN, when used in conjunction with HashCat, cracked around 51% to 73% more unique passwords than HashCat alone.

2.3 Deepfakes and Image Manipulation

Deepfakes require close attention. In addition to threatening a person's personal safety, deepfakes can cause political instability, trigger chaos, or even escalate a war conflict due to fabricated media data being used to discredit a person and damage their reputation by spreading fake photos, fake videos, or fake news articles [10]. The seriousness of the problem is evidenced by the growing number of scientific publications devoted to deepfakes from year to year (Fig. 2).

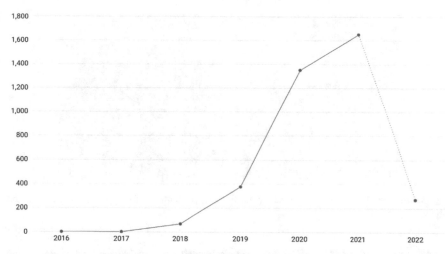

Fig. 2 Number of deepfake-related papers from 2016 to 2022, obtained from https://app.dimens ions.ai at the end of 2021 with the search keyword "deepfake" applied to full text of scholarly papers

In addition, the Defense Advanced Research Projects Agency (or DARPA; a research and development agency of the United States Department of Defense) launched the Media Forensics program that studies methods and approaches to resist fake media generated using GANs [11].

Deepfakes are usually created using a deep learning framework that is called an autoencoder. Through the encoder-decoder chain, the model is capable to replace the face of one person with another (applicable to both photo and video; the Reface app, for example).

GANs are used as a tool for generating synthesized fake human photographs and videos that can be used for malicious purposes. For instance, such realistic images can be used as profile photos of fake social media accounts (Fig. 3). The key GAN advantage is the principle behind its architecture: generator and discriminator are in a constant adversarial interplay that ensures the high photorealistic quality of synthesized data. Thus, such generated fakes are difficult to identify and debunk.

Due to the high quality of deepfakes, detection methods also need to be constantly improved. Nguyen et al. listed and summarized different methods for deepfake detection using various approaches including deep learning itself and feature extraction methods [12]. The authors note that GAN-synthesized images are probably the most difficult to identify, as they have a high level of realism and quality due to the ability of this model to learn the distribution of complex input data and produce new output data with a similar distribution.

Agarwal and Varshney suggested a method that is based on the authentication theory that considers deepfake detection as a problem of a hypothesis validation that focuses on GAN output data using a statistics framework as a baseline [13, 14]. However, when it comes to exceedingly accurate GAN that can generate fake videos or images of extremely high quality, it is hard to detect a fake by this or any

Fig. 3 "Photos" of non-existent people synthesized by the authors of this paper using pre-trained StyleGAN

other known methods. Thus, the problem of detecting deepfakes created using GANs remains unresolved.

Another GAN framework that is not related to cybersecurity directly but may have the potential for malicious application is CycleGAN. The model applies a cycle sequence loss to ensure a learning process without paired data. Its architecture allows it to map from one domain (A in Fig. 4) to another (B in Fig. 4) without pairwise alignments between the initial and focus domain [15].

Considering the specifics of this GAN framework, it is easy to imagine that the said approach can be easily applied to manipulate real images altering particular properties. Like, for instance, turning an image of horse into an image of zebra, or changing skin color. Misusing of CycleGAN can be used for creating fake images or videos of politicians or other persons and objects. The generation of fake photos and videos jeopardizes national, international, and personal security. Additionally, it is suited for attacking well-trained face recognition models.

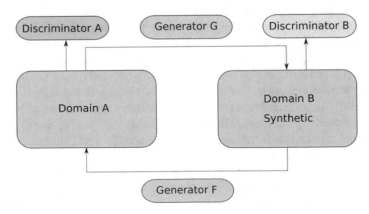

Fig. 4 CycleGAN architecture

Hsu, Lee, and Zhuang developed a deep forgery discriminator (DeepFD) that was able to successfully detect 94.7% tampered images synthesized by different GAN frameworks [16]. The algorithm is based on embedding the contrastive loss for efficient detection of synthesized images generated by GANs. The work demonstrates that there is a wide range of possibilities that allows solving the fake image detection problem. Therefore, further research is needed.

2.4 Forging Biometric Data

Real fingerprints data belongs to the category of personal information due to their sensitive nature. Synthesized by GAN fingerprints are of interest to cybersecurity. The authors of the paper [17] used a GAN modification—Adaptive Deep Convolutional Generative Adversarial Network (ADCGAN)—to create realistic fingerprint samples that appeared to be similar to real dactylograms in terms of their feature spectrum (Fig. 5).

The described approach makes it possible to use GANs for synthesizing fake fingerprints that theoretically can bypass fingerprint verification security systems, especially if a principle similar to the combination of PassGAN and HashCat is applied.

Bontrager et al. described a cognate approach that involves GAN for generation of synthetic dactylograms [18]. The work combines two GAN-based methods: the first one used an evolutionary optimization algorithm in the latent space and the second one applies gradient descent search.

In [19], the two-stage GAN has been presented: the first GAN generally "describes" fingerprints, while the second GAN applies the output data from the first one to generate fingerprint images. Then a feature extraction algorithm is applied to extract minutiae. The above methodologies can be used for malicious purposes in order to gain access to protected information, both related to national security and personal data.

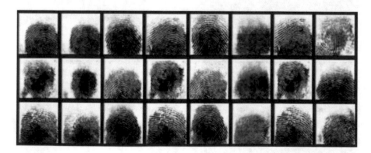

Fig. 5 Generated by ADCGAN fake fingerprints

The same goes for face recognition, as nowadays the number of different GAN-based attack types in this domain is constantly increasing, especially when it comes to cracking the DL-based face recognition systems.

2.5 Steganography

Steganography is the process of concealment information in plain sight. It's usually applied for embedding encryption in images or audio. For instance, it allows hiding text messages inside any picture or photo so no one will even guess that something is hidden there. In some cases, a second image may be encrypted in the first (Fig. 6).

Shi et al. proposed a novel approach of Secure Steganography based on GANs that was able to produce valid and safe covers for steganography [20]. The specificity of the proposed model is that the SSGAN architecture consists of one generator and two discriminators at the same time. The generator primarily estimates the graphic quality of the synthesized images for steganography, while the discriminator networks are used to evaluate their appropriateness in terms of data hiding. In addition, the said framework is based on WGAN that uses the Wasserstein distance in place of the Jensen-Shannon divergence. The suggested approach is suitable for the LSB steganography, or the least significant bit steganography method where information is written in the lowest bit of a byte.

The results demonstrated the capability of the model to be potentially used for adaptive steganography algorithms for a wide range of applications.

As for the security side, LSB steganography is hard to detect, since there's virtually an infinite number of ways it can be implemented: various bit conjunctions, different

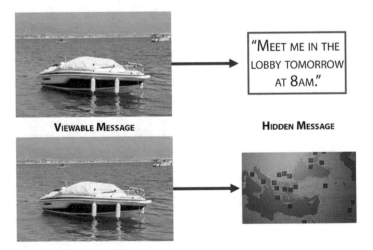

Fig. 6 Steganography visualization

bits, passwords usage, additional encryptions. Therefore, it is advisable to consider the use of deep learning methods and GAN for steganography detection.

Qian et al. proposed Gaussian-Neuron Convolutional Neural Network (or GNCNN) for steganographic "fingerprint" detection [21]. The authors presented a DL-based method for steganalysis that allows learning feature representations automatically. The developed model is able to capture the sophisticated dependencies that are beneficial for steganalysis and can learn feature representations by dint of convolutional layers.

Given the complexity of steganography procedures and given the importance of methods for detecting this type of hidden encryption, further research and advancement in this area are in high demand today.

2.6   Denial-of-Service Attacks

Denial-of-service attack (DoS attack) and distributed denial-of-service attack (DDoS attack) are one of the most common cyber threats in the field of information security. DoS attack is the situation when the malicious actor attempts to paralyze a computer or network resource and make it unavailable to its target users by transiently or permanently halting the services of a host connected to a network. During a DDoS attack, the incoming traffic flood arises from many various sources, so simply blocking one source won't stop the attack.

Usually, the target of perpetrators who perform DDoS attacks is websites and services that are hosted on high-level servers. These can be banks and payment gateways, official government websites. The malicious actors here are generally government intelligence agencies, hacktivists, vigilantes, blackmailers, etc.

Recently, DL-based network intrusion detection systems (NIDS) have reached an extremely high level in terms of detection efficiency [22]. This is exactly the domain where GANs can pose a serious threat.

Yan et al. developed DoS-WGAN that utilizes WGAN (Wasserstein generative adversarial network) with gradient penalty approach that allows evading traffic classification systems [23]. In order to disguise malicious DoS attacks as regular network traffic, DoS-WGAN automatically generates attack traces that can bypass modern NIDS and security protocols against DoS threats. The model produces attack traffic that is extremely similar to the normal one and thus cannot be distinguished by detection algorithms. The detection rate of CNN-based NIDS dropped from 97.3% to 47.6% once the researchers injected the generated DoS attack traffic into the detecting system.

Thus, we can safely conclude that this aspect of GANs requires thorough research and analysis.

2.7   Intrusion and Botnet Detection

Intrusion detection is usually carried out by special systems that monitor a network for malicious activity or policy violations and are implemented in the form of software or hardware applications. The intrusion detection systems (IDS) are responsible for identifying network attacks perpetrated by malicious traffic. Today, machine learning algorithms are used more and more in intrusion detection systems. Despite a certain level of efficiency of such systems, there are still doubts about their robustness.

Lin, Shi, and Xue in their paper presented a new framework that is based on Wasserstein GAN and contains generator, discriminator, and black-box IDS—IDSGAN [24]; the main purpose of the framework is to synthesize adversarial malevolent traffic entries aiming to attack IDS by spoofing and dodging the detection. The model demonstrated good results for different types of attacks, causing the detection rates of diverse black-box IDS architectures to drop to 0.

Another work presented by Usama et al. [25] proved the vulnerability of ML-based models to adversarial indignation that can result in an IDS malfunction by injecting a tiny, insignificant resentment in the network traffic. The researchers developed an adversarial ML-GAN attack that can efficiently avoid ML-based IDS. Additionally, it was found that GANs are applicable for the IDS inoculation so that it can respond to adversarial perturbations more effectively.

There are also successful examples of GAN applications for botnet detection. Botnets are one of the most serious threats to cybersecurity and are often used for large-scale malicious attacks. One of the main tasks today is accurate botnet detection.

Yin et al. proposed a GAN-based model for augmenting botnet detection methods—Bot-GAN. The obtained results proved that Bot-GAN is efficient and can improve regular detection systems [26]. Bot-GAN showed solid improvements in terms of detection performance, along with decreasing the false positive rate.

Table 1 briefly summarizes the domain, purpose, frameworks, tools, and description of GAN applications in cybersecurity that have been reviewed within the article.

3   Response and Forensics

3.1   Incident Response Overview

In view of the above, the most crucial element of the research is incident response and forensic methods that allow effectively counteract cyberattacks and detect digital "fingerprints" of perpetration.

In order to be able to properly detect and respond to GAN-based cyber threats, it is important to develop effective solutions that will take into account the inner machinery of GAN architecture and its sophisticated peculiarities. Given the growing popularity of machine learning and GANs, we should expect an extension of the scope

Table 1 Cybersecurity GANs generalization and areas of their application

Domain	Purpose	Frameworks & tools	Description
Malware	Malicious/defensive	MalGAN, tGAN, tDCGAN	GANs can create malware signatures & bypass ML-based detectors; also, they can classify and detect attacks
Password Guessing	Malicious	PassGAN, CrackStation, RainbowCrack, HashCat	GANs demonstrated the ability to assist in password guessing
Deepfakes & Image Manipulation	Malicious/defensive	StyleGAN, CycleGAN, DeepFD	GANs are used both for DF creation and detection
Forging Bioinformatic Data	Malicious	ADCGAN	Can be used to fake fingerprints and to bypass dactyloscopic security systems
Steganography	Malicious/defensive	SSGAN, GNCNN	GANs can produce solid covers for steganography; DL and GANs can be helpful for steganography detection
DoS/DDoS Attacks	Malicious	DoS-WGAN	GAN models allow evading traffic classification systems; they automatically generate attack traces that can bypass detectors and security protocols
Intrusion and Botnet Detection	Malicious/defensive	IDSGAN, Bot-GAN	GANs can synthesize adversarial malevolent traffic aiming to attack detectors without being identified; can efficiently avoid ML-based detectors. GAN-based models are efficient botnet detection

of this framework in the cybersecurity domain in the near future. What that means is that novel detection systems should be designed keeping machine learning and GANs in mind because this will help researchers and cybersecurity experts to get more well-prepared for new types of attacks and to better understand their identity and nature.

Incident response is a set of measures aimed at neutralizing of a malicious activity and addressing its aftermath. If an incident is not handled properly, a more destructive event such as a data leak or system failure can occur. The main goal of incident response is to restrain the incident, minimize damage, and get all processes back to normal, eliminating the negative consequences.

Incident response planning usually consists of four major steps: preparation, detection and analysis, containment-eradication-recovery, and post-incident activity. This strategy is certainly effective, but in the case of GANs, some of its components are specific and require a more comprehensive approach.

We won't review the conventional response methods mentioned above because this topic is beyond the scope of this work. The key point here is that the traditional incident response approaches should be actualized considering cutting-edge AI technologies, ML/DL methods, and GANs while applying them in practice.

As for the post-incident activity, GAN-specific digital forensics methods should be injected into traditional protocols.

3.2 Digital Forensics

The most significant task here is detecting GAN's digital "fingerprints" using forensic methods once the incident occurs so that cybersecurity experts and researchers can properly identify them, take adequate action, and study. Currently, several approaches have been considered to detect samples created by GANs.

The intrusive approach is intended to detect GAN-synthesised samples by using the discriminators, but it has several flaws, one of them is a comparatively high false detection rate. The feature extraction method is a non-intrusive technique and demonstrates solid results but only in cases where the training and testing data match.

Li et al. described the following non-intrusive detection approaches that are based on face quality assessment, inception score, and VGG features [27]; and demonstrate high performance results. Even though these methods are applied to image data, some of the principles can be extrapolated to other types of synthesized data. It is possible to train a forensics classifier on image data from one framework so that it can generalize obtained 'knowledge' by applying it to other models [28].

There are also manual methods for detecting a fake generated by GANs, they are close to the methods of conventional phototechnical examination [29]. However, these techniques mainly apply to images. Anyway, even when it comes to images generated by state-of-the-art GANs, it's still pretty hard to detect them due to a lack of reliable tools [30]. Bonettini et al. proposed detecting GAN-generated images using Benford's law to retrieve Benford-related features from a graphical sample based on divergence definitions that are different from each other [31]. According to Benford's law (or the first-digit law), in many real-life sets of numerical data (for instance, transactional data, stock market prices, country population, etc.), the leading digit is likely to be small; the conception can be described with the following Eq. (2):

$$p(d) = \log_{10}\left(1 + \frac{1}{d}\right), \tag{2}$$

where d is the first digit in base 10 ($d \in \{1, \ldots, 9\}$).

The authors conducted a set of experiments using a Random Forest classifier to examine the amount of information picked up by the features. Results demonstrated that GAN-synthesized graphical samples often violate Benford's law, hence can be distinguished from real data. Some types of CNN model, however, can generate images that are harder to identify. Furthermore, if we consider embedding principles of Benford's law into the network model, so it can strengthen GAN loss function, we can expect that the quality of the output images will be near-perfect, and counterfeit will be hard to detect.

Wang et al. analyzed existing GAN-image detection methods, dividing them into four categories: deep learning-based, physical-based, physiological-based, and performance against human vision [32]. But even despite the existing and quite effective approaches, the detection of GAN-images in a real environment is still a rather time-consuming task and requires in-depth research.

The examples above cover image-related threat examples, while many variations of GANs are used for attacks in other domains unrelated to image data. In this regard, GAN modifications aimed at malicious applications in other areas require intensified and thorough research.

In order to be more effective in preventing and responding to GAN-induced threats, it is necessary to use the methods of digital forensics and the principles of classical trace evidence science, which will allow us to effectively isolate and extract GAN traces, study potential, real cases, as well as the still hidden nuances of GAN architectures [33].

4   Conclusions

This article provides an in-depth, up-to-date survey of GAN-based applications in the information security domain. We have analyzed the state-of-the-art approaches and frameworks from different perspectives, detailed peculiarities of major approaches, and suggested response and forensics methods.

Given the current situation in the world, where cyber warfare is one of the most important tools for confronting ideologies, the importance of studying artificial intelligence [34, 35], constantly evolving methods of machine learning [36] and GANs [37] in information security problems is critical.

All reviewed GAN frameworks give us a firm understanding that we cannot trust data anymore, at least until we've got effective countermeasures and security protocols that keep up with the times. Given the ability of GANs to exploit vulnerabilities in state-of-the-art cybersecurity systems, this problem no longer lies solely in the domain of machine learning researchers. It is essential to involve information security experts in order to stay on top of the research and anticipate all possible techniques for a better response in identifying and countering cyberattacks.

Detecting GAN "footprints" in security breach incidents and identifying GAN-based threats remains challenging and requires future comprehensive research.

The researchers intend to focus their future endeavors on GAN application in DDoS attacks.

References

1. Goodfellow, I., Pouget-Abadie, J., Mirza, M., Xu, B., Warde-Farley, D., Ozair, S., Courville, A., Bengio, J. (2014). Generative Adversarial Networks. in *Proceedings of the International Conference on Neural Information Processing Systems* (NIPS), pp. 2672—2680.
2. Hu, W., Tan, Y., *Generating Adversarial Malware Examples for Black-Box Attacks Based on GAN*, [Online]. Available at: https://arxiv.org/pdf/1702.05983.pdf.
3. Guo, S., Zhao, J., Li, X., Duan, J., Mu, D., Jing, X. (2021). A black-box attack method against machine-learning-based anomaly network flow detection models, *Security and Communication Networks, 2021*(5578335), 13. doi: https://doi.org/10.1155/2021/5578335.
4. Shahpasand, M., Hamey, L., Vatsalan, D., Xue, M. (2019). Adversarial attacks on mobile malware detection. In *Proceedings of 2019 IEEE 1st International Workshop on Artificial Intelligence for Mobile* (AI4Mobile), pp. 17–20. doi: https://doi.org/10.1109/AI4Mobile.2019. 8672711.
5. Kargaard, J., Drange, T., Kor, A., Twafik, H., Butterfield, E. (2018). Defending IT systems against intelligent malware. In *Proceedings of 2018 IEEE 9th International Conference on Dependable Systems, Services and Technologies* (DESSERT) pp. 411–417. doi: https://doi. org/10.1109/DESSERT.2018.8409169.
6. Taheri, R., Shojafar, M., Alazab, M., & Tafazolli, R. (2021). Fed-IIoT: a robust federated malware detection architecture in industrial IoT. *Proceedings of IEEE Transactions on Industrial Informatics, 17*(12), 8442–8452. https://doi.org/10.1109/TII.2020.3043458
7. Kim, J., Bu, S., & Cho, S. (2017). Malware detection using deep transferred generative adversarial networks. *Proceedings of ICONIP, Part I, LNCS, 10634*, 556–564. https://doi.org/10. 1007/978-3-319-70087-8_58
8. Kim, J., Bu, S., & Cho, S. (2018). Zero-day malware detection using transferred generative adversarial networks based on deep autoencoders. *Information Sciences, 460–461*, 83–102. https://doi.org/10.1016/j.ins.2018.04.092
9. Hitaj, B., Gasti, P., Ateniese, G., Perez-Cruz, F. (2019). *PassGAN: A Deep Learning Approach for Password Guessing*, 2019, [Online]. Available at: https://arxiv.org/abs/1709.00440.
10. Striuk, O., Kondratenko, Y. (2021). Generative adversarial neural networks and deep learning: successful cases and advanced approaches, *International Journal of Computing* 339—349. doi: https://doi.org/10.47839/ijc.20.3.2278.
11. W. Knight, *The Defense Department has produced the first tools for catching deepfakes*, 2018, [Online]. Available at: https://www.technologyreview.com/2018/08/07/66640/the-defense-dep artment-has-produced-the-first-tools-for-catching-deepfakes/.
12. Nguyen, T. T., Nguyen, Q. V. H., Nguyen, D. T., Nguyen, D. T., Huynh-The, T., Nahavandi, S., Nguyen, T. T., Pham, Q.-V., Nguyen, C. M. (2022). *Deep Learning for Deepfakes Creation and Detection: A Survey*, [Online]. Available at: https://arxiv.org/abs/1909.11573.
13. Agarwal, S., Varshney, L. R. (2019). *Limits of Deepfake Detection: A Robust Estimation Viewpoint* [Online]. Available at: https://arxiv.org/abs/1905.03493.
14. Maurer, U. M. (2000). Authentication theory and hypothesis testing. *Proceedings of IEEE Transactions on Information Theory, 46*(4), 1350–1356. https://doi.org/10.1109/18.850674

15. Zhu, J.-Y., Park, T., Isola, P., Efros. A. A. (2017). *Unpaired Image-to-Image Translation using Cycle-Consistent Adversarial Networks*, [Online]. Available at: https://arxiv.org/abs/1703.10593.
16. Hsu, C.-C., Lee, C.-Y., Zhuang, Y.-X. (2018) *Learning to Detect Fake Face Images in the Wild* [Online]. Available at: https://arxiv.org/abs/1809.08754.
17. Struik, O., Kondratenko, Y. (2021). Adaptive deep convolutional GAN for fingerprint sample synthesis. In *Proceedings of 2021 IEEE 4th International Conference on Advanced Information and Communication Technologies* (AICT), pp. 193–196. doi: https://doi.org/10.1109/AICT52120.2021.9628978.
18. Bontrager, P., Roy, A., Togelius, J., Memon, N., Ross, A. (2017). *DeepMasterPrints: Generating MasterPrints for Dictionary Attacks via Latent Variable Evolution* [Online]. Available at: https://arxiv.org/abs/1705.07386.
19. Kim, H., Cui, X., Kim, M.-G., Nguyen, T. H. B. (2019). Fingerprint generation and presentation attack detection using deep neural networks. In *Proceedings of 2019 IEEE Conference on Multimedia Information Processing and Retrieval* (MIPR), 375—378. doi: https://doi.org/10.1109/MIPR.2019.00074.
20. Shi, H., Dong, J., Wang, W., Qian, Y., Zhang, X. (2017). *SSGAN: Secure Steganography Based on Generative Adversarial Networks*, 2017, [Online]. Available at: https://arxiv.org/abs/1707.01613
21. Qian, Y., Dong, J., Wang, W., Tan, T. (2015). Deep learning for steganalysis via convolutional neural networks. In *Proceedings, Media Watermarking, Security, and Forensics, 9409*. https://doi.org/10.1117/12.2083479.
22. Vinayakumar, R., Soman, K. P., Poornachandran, P. (2017). Applying convolutional neural network for network intrusion detection. In *Proceedings of 2017 International Conference on Advances in Computing, Communications and Informatics* (ICACCI), pp. 1222–1228. doi: https://doi.org/10.1109/ICACCI.2017.8126009.
23. Yan, Q., Wang, M., Huang, W., Luo, X., & Yu, F. R. (2019). Automatically synthesizing dos attack traces using generative adversarial networks. *International Journal of Machine Learning and Cybernetics, 10*, 3387–3396. https://doi.org/10.1007/s13042-019-00925-6
24. Lin, Z., Shi, Y., Xue, Z. (2018). *IDSGAN: Generative Adversarial Networks for Attack Generation against Intrusion Detection*, 2018, [Online]. Available at: https://arxiv.org/abs/1809.02077.
25. Usama, M., Asim, M., Latif, S., Qadir, J., & Ala-Al-Fuqaha. Generative adversarial networks for launching and thwarting adversarial attacks on network intrusion detection systems. In *Proceedings of 15th International Wireless Communications & Mobile Computing Conference* (IWCMC), pp. 78—83, doi: https://doi.org/10.1109/IWCMC.2019.8766353.
26. Yin, C., Zhu, Y., Liu, S., Fei, J., & Zhang, H. (2018). An enhancing framework for botnet detection using generative adversarial networks. In *Proceedings of International Conference on Artificial Intelligence and Big Data* (ICAIBD), pp. 228—234, doi: https://doi.org/10.1109/ICAIBD.2018.8396200.
27. Li, H., Chen, H., Li, B., Tan, S. (2018). Can forensic detectors identify GAN generated images?" In *Proceedings of Asia-Pacific Signal and Information Processing Association Annual Summit and Conference* (APSIPA ASC), pp. 722–727, doi: https://doi.org/10.23919/APSIPA.2018.8659461.
28. Wang, S.-Y., Wang, O., Zhang, R., Owens, A., Efros, A. A. (2019) *CNN-generated Images are Surprisingly Easy to Spot... For Now*, [Online]. Available at: https://arxiv.org/abs/1912.11035.
29. McDonald, K. (2018). *How to Recognize Fake AI-generated Images* [Online]. Available at: https://kcimc.medium.com/how-to-recognize-fake-ai-generated-images-4d1f6f9a2842
30. Gragnaniello, D., Cozzolino, D., Marra, F., Poggi, G., Verdoliva, L. (2021). Are GAN generated images easy to detect? A critical analysis of the state-of-the-art. In *Proceedings of 2021 IEEE International Conference on Multimedia and Expo* (ICME), Shenzhen, China, pp. 1–6, doi: https://doi.org/10.1109/ICME51207.2021.9428429.
31. Bonettini, N., Bestagini, P., Milani, S., Tubaro, S. (2020). *On the Use of Benford's Law to Detect GAN-generated Images*, [Online]. Available at: https://arxiv.org/abs/2004.07682.

32. Wang, X., Guo, H., Hu, S., Chang, M.-C., Lyu, S. (2022) *GAN-generated Faces Detection: A Survey and New Perspectives (2022)*, [Online]. Available at: https://arxiv.org/abs/2202.07145.

33. Striuk, O., Kondratenko, Y., Sidenko, I., Vorobyova, A. (2020). Generative adversarial neural network for creating photorealistic images. In *Proceedings of 2020 IEEE 2nd International Conference on Advanced Trends in Information Theory* (ATIT), pp. 368–371, doi: https://doi.org/10.1109/ATIT50783.2020.9349326.

34. Kondratenko, Y., Atamanyuk, I., Sidenko, I., Kondratenko, G., Sichevskyi, S. (2022). Machine learning techniques for increasing efficiency of the robot's sensor and control information processing, *Sensors, 22*(3), 1062. doi: https://doi.org/10.3390/s22031062.

35. Kondratenko, Y.P., Kuntsevich, V.M., Chikrii, A.A., Gubarev, V.F. (Eds.) (2021). Advanced Control Systems: Theory and Applications. Series in Automation, Control and Robotics; River Publishers: Gistrup. ISBN: 9788770223416.

36. Duro, R.J., Kondratenko, Y.P. (Eds.) (2015). Advances in Intelligent Robotics and Collaborative Automation. Series in Automation, Control and Robotics; River Publishers: Aalborg. ISBN: 9788793237032.

37. Striuk, O., Kondratenko, Y. (2023). Implementation of generative adversarial networks in mobile applications for image data enhancement. *Journal of Mobile Multimedia, 19*(03), 823–838. https://doi.org/10.13052/jmm1550-4646.1938

Printed in the United States
by Baker & Taylor Publisher Services